MW01010536

Study Guide to Accompany

Evolution

Making Sense of Life Second Edition

Alison E. H. Perkins

Roberts and Company, Greenwood Village, Colorado

Study Guide to Accompany
Evolution: Making Sense of Life, Second Edition

Roberts and Company Publishers, Inc.
4950 South Yosemite Street, F2 #197
Greenwood Village, CO 80111 USA
Tel: (303) 221-3325
Fax: (303) 221-3326
Email: info@roberts-publishers.com
Internet: www.roberts-publishers.com

Photo credits: p. 107, Syed Zillay Ali/Wikimedia Commons; p. 167, Douglas J. Emlen; p. 195, gvictoria/Shutterstock.com; p. 197, iStockphoto.com/Steve Byland/123rf; p. 208, H. Douglas Pratt.

Manufactured in the United States

ISBN: 978-1-936221-85-1
==
10 9 8 7 6 5 4 3 2 1

adaptive radiation

CONTENTS

INTRODUCTION TO THE STUDY GUIDE

This study guide is a tool to help you understand what you've learned reading *Evolution: Making Sense of Life*, Second Edition, by Carl Zimmer and Doug Emlen. It's designed to be engaging, using games and exercises and online resources to broaden your experiences. You can dive in as deeply as you choose, doing some exercises and skipping others, and you can test your comprehension when you're done.

Check Your Understanding

The *Check Your Understanding* questions are designed to remind you of important concepts—concepts that the current chapter builds on. The answers offer additional insight to the concepts being reviewed, so if you struggle too much, pop back to previous chapters and revisit the concepts.

Learning Objectives for the Chapter

Use the table provided to add important definitions and notes next to each learning objective for each chapter to help guide your understanding.

Learning Objective	Important Definitions	Notes
Learning objectives allow you to identify important ideas presented in the text.	biological evolution	change in the inherited traits of a pop. over time
Learning objectives prompt you to reach certain goals with your thinking.		three questions biological evolution can potentially address
Learning objectives help identify the skills you should develop as a learner.		evidence for understanding evolution

Identify Key Terms

The chapters in this guide introduce and discuss important concepts related to the theory of evolution. Here, you are challenged to match key terms with their definitions. These key terms are also important to developing concept maps and linking concepts in the next section.

Link Concepts

Each chapter includes a tool, either a concept map or another visual activity, to help you organize part of the chapter in your mind. Concept maps can be difficult to "get," but, really, there is no "right" answer; concept maps are works in progress. The idea is to help you visually link ideas outlined in the chapter and gradually build an understanding of how those ideas go together. The arrows used to link the concepts don't necessarily mean that one concept "leads to" another—simply that one concept *is related to* another in some way. You may prefer to think of the links in the opposite direction, or maybe you have a different way of linking the concepts altogether. The important thing is to think about how concepts you've learned are linked and *why*, instead of just jumbling everything up. Talking about concept maps with classmates or friends can help you develop your ideas and improve your understanding.

The first concept map is outlined, showing one way of thinking about how the relationships are delineated, and the key concepts that need to be considered are listed to the side. Take it to the next step and come up with your own concept map for another part of the chapter. Concept maps can be great tools for visual learners!

Key Concepts

Another way to test your mastery of the subject is by using the key concept statements at the end of most subsections within the chapters. Fill in the blanks where you can, then return to the chapters and see how well you did.

Interpret the Data

Scientists rely on visual representations of data to help others understand their evidence—to tell their stories. But charts and diagrams can be difficult to grasp, and interpreting graphs takes practice. It's often too easy to read the caption and skip over actually evaluating the data presented. Check to make sure you understand the graph from the textbook presented in this section. You'll gain insight into the kinds of evidence scientists gather and how scientists use that evidence.

Games and Exercises

The study guide includes some hopefully fun, but definitely explanatory, games that demonstrate the principles important to understanding evolution. Some can be done alone, and some require the help of friends or classmates. The games often involve jelly beans, M&Ms, or money, so don't eat or spend until the end of the book!

Explore!

The Internet is an incredible resource for videos, data, analyses—and baloney—about evolutionary biology. Some of the resources included in the study guide offer stunning visuals (PBS *Evolution*, *NOVA*, National Geographic) and even YouTube videos (although beware—YouTube and the Internet are full of misinformation, too). For cutting edge scientific studies, check out the links to *Science* magazine from the American Association for the Advancement of Science (AAAS). The links to *Science*Now are recent news stories that provide readable summaries and explanations of the current state of scientific research. Check the AAAS website regularly to find even more recent examples of the work scientists are doing.

The QR Codes

QR (short for "quick response") codes are essentially bar codes that can be read by imaging devices, such as cameras. If your smartphone has a camera, there's likely an app available that you can use to read the QR codes.

The QR codes used in the study guide use the URLs to take you to the websites listed in the text. The codes were generated using GoQR.ME, created by Andreas Haerter and Andreas Wolf and should be completely free of advertising. "Like" them!

Overcoming Misconceptions

The theory of evolution involves some abstract and complex concepts, and these concepts can be particularly difficult to grasp. Some misconceptions result simply because the ideas were never explained very well, and others because they seem to make sense—even though they are misconceptions. Other misconceptions are actively played up in the media, on the Internet, and by sources motivated to discredit science. Think carefully about the common misconceptions outlined in each chapter, and make sure you understand why they are misconceptions.

Go the Distance: Examine the Primary Literature

Scientists rely on primary literature for communicating their results. Thousands and thousands of research papers are published every year, and sifting through all that evidence can feel overwhelming. Check out the research paper highlighted for each textbook chapter and see if you can answer the questions.

Delve Deeper

You can really test your understanding with some extra short answer questions. They are designed to help you put concepts together and to address some of the confusion that may arise as you learn more and more about the theory of evolution. See how many you can tackle successfully.

Test Yourself

The extra multiple choice questions at the end of each study guide chapter can be a great quick assessment of your understanding of ideas presented in the text that you may have overlooked.

Contemplate

Sometimes it's important to just think about stuff, and the questions posed in this section give you the opportunity to get a little creative with your thinking. Some questions really don't have answers, but share your ideas with classmates and friends. You might be surprised at what you can come up with.

Answers

Of course answers to almost everything can be found at the end of the study guide. The Check Your Understanding questions go a little further and help you out if you're unclear about the right answer and why. If you are still uncomfortable with the concepts, you should probably brush up before moving on.

The answers to the Key Concepts can be found at the end of the relevant sections in the text.

There are no right answers for the concept maps, so don't expect to find an answer to the challenge to develop your own map. Nor will you find answers to questions posed in the Contemplate section. You should talk to your friends and classmates about their ideas on these two sections—you'd be surprised how informative those conversations can be.

Similarly, the Games and Exercises may or may not give you answers consistently. That's the beauty of data and evidence and interpretation. Many experiments need to be conducted over and over and over until a trend emerges. Sometimes random processes are more influential than experimental design, and sometimes experimental error affects results. Talk about the results you get with your classmates or your friends—compare notes, procedures, and ideas—and then think about what it means to do science.

ACKNOWLEDGMENTS

Understanding evolution is more important today than it has ever been. New strains of viruses are killing more and more people, hospitals and farmers are battling antibiotic resistance, and climate change is altering the selective environment for most organisms on the planet.

Carl Zimmer and Doug Emlen have written an incredible textbook outlining the basic principles of the theory of evolution. They've shared stories about where we are with the science and how we got here. But understanding the theory can be difficult. The underlying principles can be hard to grasp, and the evidence supporting the theory is growing at a breakneck pace! This study guide is designed to add to concepts presented in the textbook, to help readers strengthen their knowledge, and most importantly, to offer opportunities to explore some of the evidence of evolution more deeply.

In fact, the study guide links readers with some of the tremendous resources that evolution and science educators are developing. So many people are thinking about how to help students of all ages understand the complexities of evolutionary theory and the scientific endeavor. They are developing curricula, websites, and interactives, and this guide will introduce you to some of the amazing contributions from the American Association for the Advancement of Science (AAAS), Janis Antonovics and Doug Taylor of the Biology Department at University of Virginia, Ball State University Electronic Field Trips, Carol Brewer, Cornell Lab of Ornithology, Darren Fix (ScienceFix.com), Larry Flammer and the Evolution and the Nature of Science Institutes (ENSI), Judy Parrish, the Smithsonian, the Society for the Study of Evolution, and Tracy Tomm (ScienceSpot.net).

I want to especially thank Doug Emlen. He was invaluable in the development of the study guide. He reviewed everything, in spite of hectic schedules, and he offered incredible insight, support, mentorship, and, of course, friendship. I'm so glad to have worked with both him and Carl.

A number of reviewers provided valuable insight and expertise crafting the study guide. Bruce Cochrane, Miami University, Devin M. Drown, University of Alaska Fairbanks-Fairbanks, and Maheshi Dassanayake, Louisiana State University, were especially instrumental to the development. Their thoughtful insight contributed to the entire study guide.

The guide was written jointly with the textbook and as such many of our readers from the text also read the study guide chapters. We want to thank them all.

Windsor Aguirre, DePaul University
Lisa Belden, Virginia Tech University
Stewart Berlocher, University of Illinois at Urbana–Champaign
Annalisa Berta, San Diego State University
Gregory Bole, University of British Columbia
Jeffrey L. Boore, University of California, Berkeley
Brent Burt, Stephen F. Austin State University
Nancy Buschhaus, University of Tennessee at Martin
Douglas Causey, University of Alaska Anchorage
Robert Cox, University of Virginia
Robert Dowler, Angelo State University
Abby Drake, Skidmore College
David Fastovsky, University of Rhode Island
Charles Fenster, University of Maryland
Caitlin Fisher-Reid, Bridgewater State University
David Fitch, New York University
Jennifer Foote, Algoma University
Anthony Frankino, University of Houston
Barbara Frase, Bradley University
Nicole Gerlach, University of Florida
Jeff Good, University of Montana

Charles Goodnight, University of Vermont
Neil Greenspan, Case Western Reserve University
T. Ryan Gregory, University of Guelph
David Hale, United States Air Force Academy
Benjamin Harrison, University of Alaska Anchorage
Sher Hendrickson-Lambert, Shepherd University
David Hoferer, Judson University
Luke Holbrook, Rowan University
Elizabeth Jockusch, University of Connecticut
Charles Knight, Cal Poly San Luis Obispo
Patrick Krug, California State University
Simon Lailvaux, University of New Orleans
David Lampe, Duquesne University
Hayley Lanier, University of Wyoming at Casper
Kari Lavalli, Boston University
Amy Lawton-Rauh, Clemson University
Brian Lazzarro, Cornell University
Matthew Lehnert, Kent State University Stark
Kevin Livingstone, Trinity University
John Logsdon, University of Iowa
Patrick Lorch, Kent State University
J. P. Masley, University of Oklahoma
Lauren Mathews, Worcester Polytechnic Institute
Rodney Mauricio, University of Georgia
Joel McGlothlin, Virginia Tech University
Steve Mech, Albright College
Matthew Miller, Villanova University
Nathan Morehouse, University of Pittsburgh
James Morris, Brandeis University
Brian Morton, Barnard College
Barbara Musolf, Clayton State University
Mohamed Noor, Duke University
Steve O'Kane, University of Northern Iowa
Brian O'Meara, University of Tennessee
Cassia Oliveira, Lyon College
Daniel Pavuk, Bowling Green State University
Rob Phillips, California Institute of Technology
Marcelo Pires, Saddleback College
Patricia Princehouse, Case Western Reserve University
Sean Rice, Texas Tech University
Christina Richards, University of South Florida
Ajna Rivera, University of the Pacific
Antonis Rokas, Vanderbilt University
Sean Rogers, University of Calgary
Cameron Siler, University of Oklahoma
Sally Sommers Smith, Boston University
Chrissy Spencer, Georgia Tech
Joshua Springer, Purdue University
Christina Steel, Old Dominion University
Judy Stone, Colby College
Thomas Turner, University of California, Santa Barbara
Steve Vamosi, University of Calgary
Matthew White, Ohio University
Christopher Wills, University of California, San Diego
Peter Wimberger, University of Puget Sound
Christopher Witt, University of New Mexico
Lorne Wolfe, Georgia Southern University
Danielle Zacherl, California State University, Fullerton
Robert Zink, University of Minnesota

Most importantly, I am so grateful to Julianna Scott Fein for seeing this whole project through. She went above and beyond, reviewing, checking, and gently needling. Designer and compositor, Danielle Foster, was so patient, and her thoughtful contributions to the organization and structure were so helpful. Emiko Paul always impresses me. She's got such an eye for visually presenting scientific information, and she added significantly to the study guide. And Ben Roberts has the vision and patience for us all.

REFERENCES

Brennan, P. L. R., R. O. Prum, K. G. McCracken, M. D. Sorenson, R. E. Wilson, et al. 2007. Coevolution of male and female genital morphology in waterfowl. *PLoS ONE* 2 (5):e418.

Brewer, C. A. 2004. Near Real-Time Assessment of Student Learning and Understanding in Biology Courses. *BioScience* 54 (11): 1034–39.

Burger, J. M. S., M. Kolss, J. Pont, and T. J. Kawecki. 2008. Learning ability and longevity: A symmetrical evolutionary trade-off in *Drosophila. Evolution* 62 (6):1294–1304.

Coyne, J. A., and H. A. Orr. 2004. *Speciation.* Sinauer Associates, Sunderland, MA. 545 pp.

Ding, L., T. J. Ley, D. E. Larson, C. A. Miller, D. C. Koboldt, et al. 2012. Clonal evolution in relapsed acute myeloid leukaemia revealed by whole-genome sequencing. *Nature* 481 (7382):506–10.

Eterovic, A., and C. M. D. Santos. 2013. Teaching the role of mutation in evolution by means of a board game. *Evolution: Education and Outreach* 6:22.

Flynn, J. J., J. A. Finarelli, S. Zehr, J. Hsu, and M. A. Nedbal. 2005. Molecular phylogeny of the Carnivora (Mammalia): Assessing the impact of increased sampling on resolving enigmatic relationships. *Systematic Biology* 54 (2): 317–37.

Gatesy, J., and M. O'Leary. 2001. Deciphering whale origins with molecules and fossils. *Trends in Ecology and Evolution* 16: 562–70.

Gregory, T. 2008. Understanding evolutionary trees. *Evolution: Education and Outreach* 1:121–37.

Herrmann, E., C. Josep, H.-L. Maráa Victoria, H. Brian, and T. Michael. 2007. Humans have evolved specialized skills of social cognition: The Cultural Intelligence Hypothesis. *Science* 317 (5843):1360–66. doi: 10.1126/science.1146282.

Losos, J. B., and R. E. Ricklefs. 2009. Adaptation and diversification on islands. *Nature* 457:830–36.

Meir, E., J. Perry, J. C. Herron, and J. Kingsolver. 2007. College students' misconceptions about evolutionary trees. *The American Biology Teacher* 69(7):e71–e76.

Seife, C. 2010. *Proofiness: The Dark Arts of Mathematical Deception.* Viking. 295 pp.

Tishkoff, S. A., F. A. Reed, F. R. Friedlaender, C. Ehret, A. Ranciaro, et al. 2009. The genetic structure and history of Africans and African Americans. *Science* 324 (5930):1035–44.

1 The Whale and the Virus

How Scientists Study Evolution

Check Your Understanding

1. Because evolution is a theory, that means:
   a. It is a guess or a hunch.
   b. It has very little evidence to support it.
   c. Scientists may or may not believe in it.
   d. It has never been observed.
   e. None of the above
2. Why is understanding evolution important?
   a. Understanding evolution can help us understand biodiversity issues associated with deforestation and global warming.
   b. Understanding evolution can help us understand the evolution of antibiotic resistance and cancer.
   c. Understanding evolution can help us understand our own genetic makeup and how it affects our lives.
   d. All of the above
3. Which of the following statements about evolution is true?
   a. Once biologists find all the missing links, they will be able to understand evolution.
   b. Evolution is a process that leads to more and more complex organisms.
   c. Evolution is entirely random.
   d. Some forms of life are higher on the ladder than other forms of life.
   e. None of the above is a true statement.

Learning Objectives for Chapter 1

Add important definitions and notes next to each learning objective for this chapter to help guide your understanding.

Learning Objective	Important Definitions	Notes
Define biological evolution, and pose three questions biological evolution can potentially address.		
Using evidence from fossil whales, demonstrate how lineages change through time.		
Identify the characteristics of viruses that make them difficult to control.		
Describe three lines of evidence that scientists use for understanding evolution.		

Identify Key Terms

Match terms and definitions by filling in the blank to the right of the term with the appropriate letter.

1. **Biological evolution** c
2. **Genetic drift** j f
3. **Homologous traits** e
4. **Lineage** g
5. **Mutation** d
6. **Natural selection** b
7. **Phenotypes** f j
8. **Phylogeny** h
9. **Synapomorphy** i
10. **Viral reassortment** a

a. Occurs when genetic material from different strains gets mixed into new combinations within a single individual

b. A mechanism that can lead to evolution, whereby differential survival and reproduction of individuals cause some of them to survive and reproduce more effectively than others

c. Any change in the inherited traits of a population that occurs from one generation to the next (i.e., over a time period longer than the lifetime of an individual in the population)

d. Any change to the genomic sequence of an organism

e. Characteristics that are similar in two or more species because they are inherited from a common ancestor

f. Evolution arising from random changes in the genetic composition of a population from one generation to the next

g. A chain of ancestors and their descendants. It may be the successive generations of organisms in a single population, the members of an entire species during an interval of geological time, or a group of related species descending from a common ancestor

h. A visual representation of the evolutionary history of populations, genes, or species

i. A derived form of a trait that is shared by a group of related species (i.e., one that evolved in the immediate common ancestor of the group and was inherited by all its descendants)

j. Measurable aspects of organisms, such as morphology (structure), physiology, and behavior. Genes interact with other genes and with the environment during the development of the phenotype

Link Concepts

Fill in the bubbles with the appropriate terms.

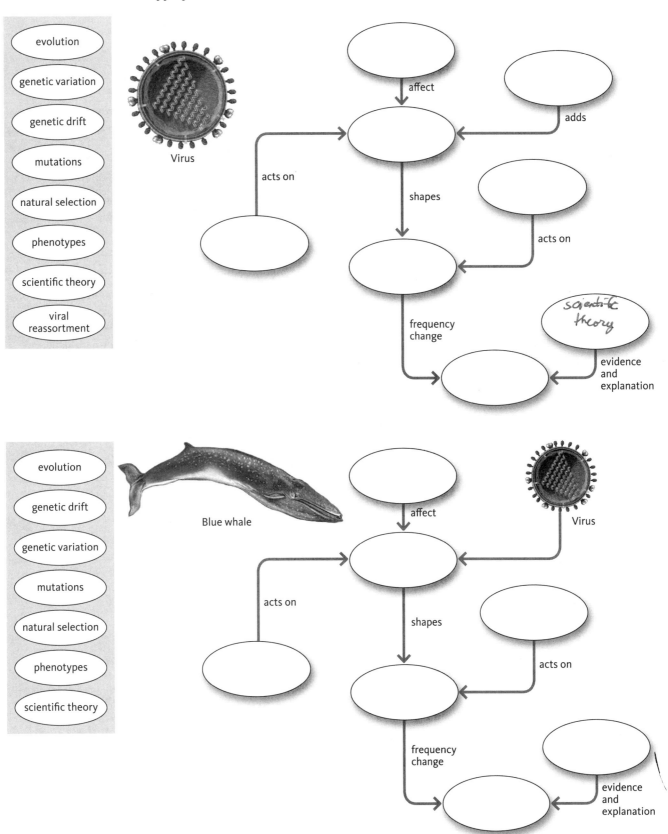

- evolution
- genetic variation
- genetic drift
- mutations
- natural selection
- phenotypes
- scientific theory
- viral reassortment

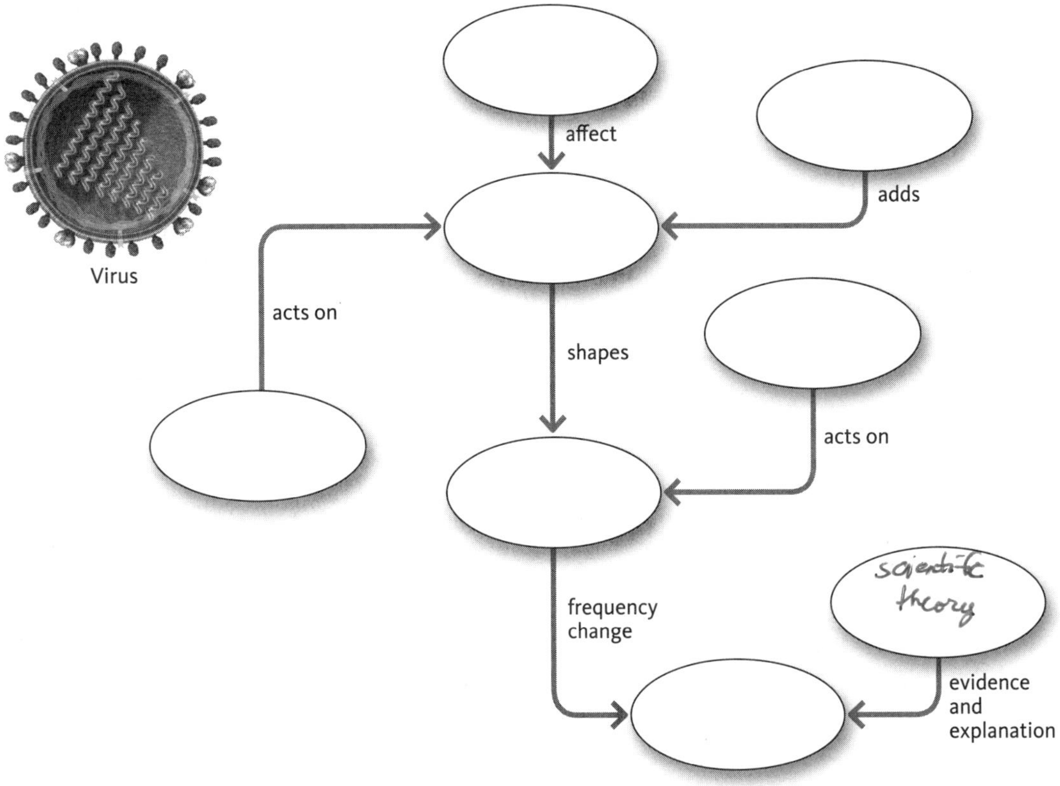

- evolution
- genetic drift
- genetic variation
- mutations
- natural selection
- phenotypes
- scientific theory

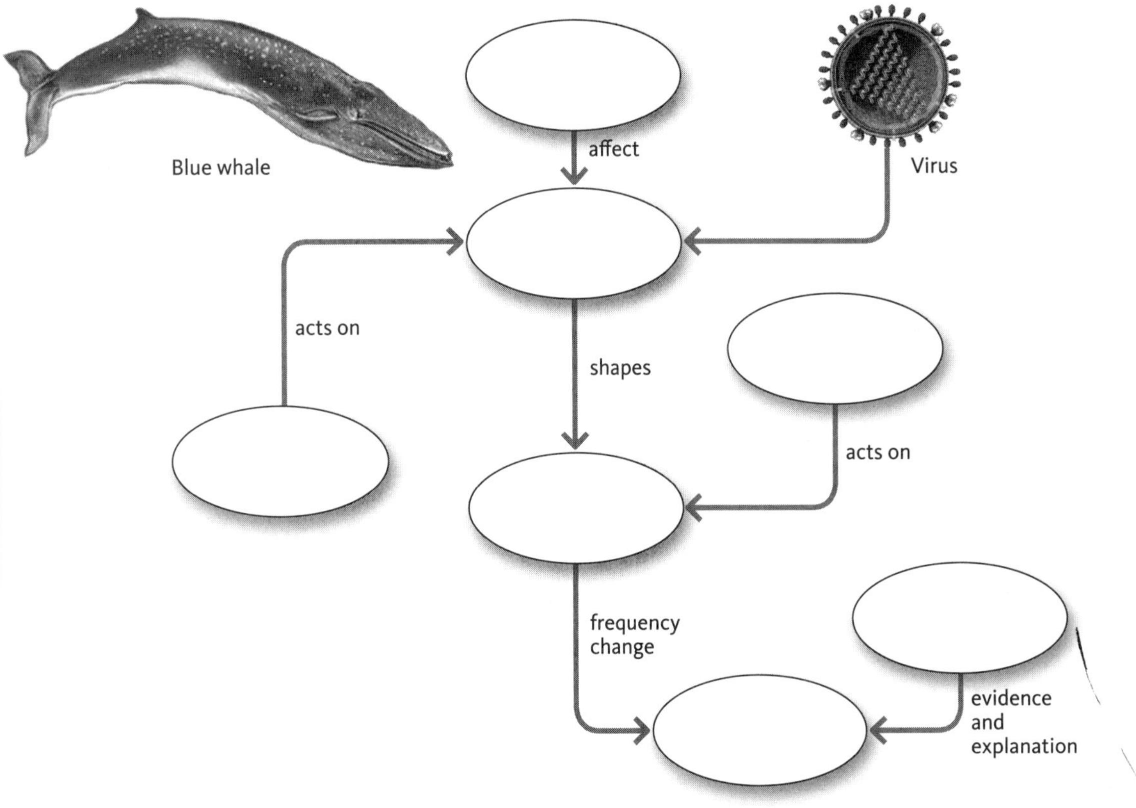

Key Concepts

Fill in the blanks for the key concepts from the chapter.

By __________ evolution, we can understand why the __________ is the way it is.

1.1 Whales: Mammals Gone to Sea

__________ and __________ lineages, evolving independently, __________ on body forms that are superficially similar.

Ambulocetus is a fossil whale with __________. This animal had traits that were intermediate between __________ whales and their terrestrial __________.

Scientists use different lines of evidence to study evolution. The __________ of fossil whales documents a transition from __________ to estuaries to the open __________ —the __________ transition documented in the changing shape of their skeletons.

As further __________ of their evolution from __________ mammalian ancestors, whales began to develop __________.

1.2 Viruses: The Deadly Escape Artists

__________ favors new variants of influenza viruses that can __________ detection (or destruction) by the __________.

Reassortment of the __________ is terrifying because it could blend dangerous elements of pig or bird flu with the infectious potential of human flu, instantaneously generating a strain that is both __________ *and* __________.

By monitoring bird flu, __________ can catch the evolution of new flu __________ as they occur.

Interpret the Data

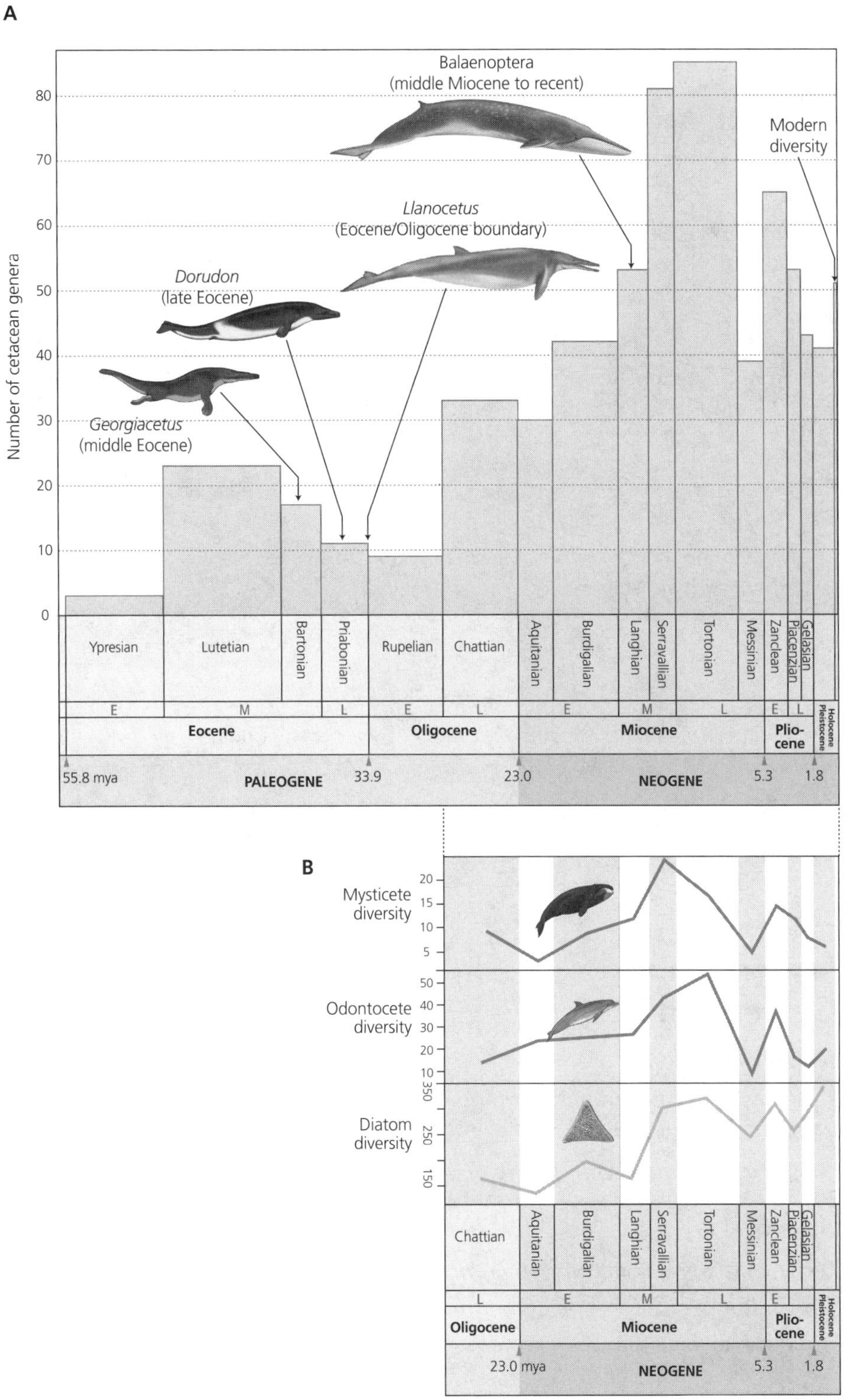

Figure 1.13 shows that the rise in whale diversity starting in the Oligocene may have been made possible by a rise in the diversity of diatoms, algae-like organisms that can support an ecological food web (Marx and Uhen 2010).

- When did the diversity of baleen whales (Mysticetes) peak?

- When did the diversity of toothed whales (Odontocetes) peak?

- When was the diversity of diatoms high?

- From this graph, what might you predict about the diversity of toothed whales (Odontocetes) in the future? What is your prediction based on?

- Can you predict the same about the diversity of baleen whales (Mysticetes)? Why or why not?

- What factors might affect the accuracy of your predictions?

Games and Exercises

Evolution in Action

This interactive game from PBS *NOVA* allows you to see how random mutations can lead to the evolution of a population of virtual creatures in different environments.

http://www.pbs.org/wgbh/nova/evolution/evolution-action.html

Whale Kiosk

This interactive site can help you understand the evidence for whale evolution created by Lara Sox-Harris.

http://www.indiana.edu/~ensiweb/lessons/whalekiosk.html

Guess the Embryo

This interactive game from PBS *NOVA* uses images of different embryos and the development of species to help show that embryos of different species can appear startlingly similar to one another, and the embryos of many species may or may not resemble their adult forms.

http://www.pbs.org/wgbh/nova/evolution/guess-embryo.html

Explore!

Return to the Water: Inside the Blue Whale

From the "Life of Mammals—Return to the Water" episode, produced by the BBC, this video takes you inside the blue whale, exploring the anatomy of the blue whale and its adaptations to life in the water.

http://www.bbc.co.uk/programmes/p004t035

Evolution: Great Transformations

This video from the PBS *Evolution* Library and WGBH shares the fossil evidence that supports the evolutionary transformation of whales and the characteristics that align whales with mammals not fish or sharks.

http://www.pbs.org/wgbh/evolution/library/03/4/l_034_05.html

Fellowship of the Whales: Cooperative Feeding in Humpback Whales

PBS *Nature* provides fantastic footage of cooperative hunting in humpback whales.

http://www.pbs.org/wnet/nature/episodes/fellowship-of-the-whales/video-cooperative-feeding/5324/

You can watch the full episode of the "Fellowship of the Whales" here:

http://www.pbs.org/wnet/nature/episodes/fellowship-of-the-whales/introduction/5263/

Dolphin Language: Two Dolphins Communicate with Each Other

This YouTube video shows trainers working with two captive dolphins as they communicate and learn a new trick.

http://www.youtube.com/watch?v=YSjqEopnC9w

New Whale Species Unearthed in California Highway Dig

Science Now reports the discovery of four new fossil species that offer new ideas about when toothed baleen whales went extinct.

http://news.sciencemag.org/sciencenow/2013/02/new-whale-species-unearthed-in-c.html

Overcoming Misconceptions

Outside of science, evolution is often characterized as something it is not. Descent with modification can, and has, produced an amazing diversity of life, but this simple process does not necessarily fit the way we *want* to look at nature.

Peaceful Balance

Evolution is not about creating a peaceful balance among organisms. Organisms eat other organisms, and that may sometimes be gruesome in our eyes. The same processes that give rise to exquisite adaptations, also give rise to the remarkable capabilities of predators for finding and killing prey. Parasites may also have amazing adaptations, such as the capacity to manipulate individual molecules within their hosts that allow them to devour their hosts from the inside out. Predation and parasitism may seem cruel, but predators and parasites are not evil. These organisms are simply responding to the same selective pressures that all organisms face, driving both the diversification and extinction of life.

Adaptations Species "Need"

The match between species and their environments can be striking, but this match did not come about because the species needed specific adaptations. Evolution cannot determine need—after all, one organism may "need" to live and breed, but another organism may "need" to kill and eat the first organism. Evolution is based on simple processes, such as mutations that become more or less common over the course of generations. A characteristic in one individual that performs better than one in another individual in a particular environment should result in more offspring that have the genetic instructions that produced the better version of the adaptation in the next generations—better and better adaptations become more and more common.

Perfect Adaptations

As impressive as some adaptations may be, they are far from perfect. Adaptations do not evolve from scratch; evolution modifies what already exists. Because beneficial mutations are limited, new forms evolve under tight constraints. Plus, mutations can have several different effects at once, so evolution also involves trade-offs. Many of the diseases we face may result from these tradeoffs.

Which of the following is a true statement?

a. Mutations always cause the improvement of a trait.

b. Having bigger brains gave both advantages and disadvantages to human ancestors.

c. The ancestors to whales needed more food than could be found on land, so they evolved features that allowed them to survive in the water.

d. Natural selection always favors viruses that infect only the weakest individuals in a population.

Go the Distance: Examine the Primary Literature

Hans Thewissen and his colleagues (2006) examined how genes affect development of hind limbs in modern dolphins (yes, dolphin embryos begin to grow legs). They construct a hypothesis about both the small changes (genes) and large changes (species) in their evolutionary history.

- What evidence did they use?

- How did they develop their hypothesis about whale evolution?

Thewissen, J. G. M., M. J. Cohn, L. S. Stevens, S. Bajpai, J. Heyning, and W. E. Horton, Jr. 2006. Developmental Basis for Hind-Limb Loss in Dolphins and Origin of the Cetacean Body Plan. *Proceedings of the National Academy of Sciences* 103 (22): 8414–8418. http://www.pnas.org/content/103/22/8414.full.

Delve Deeper

1. How do scientists use evidence to explain events that happened in the past?

2. Does evolution produce a peaceful balance in the natural world?

Test Yourself

1. Whales are most closely related to which group of animals?
   a. Camels
   b. Fish
   c. Seals
   d. They are equally related to all of the above.
2. According to Figure 1.8, which shared derived character (synapomorphy) links modern whales and the fossil whale *Indohyus*?
   a. A nasal opening that is shifted backwards
   b. A large powerful tail
   c. The presence of an involucrum
   d. All of the above
   e. None of the above
3. Which trait would you consider the most important in determining whether a fossil should be considered a cetacean or not?
   a. The teeth, because modern whales have peg-like teeth but *Dorudon*, one of the earliest cetacean fossils completely adapted to life in the water, had diverse teeth
   b. The involucrum, because it has a unique from in cetaceans not found in other mammals
   c. The astragalus, because if cetaceans evolved from artiodactyls, then early cetaceans should have double-pulley astragali
   d. The presence of flippers, because the earliest whale fossils had flippers much like those of modern whales
   e. The absence of legs, because cetaceans do not have legs
4. What characters of *Dorudon* are similar to modern whales?
   a. Presence of flippers
   b. A long vertebral column
   c. The shape of the involucrum
   d. All of the above are characters shared by *Dorudon* and modern whales.
   e. None of the above is a character shared by *Dorudon* and modern whales.

5. Why was finding the fossils of *Pakicetus* so important in understanding whale evolution?
    a. Because the fossils made the researchers famous
    b. Because *Pakicetus* shared many whale traits, but it lived on land
    c. Because *Pakicetus* was not really a whale because it lived on land
    d. Because the fossils weren't old enough to be considered the common ancestor of whales
6. Which is more important to the evolution of viruses, mutations or rapid reproductive potential?
    a. Mutations, because they are harmful to viruses and reduce reproductive potential
    b. Mutations, because all mutations benefit the virus
    c. Both, because mutations add variation on which natural selection can act through reproductive potential
    d. Rapid reproductive potential, because viruses that reproduce more rapidly can spread more rapidly
    e. Neither, because the ability to invade a host cell is the most important factor affecting virus evolution
7. What mechanism do scientists suggest led to the appearance of the H7N9 flu virus?
    a. Eating chicken
    b. Reassortment
    c. Natural selection
    d. The evolution of hemagglutinin
    e. None of the above
8. What do scientists understand about the evolutionary history of H7N9 and other emerging viruses?
    a. That through reassortment, a variety of different strains can combine to produce one deadly virus strain
    b. That even within a host, strains resulting from reassortment continue to evolve
    c. That a single mutation to a virus strain may lead to a highly infectious virus
    d. That a virus strain can circulate among some animal hosts for years before a mutation allows it to jump to humans
    e. All of the above
9. What is genetic drift?
    a. A random process that changes the genetic composition of a population from one generation to the next
    b. An unimportant process in evolutionary biology
    c. A random change in amino acid sequences
    d. A random process that gives rise to genetic variation within a population from one generation to the next
    e. A statistical anomaly that results when gene frequencies change over time

Contemplate

Could a mutation that is detrimental to individuals persist in a population? How might that occur?	Are embryos miniature versions of the animals they eventually become?

2 From Natural Philosophy to Darwin

A Brief History of Evolutionary Ideas

Check Your Understanding

1. How would you define biological evolution?
    a. A gradual process in which something changes into a different and usually more complex or better form
    b. Any change in the frequency of heritable traits within a population from one generation to the next
    c. A process of slow, progressive change
    d. Any kind of change over time
    e. A theory that humans have their origin in other types of animals, such as apes, and that the distinguishable differences between humans and apes are due to modifications in successive generations
2. Why are mutations important in evolution?
    a. Because mutations are always deleterious, and organisms with these deleterious mutations do not survive and reproduce
    b. Because mutations only occur in viruses
    c. Because mutations are random, and evolution is a random process
    d. Because mutations create the variation among individuals on which other mechanisms of evolution can act
3. What kinds of evidence have scientists used to study the evolution of whales?
    a. Evidence from DNA
    b. Evidence from fossils, such as their anatomical traits
    c. Evidence from geology
    d. Evidence from living species
    e. All of the above

Learning Objectives for Chapter 2

Add important definitions and notes next to each learning objective for this chapter to help guide your understanding.

Learning Objective	Important Definitions	Notes
Identify early naturalists and their contributions to evolutionary theory.		
Analyze the role the fossil record played in the development of the concept of evolution.		
Explain how Darwin's observations of nature led to the inferences he developed regarding natural selection.		
Give three examples of homologies, and explain why they are homologies rather than analogies.		

Identify Key Terms

Match terms and definitions by filling in the blank to the right of the term with the appropriate letter.

1. **Adaptations** c
2. **Extinction** e
3. **Genetic drift** g
4. **Homologous traits** k
5. **Hypothesis** h
6. **Natural selection** f
7. **Paleontology** d
8. **Stratigraphy** j
9. **Theory** b
10. **Taxa (singular, taxon)** a
11. **Taxonomy** i
12. **Uniformitarianism** l

a. Groups of organisms that a taxonomist judges to be cohesive units, such as species or orders

b. An overarching set of mechanisms or principles that explain a major aspect of the natural world

c. Inherited aspects of an individual that allow it to outcompete other members of a population that lack the trait (or that have a slightly different version of the trait). They are traits that have evolved through the mechanism of natural selection

d. The study of prehistoric life

e. The permanent loss of a species. It is marked by the death or failure to breed of the last individual

f. A mechanism that can lead to evolution, whereby differential survival and reproduction of individuals cause some of them to survive and reproduce more effectively than others

g. Evolution arising from random changes in the genetic composition of a population from one generation to the next

h. Tentative explanations or ideas that are grounded in evidence

i. The science of describing, naming, and classifying species of living or fossil organisms

j. The study of layering in rock

k. Characteristics that are similar in two or more species because they are inherited from a common ancestor

l. The idea that the natural laws observable around us now are also responsible for events in the past. One part of this view, for example, is the idea that the Earth had been shaped by the cumulative action of gradual processes like sediment deposition and erosion

Link Concepts

Fill in the bubbles with the appropriate terms.

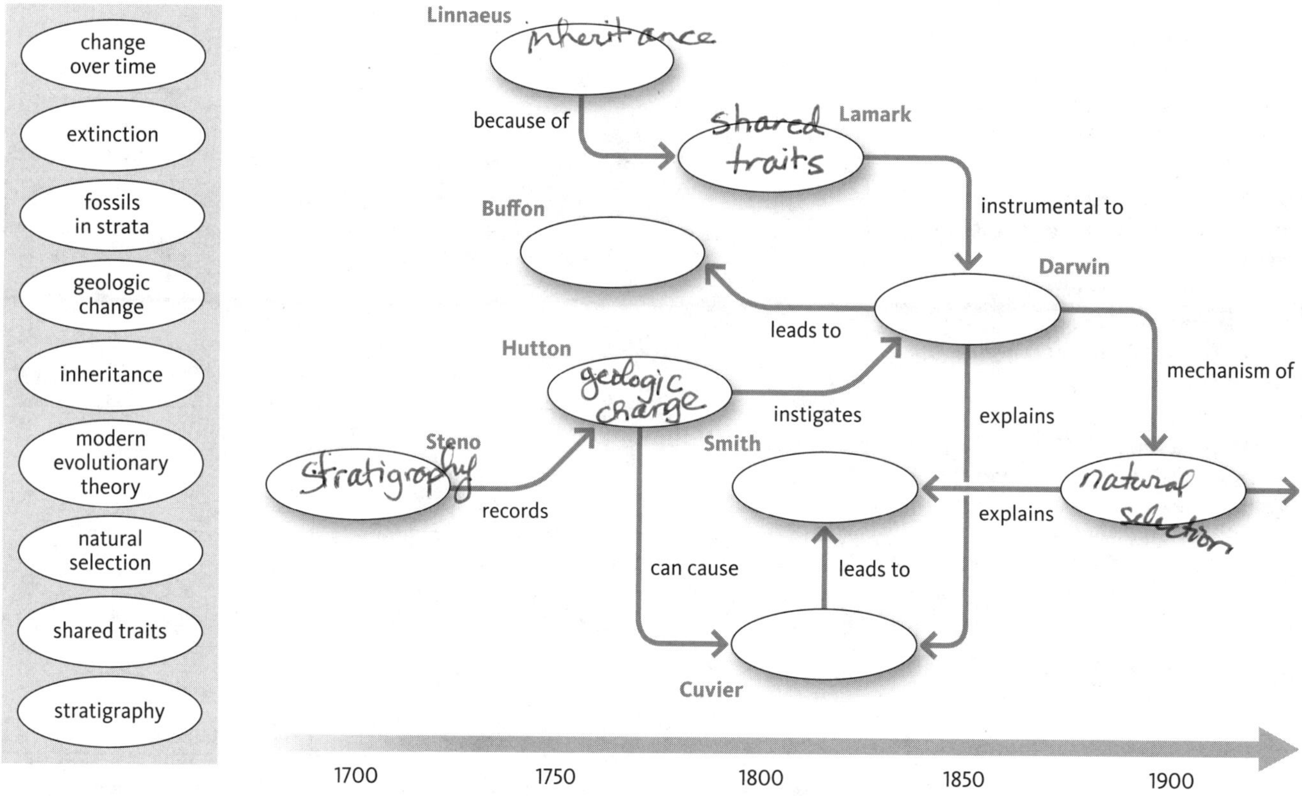

Key Concepts

Fill in the blanks for the key concepts from the chapter.

2.1 Nature before Darwin

Linnaeus is considered the father of modern taxonomy because his system for grouping organisms into a nested hierarchy is still in use today (although many of the groupings he proposed are not).

Stratigraphy is the study of layers in rock (stratification). Nicolaus Steno, the father of geology, pioneered the use of stratigraphy as a method for analyzing the past.

William Paley proposed that the mechanical structure of animal organs provided proof for the existence of a Divine Creator.

2.2 Evolution before Darwin

Early naturalists contributed to evolutionary theory. George-Louis Buffon proposed that new adaptations of a species could arise in response to new habitats. However, he did not believe that species could arise this way.

Some of the first compelling evidence for ____________ came from research conducted by Georges Cuvier, a pioneer in comparative ____________ and ____________ and an ardent anti-evolutionist.

James Hutton envisioned a world with a deep history shaped by gradual transformations of ~~geology~~ Earth through imperceptibly slow changes

The first geological map of ____________ and ____________ was developed by William Smith, an English geologist and land surveyor.

Jean-Baptiste Lamarck was an early pioneer of evolution as a process that obeyed natural laws.

2.3 The Unofficial Naturalist

Charles Lyell, who argued that Earth's history were the result of change, had a strong influence on the young Charles Darwin.

Comparisons between embryos can reveal similarities not evident in adulthood.

Thomas Malthus proposed that the increase of the human population is necessarily ~~monitored~~ controlled by the means of subsistence. His argument greatly influenced Darwin in his development of natural selection.

Darwin recognized that nature provided a reason that could explain how and why organisms evolved.

Alfred Russel Wallace also arrived at the idea of natural selection as a mechanism for evolution.

Interpret the Data

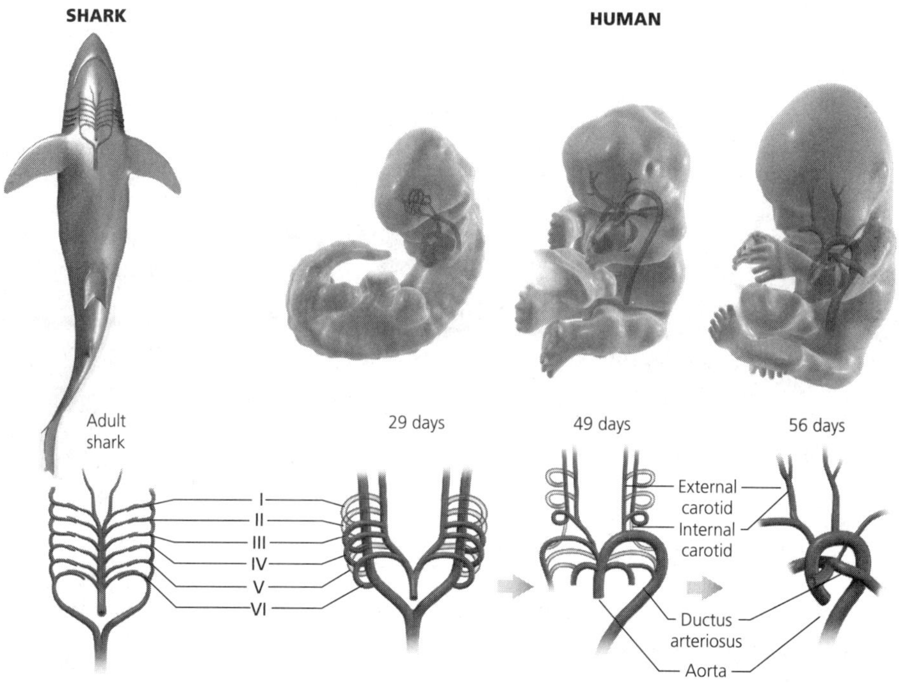

Figure 2.18 shows the arrangement of branching blood vessels in adult sharks and human embryos at various stages of development.

- At which of the stage of human development pictured does the pattern of blood vessels look most similar to the pattern found in sharks?

  29 days

- If scientists were only to examine human embryos at 56 days, would the homologies be apparent? no

- What does this homology indicate about the relationship between fishes and humans? came from a common ancestor

Games and Exercises

How Good a Predator Are You?

Survival and reproduction are not random. An individual with one version of a trait that helps it to thrive in its environment, such as better camouflage, may contribute more offspring to the next generation than other individuals—and those offspring are more likely to have that version of the trait. You can test how well certain individuals with certain traits survive in certain environments by pretending to be a predator to see how well you do spotting your prey. It's pretty easy to test your and your friends' abilities as predators—you need individuals that vary in color and a background that serves as camouflage for some individuals and not others.

Go to the craft store and buy some red, white, and green pipe cleaners. Twist 2–3 pipe cleaners together using some pipe cleaners that are the same color and some of different colors until you have 20 "animals." Take the animals outside and place them in bushes, on tree branches, on the ground, etc., along a stretch of pathway. [This activity is adapted from "Pipe Cleaner Animal Camouflage" by T. J. Fontaine from http://www.bioed.org/ECOS/inquiries/inquiries/FCCamo.pdf.]

Invite a couple friends to walk the stretch of pathway, at a moderately slow pace, and list the colors of animals they can find. Can they find all 20? If not, which colors were they more likely to find? Which were harder to find?

You can also do a similar experiment with green and plain colored toothpicks scattered in the grass in your yard in an exercise designed by Don Dunton, Fred Fisher, and Larry Flammer at ENSI (http://www.indiana.edu/~ensiweb/lessons/ns.st.wm.html). Spread equal numbers of green and plain toothpicks randomly in the grass. Give yourself and your friends 1 minute to collect as many as you can, and then determine the proportion of green versus plain toothpicks you each located. Did the green trait affect the survival of some individual more than the plain trait?

Alternatively, you can play Nowhere to Hide, an interactive online game that illustrates the same point.

http://sciencenetlinks.com/tools/nowhere-to-hide/.

From the Science Channel, journey with Charles Darwin as he explains natural selection and play an interactive selection game.

http://science.discovery.com/games-and-interactives/charles-darwin-game.htm

Explore!

A Film about Carl Linnaeus

Learn more about Carl Linnaeus and his influence on modern-day taxonomy in this biography produced by the Natural History Museum in London.

http://www.youtube.com/watch?v=Gb_IO-SzLgk

Fossil Great White Shark from Antwerp

This short video shows a tongue stone and the discovery of a fossil great white shark, *Carcharodon carcharias,* from the Pliocene, in Port of Atwerp, Belgium.

http://www.wat.tv/video/fossil-great-white-shark-from-4thu5_2htaj_.html

Time and Death: The Secrets of Evolution with Sagan, Cuvier, Darwin, Eiseley, and Barlow

Connie Barlow gives a really interesting lecture about the history of Carl Sagan's claim that "the secrets of evolution are time and death." The video begins with the insight of Georges Cuvier that species have indeed gone extinct in the past.

http://www.youtube.com/watch?v=mnTAzhLIEIg

Darwin's Dangerous Idea

From PBS, this short video introduces you to Charles Darwin and his voyage on the *Beagle.* Darwin uses the scientific process to amass evidence contrary to current ways of thinking, ultimately leading to publication of *On the Origin of Species.*

http://www.pbs.org/wgbh/evolution/darwin/index.html

Overcoming Misconceptions

Opponents of evolution often make many misleading claims about evolution, BUT:

- Evolutionary Biologists Do Not Study How Life Began

  People that oppose evolution often state that "evolutionary biologists have not discovered how life began." Evolutionary biologists do not study how life began but how life diversified *after* it began; other scientific fields of study are asking fruitful questions about how life began.

- Our Planet Is Not a Closed System

  People that oppose evolution often claim that "evolution violates the second law of thermodynamics." The second law of thermodynamics only holds in closed systems, but our planet receives outside energy from the sun—it is not a closed system.

- Evolution Is Not an Entirely Random Process

  People who oppose evolution state that "it is statistically impossible for evolution to have produced complex molecules like hemoglobin or organs like the eye," but that statement is based on the presumption that random processes led to a complex adaptation. Mutation is random, and genetic drift is random. But evolution through natural selection is not a random process, occurring because mutations with beneficial effects spread through populations. Complex adaptations evolve as beneficial mutations accumulate over time.

- Mutations Can Easily Provide the Raw Material for Innovations

  People that oppose evolution claim that "natural selection cannot produce innovations, because it simply favors some pre-existing variants over others." Mutations can easily provide the raw material for innovations—genes can be duplicated and mutations can lead to different responses or responses to different signals, any of which can produce new gene networks that can take on entirely new functions and even produce new structures.

- Gaps in the Fossil Record Should Be Expected

  People that oppose evolution argue that evolution fails as an explanation because of "absence of fossils that document major transitions." Paleontologists are actively exploring the fossil record, adding to our understanding of the relationships among taxa. But fossilization is not a perfect process—not everything that dies is preserved.

- Evolution Can Be Witnessed

  People that oppose evolution often confound the kinds of evidence used in science. They argue that scientists have not witnessed evolution *directly*. But scientists have witnessed evolution. Even using a very restrictive definition of "directly" meaning "to see with one's own eyes," scientists (and other careful observers, such as animal breeders) *commonly* witness changes in gene frequencies. Biological evolution is "any change in the inherited traits of a population that occurs from one generation to the next" (see Chapter 1). Scientists have studied changes in gene frequency both in the lab and in the wild (see Chapter 8).

  Scientists don't limit themselves to this narrow definition of "direct" evidence though. If we did, we'd be nowhere. For example, scientists only recently "directly" observed the Earth moving around the sun. Copernicus proposed that the planets revolved around a relatively stationary sun in the second century based entirely on evidence he could not witness "directly." And until a rover landed on Mars, scientists had very little "direct" evidence that Mars was a planet at all (nor direct evidence for any other planet). In science, direct evidence is evidence that comes *directly* from specific research results, and indirect evidence is other relevant information, observations, and resources. Both are common in evolutionary biology. And when combined with the myriad of other independent lines of evidence, scientists accept the theory of evolution as the best explanation for the diversity of life on Earth.

The Role of Debate in Science

Debate is a critically important component of science! Scientists constantly evaluate and criticize evidence. Eventually they come to a consensus about some important questions—such as how heredity makes natural selection possible—so they don't need to keep revisiting those questions. They delve deeper into aspects about the theory of evolution and continue on with the debate. But it's important to recognize that scientists aren't debating whether or not evolution occurs; they debate the nuances of how it occurs—for example, whether genetic drift or selection is more important in driving the evolution of particular characteristics or of entire populations under specific conditions or in particular environments.

What Do You Know about the Great Chain of Being and the March of Progress?

Evolution does not produce any kind of hierarchy in life—life does not proceed from simple to complex with humans holding the top spot. Think about it this way. Your family tree was not a chain of being with you as the final outcome:

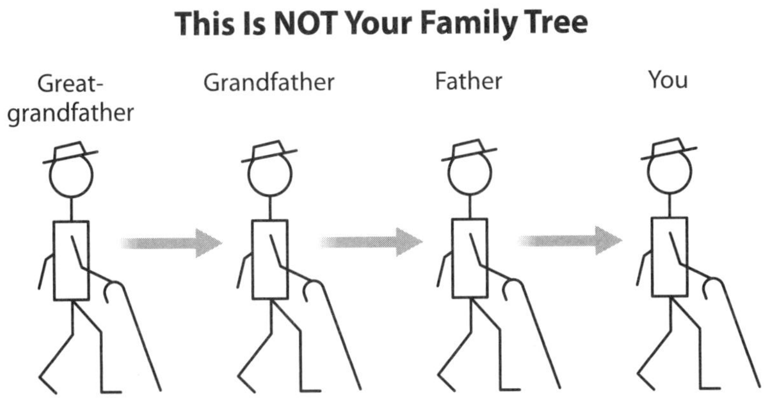

CREATED BY MATTHEW BONNAN OF MACOMB, ILLINOIS FOR A STICK-FIGURE CARTOON CONTEST IN THE SUMMER OF 2010 BY THE FLORIDA CITIZENS FOR SCIENCE.

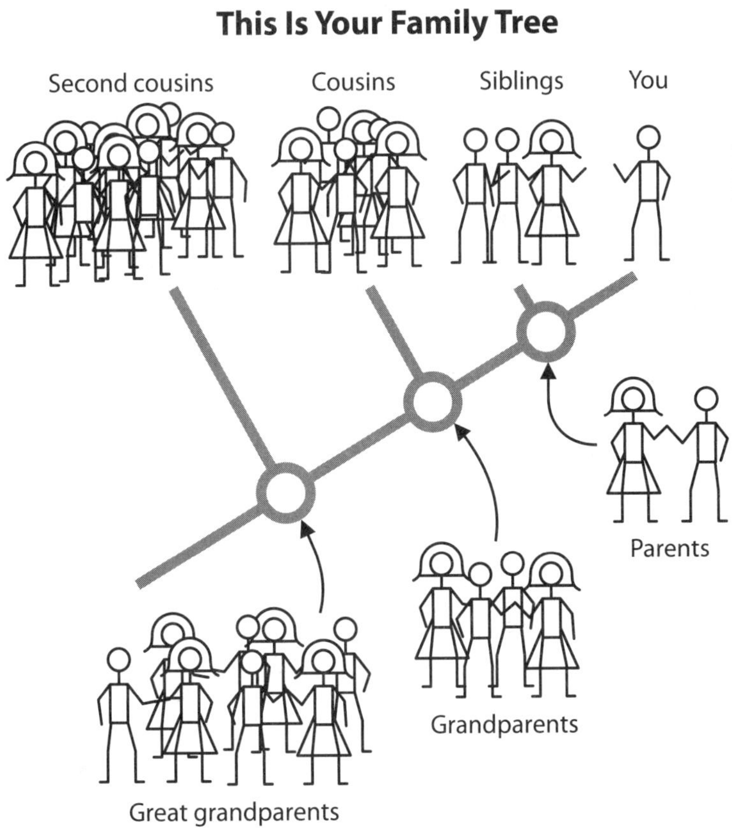

CREATED BY MATTHEW BONNAN OF MACOMB, ILLINOIS FOR A STICK-FIGURE CARTOON CONTEST IN THE SUMMER OF 2010 BY THE FLORIDA CITIZENS FOR SCIENCE.

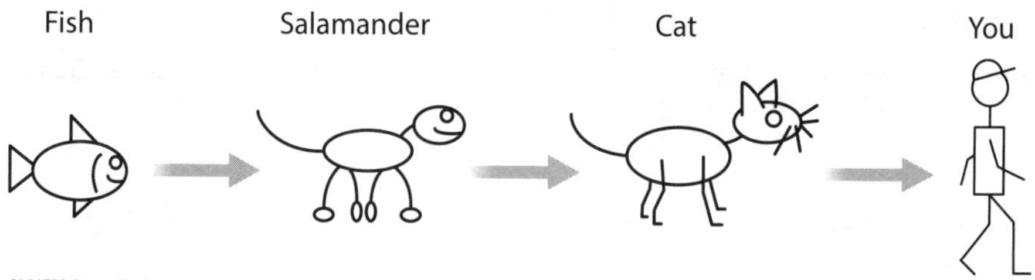

CREATED BY MATTHEW BONNAN OF MACOMB, ILLINOIS FOR A STICK-FIGURE CARTOON CONTEST IN THE SUMMER OF 2010 BY THE FLORIDA CITIZENS FOR SCIENCE.

So Why Isn't This Evolution?

Using the example above, redraw the relationships among fish, salamanders, cats, and you:

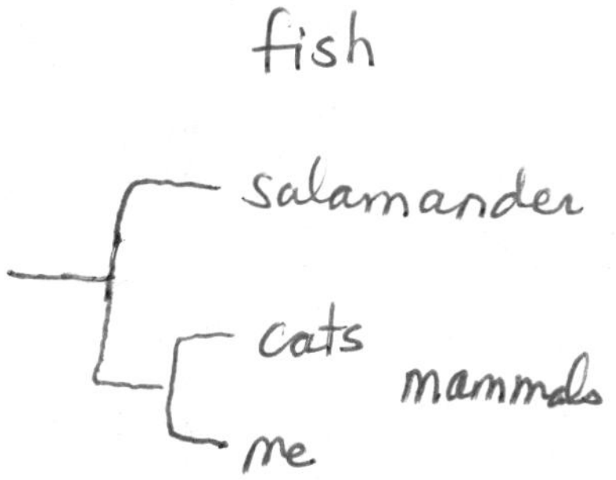

Which of the following is a misconception about evolution?

a. The fossil record will never be a complete record of life on Earth. t
b. Evolution is entirely random.
c. Evolutionary biologists are searching for the beginning of life. f
d. Scientists have never observed evolution directly.

Go the Distance: Examine the Primary Literature

In his essay "Darwin's Enduring Legacy," Kevin Padian outlines ten of the major contributions Charles Darwin made to the study of evolution.

- List three major concepts Darwin was unaware of as he developed his ideas.

- According to Padian, how important is scientific testing to Darwin's contributions to evolutionary theory? Why?

Padian, K. 2008. Darwin's Enduring Legacy. *Nature* 451:632-634.doi:10.1038/451632a. http://www.nature.com/nature/journal/v451/n7179/full/451632a.html

Delve Deeper

1. What is the difference between how scientists use the word *theory* and how it is used in everyday language?

2. Consider the following discussion question and fill in the boxes.

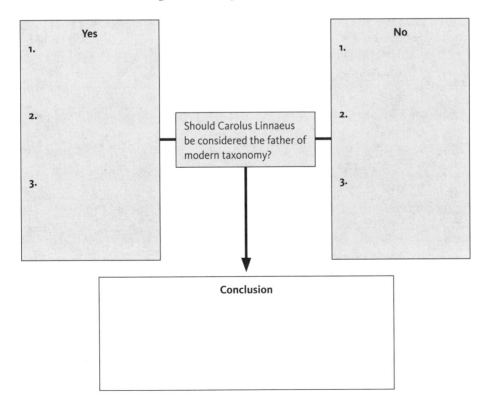

Yes

1.

2.

3.

Should Carolus Linnaeus be considered the father of modern taxonomy?

No

1.

2.

3.

Conclusion

Test Yourself

1. In which order would humans be classified according to Linnaeus?
   a. Hominidae
   b. Homo
   c. Animalia
   d. Mammalia
   e. Primates
2. Who first proposed that life could change over time?
   a. Charles Darwin
   b. Georges Buffon
   c. Carl Linnaeus
   d. Alfred Russel Wallace
3. What evidence did James Hutton use to reason that the Earth must be vastly old?
   a. The slow transformation of landscapes caused by erosion
   b. The rich diversity of fossil specie
   c. William Smith's maps of fossil strata
   d. Volcanoes that radically transformed the Scottish landscape
   e. The patterns of fossils of different ages found throughout England
4. Why did Georges Cuvier reject Lamarck's idea that species were not fixed?
   a. Because he and Lamarck were competitors at the National Museum of Natural History in Paris
   b. Because Lamarck did not believe in extinction
   c. Because his experience indicated that huge gaps existed among major groups of animal fossils with no intermediate fossils
   d. Because his experience with fossils indicated that all species were created at the same time
   e. Because he believed the Earth was not old enough for species to change
   f. Because he was a creationist (although he thought new creations occurred after catastrophic extinctions)
5. How do you define a scientific theory?
   a. A guess based on a few facts that scientists try to prove as correct
   b. An educated guess based on some experience that allows scientists to test evidence
   c. A set of laws that define the natural world
   d. A testable set of mechanisms or principles that explain a major aspect of the natural world
   e. A belief that scientists try to prove as fact
6. Why did Thomas Malthus' arguments in *An Essay on the Principle of Population* inspire Darwin and Wallace's concept of natural selection?
   a. Because Darwin and Wallace realized that poor people could not get enough food no matter how hard they tried
   b. Because Darwin and Wallace believed that populations get what they need, so even humans would not live up to their full reproductive potential
   c. Because Darwin and Wallace needed some support for their ideas about human populations and social engineering
   d. Because Darwin and Wallace realized that one generation of any organism can produce far more individuals then can actually survive and reproduce in the next

7. Is natural selection the only mechanism that can lead to biological evolution?
   a. Yes, because natural selection operates on the needs of organisms in their environment
   b. Yes, because Charles Darwin invented natural selection as part of his theory of evolution
   c. No, because evolution is entirely random
   d. No, because evolution can occur through other mechanisms such as genetic drift

Contemplate

Why did scientists' understanding of nature change as they learned more about the natural world?	**Why were Thomas Malthus' ideas about population growth so important to Darwin's and Wallace's ideas about natural selection?**

3 What the Rocks Say

How Geology and Paleontology Reveal the History of Life

Check Your Understanding

1. Why is the study of stratigraphy important to understanding evolutionary theory?
   a. Because stratigraphy provides evidence for the relative age of fossils
   b. Because Nicolaus Steno recognized that rocks occurred in layers even though he didn't believe in evolution
   c. Because stratigraphy characterizes the process of fossilization
   d. All of the above
   e. None of the above
2. What was Georges Cuvier's contribution to evolutionary theory?
   a. He strongly objected to Jean-Baptiste Lamarck's ideas about life evolving from simple to complex, drumming Lamarck out of the scientific establishment.
   b. He invented paleontology.
   c. He recognized that some fossils were both similar to and distinct from living species, and many fossil animals no longer existed.
   d. He was the first to organize and map strata according to a geological history.
   e. He rejected the idea that species evolved, instead believing that life's history was a series of appearances and extinctions of species.
3. What is a scientific theory?
   a. A belief that scientists try to prove as fact
   b. A set of laws that define the natural world
   c. A set of mechanisms or principles that explain a major aspect of the natural world
   d. A guess based on a few facts that scientists try to prove as correct
   e. An educated guess based on some experience that allows scientists to test evidence

Learning Objectives for Chapter 3

Add important definitions and notes next to each learning objective for this chapter to help guide your understanding.

Learning Objective	Important Definitions	Notes
Explain the role of debate in science.		
Describe how radioactive elements are used to determine the age of rocks.		
Discuss patterns of fossilization and how those patterns relate to the fossil record.		
Analyze how behaviors observed today can be used to understand plants and animals of the past.		
Based on your understanding of isotopes and biomarkers, design an experiment to examine the habitats of early herbivores.		
Explain why different lines of evidence are important in examining Earth's history.		
Describe the earliest forms of life on Earth.		
Describe the origins of multicellular life.		
Evaluate the contributions Ediacaran and trilobite fossils have made to our understanding of animal evolution.		
Define tetrapods and analyze their significance to human evolution.		
Develop an explanation for the patterns of diversity observed in animal and plant species alive today.		

Identify Key Terms

Match terms and definitions by filling in the blank to the right of the term with the appropriate letter.

1. **Archaea** q
2. **Bacteria** e
3. **Biomarkers** a
4. **Burgess Shale** g
5. **Chordates** k
6. **Ediacaran fauna** m
7. **Eukarya** f
8. **Hominins** o
9. **Lagerstätten (singular, Lagerstätte)** l
10. **Notochords** n
11. **Prokaryotes** d
12. **Radiometric dating** i
13. **Stromatolites** c
14. **Synapsids** h
15. **Teleosts** p
16. **Tetrapods** b
17. **Trilobites** j

a. Molecular evidence of life in the fossil record. They can include fragments of DNA, molecules such as lipids, or specific isotopic ratios.

b. Vertebrates with four limbs (or, like snakes, descended from vertebrates with four limbs). Living members include mammals, birds, reptiles, and amphibians.

c. Layered structures formed by the mineralization of bacteria

d. Microorganisms lacking a cell nucleus or any other membrane-bound organelles. These microorganisms comprise two evolutionarily distinct groups, the bacteria and the archaea.

e. One of the two prokaryotic domains of life. This domain includes organisms such as *E. coli* and other familiar microbes.

f. The third domain of life, characterized by traits that include membrane-enclosed cell nuclei and mitochondria. This domain includes animals, plants, fungi, and protists (a general term for single-celled members).

g. A Lagerstätte in Canada that preserved fossils from the Cambrian period

h. A lineage of tetrapods that emerged 300 million years ago and gave rise to mammals. The lineage can be distinguished from other tetrapods by the presence of a pair of openings in the skull behind the eyes, known as the temporal fenestrae.

i. A technique that allows geologists to estimate the precise ages at which one geological formation ends and another begins

j. Extinct marine arthropods that diversified during the Cambrian period and gradually died out during the Devonian period

k. Members of a diverse phylum of animals that includes the vertebrates, lancelets, and tunicates. As embryos, chordates all have a notochord (a hollow nerve cord), pharyngeal gill slits, and a post-anal tail. Many present-day animals lose or modify these structures as they develop into adults.

l. Sites with an abundant supply of unusually well-preserved fossils—often including soft tissues—from the same period of time

m. A group of animal species that existed during the period just before the Cambrian, between 575 and 535 million years ago. These animals included diverse species that looked like fronds, geometrical disks, and blobs covered with tire tracks.

n. Flexible, rod-shaped structures found in the embryos of all chordates. These structures served as the first "backbones" in early chordates, and in extant vertebrates the embryonic structure becomes part of the vertebral column.

o. A group including humans as well as all species more closely related to humans than to chimpanzees. Within this group, humans are the only surviving members.

p. A lineage of bony fish that comprises most living species of aquatic vertebrates, including goldfish, salmon, and tuna. They can be distinguished from other fishes by unique traits, such as the mobility of an upper jawbone called the premaxilla.

q. One of the two prokaryotic domains of life. Members of this domain superficially resemble bacteria, but they are distinguished by a number of unique biochemical features.

Link Concepts

Domains of Life
1) Archaea (prokaryote)
2) Bacteria (prokaryotes)
3)

Deep time is pretty abstract; it's hard to imagine just how long 1 million years ago is, let alone 4.567 billion years! If you could count one number every second, it would take just over 11.5 days to count to a million (that's without eating, sleeping, or taking any breaks). It would take almost 32 *years* (31 years and 255 days to be exact) to count to a billion! At that rate, you couldn't even count to 4.567 billion years in a lifetime.

It's easier to get an idea of the *relative* age of events in the Earth's past. Try this demonstration using a 1000 sheet roll of toilet paper. (You can use a roll with fewer sheets—just divide the number in the "Years before Present" column by the total number of sheets in the roll.) The roll represents the entire age of the Earth, and you can mark important events (such as the oldest known zircons or the first multicellular organisms) on different sheets as they are rolled out. The more room you have to spread out, the more you'll get a feel for exactly how short a time humans have been around on this planet!

Event	Years before Present	Number of Sheets	Rounded # of Sheets
Origin of Earth	4,567,000,000	1,000.000	1
Oldest known zircons	4,404,000,000	964.309	964
Rocks containing carbon	3,700,000,000	810.160	810
Archaea (biomarker)	3,500,000,000	766.367	766
Stromatolites	3,450,000,000	755.419	
Cyanobacteria	2,600,000,000	569.302	
Multicellularity	2,100,000,000	459.820	
Eukarya	1,800,000,000	394.132	
Algae	1,600,000,000	350.339	
Red algae	1,200,000,000	262.755	
Green algae	750,000,000	164.222	
Sponges	650,000,000	142.325	
Worm-like fossil traces	585,000,000	128.093	
Ediacaran fauna	575,000,000	125.903	
Cambrian period	542,000,000	118.677	
Chordates	515,000,000	112.765	
Oldest invertebrate traces	480,000,000	105.102	

Event	Years before Present	Number of Sheets	Rounded # of Sheets
Oldest plant fossils	475,000,000	104.007	
Oldest insect fossils	400,000,000	87.585	
Wattieza fossil	385,000,000	84.300	
Millipede fossil	428,000,000	93.716	
Oldest vertebrate traces	390,000,000	85.395	
Oldest tetrapods	370,000,000	81.016	
Synapsids	320,000,000	70.068	
Permian extinction	252,000,000	55.178	
First dinosaurs	230,000,000	50.361	
Mammals & birds	150,000,000	32.844	
Oldest flowering plants	132,000,000	28.903	
Grasses	70,000,000	15.327	
Cretaceous extinction	66,000,000	14.451	
Primates	50,000,000	10.948	
Ambulocetus	49,000,000	10.729	
Dorudon	40,000,000	8.758	
Earliest ape fossils	20,000,000	4.379	
Sahelanthropus	7,000,000	1.533	
Human-chimpanzee split	6,600,000	1.445	
Australopithecus afarensis	3,850,000	0.843	
Oldest stone tools	2,600,000	0.569	
Homo erectus	1,890,000	0.414	
Homo heidelbergensis	600,000	0.131	
Homo neanderthalensis	400,000	0.088	
Humans	200,000	0.044	
Present	—	—	

A Note on Significant Digits

No, not your fingers and toes. Scientists talk about significant figures because the digits in a number can carry meaning—they reflect the *precision* of the number. Accuracy and precision are two different concepts—accuracy refers to how close the measure is to the true number, and precision refers to how close the measured values are to each other. So, if, as a scientist, you wanted to know how long it takes to get to a class from your dorm, you might measure the number of minutes it takes for some of your friends to walk there. Say you found that for six of your friends, it took 12, 15, 13, 12, 14, and 13 minutes. Those measures could be accurate, precise, or both—information that could be really important for an 8 am class! If, on average, it actually took 8 minutes to get to class, you could say you're measures probably were not accurate (your measured values did not match with the true number). BUT they were precise, since they were all very close to each other. (Why might you're measures be different from the true value?)

So, significant digits reflect what scientists can know about precision in their measurements. In his book on how numbers can fool us, Charles Seife (2010) makes a point about significant digits with a great story about the age of a dinosaur skeleton. A museum guard claimed the dinosaur on exhibit was 65,000,038 years old because a scientist told him the skeleton was 65,000,000 years old when he was hired 38 years ago. The joke of course is that the scientist didn't know the *exact* age of the skeleton to be 65,000,000 years—not a year longer. The digits in the number reflect the accuracy with which the estimate could be made—signified by all of the zeros. In essence, the scientist rounded the number to the most accurate number of significant digits = 2 (6 and 5). [Note that this is just a story about how people generally don't understand numbers. Scientists are constantly developing tools that provide much more precise information about the ages of things like skeletons.]

Because significant digits are a reflection of precision, they depend largely on the number of significant digits in the data you are working with. So, when you average the number of minutes it takes to walk to class, you have to think about how precise your measurements actually are. Did you measure the time with a stopwatch app, or just the clock on your phone? Would one of those give you a more precise reading about the number of *seconds* it took—not just the minutes? Did you just happen to look at a clock on your way out the door and the clock on the wall in the classroom? How precise would that make your measure?

Nowadays, computers quickly calculate the outcomes of equations to as many digits as can fit in a given space. So, you can get the feeling that your measures are really good—on average it took 13.166666666666700 minutes to walk to class. You have to decide which digits actually are significant. And, of course, there are some rules:

In a number, the non-zero digits are considered significant, as are zeros appearing between two non-zero digits. Leading zeros are not.

Tricky part: The zeros at the end of number containing a decimal point are significant, so even though they may not be necessary, they tell everyone exactly how many significant digits there are. If the number doesn't contain a decimal, the significance of the zeros at the end isn't necessarily obvious. Different scientists may use different means for identifying the last significant zero (a bar over or under that zero, by changing the measurement unit, or simply stating how many significant figures there are).

Practice with these numbers. How many significant digits are there?

	Measurement	Significant Digits
Length of a year	365	
Length of a year	365.25	
Number of pixels per inch with the iPhone 6 Plus	406	
The average engagement rate of Vine videos	0.0206	
The Blood Alcohol Content equivalent to the reaction time delay of a driver using a cellphone, handheld, or hands-free device	0.08	
Age of the Earth	4,568,000,000	
Age of the Earth	4.568 billion	

- How many significant digits should you use to measure the amount of time it takes to walk to class?

Key Concepts

Fill in the blanks for the key concepts from the chapter.

3.1 The Great Age-of-the-Earth Debate

Nineteenth-century scientists ____________ the age of the Earth. Early estimates were based on wrong assumptions about the ____________ of the planet's interior.

3.2 A Curious Lack of Radioactivity

Many elements have both ____________ and isotopes, or radioactive, isotopes.

Unstable isotopes have a rate of decay.

Isotopes with high decay probabilities decay high, and those with low probabilities decay low.

3.3 A Vast Museum

The fossil record will never be complete because most organisms don't fossilize.

3.4 Bringing Fossils to Life

Technology allows scientists to gain new insights into the ____________, ____________, and ____________ of ____________ species by examining their fossils.

3.5 Traces of Vanished Biology

Isotopes and biomarker molecules carry information about the ____________.

3.6 Reading the Record: First Life

Scientists use diverse methods and independent ____________ to reconstruct the history of Earth.

Potential signs of life date back as far as ____________ billion years ago. The oldest known fossils that are generally accepted are ____________ billion years old.

3.7 The Rise of Life

The earliest signs of life are microbial, and ____________ still constitute ____________ of the world's biomass and genetic diversity.

3.8 Life Gets Big

The transition to ____________ life began at least 2.1 billion years ago, but multicellularity evolved independently in a number of ____________.

3.9 The Dawn of the Animal Kingdom

Although early ____________ fossils were highly diverse and had unique body plans, only a fraction of them share traits with living species. Nearly all Ediacaran species __vanished__ from the fossil record within 40 million years of their appearance.

Nearly all living animal ____________, including chordates, evolved during the ____________ period.

3.10 Climbing Ashore

The __emergence__ from life in the oceans to life on land marked another __epoch__ in the fossil record.

era

3.11 Recent Arrivals

Many of the most ____________ animal and plant species alive today have undergone relatively ____________ adaptive radiations.

Interpret the Data

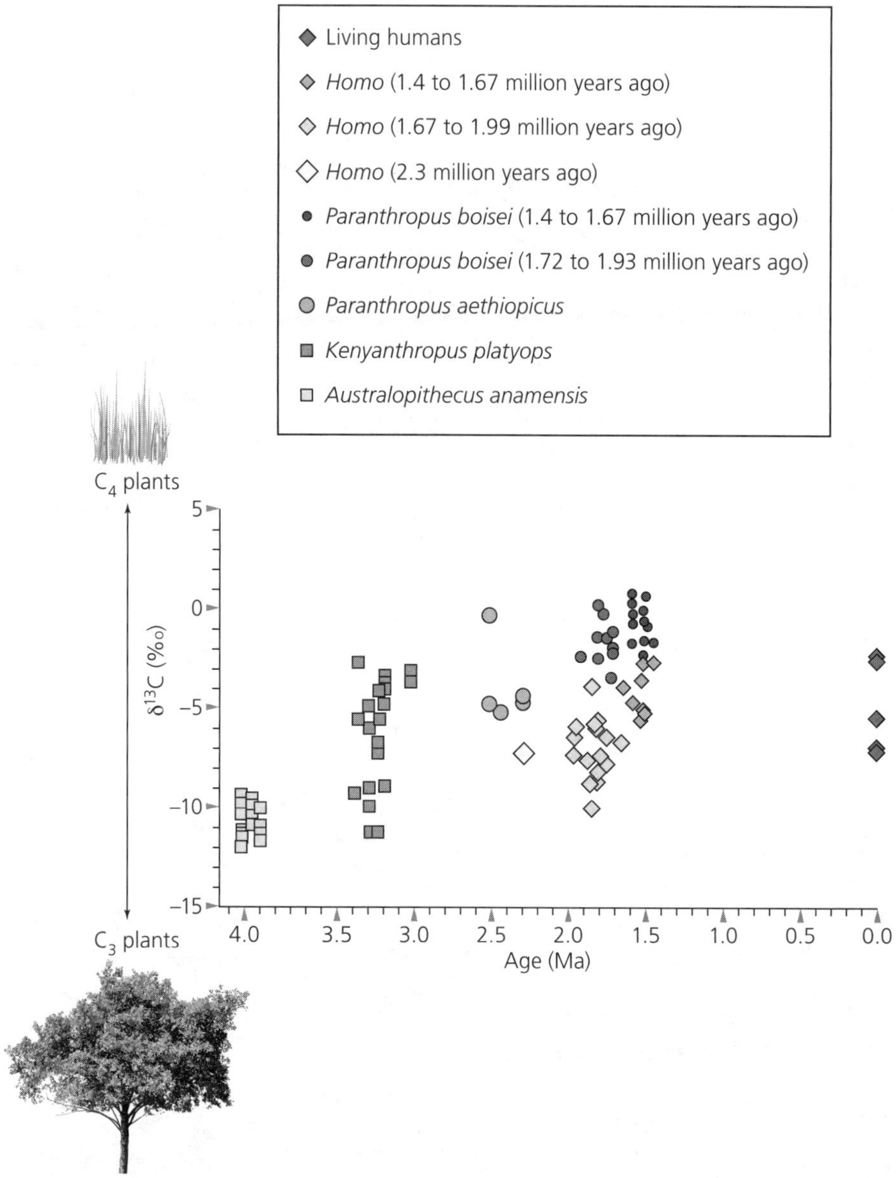

Scientists can use isotopes in fossils to understand some of the characteristics of the habitats where the organisms lived. To gain insight to the lives of ancient relatives of humans, Thule Cerling and his colleagues examined the amount of carbon isotopes found in the teeth of different hominins. They plotted the ratio of ^{13}C to ^{12}C ($\delta^{13}C$) for different aged specimens from the Turkana Basin in Africa (Figure 3.14).

- Which species had the lowest ratio of ^{13}C to ^{12}C, and when did it live?

- What is the range of $\delta^{13}C$ values for the genus *Homo* (our genus)?

- If the $\delta^{13}C$ values generally range between –24‰ and –32‰ for plants using the C3 pathway, and between –10‰ and –14‰ for plants using the C4 pathway, what can you surmise about the diets of species with low and high ratios?

- What can you determine about the diet of our genus?

Games and Exercises

A Sweet "Half Life"

Understanding decay rates can be difficult because the exact time that a certain nucleus will decay can't be predicted. Decay rates can be measured, however, and you can get a feel for how those measurements are determined by examining the "decay rates" of M&Ms or Skittles (or any candy with two "sides").

Use the blank side of the candy to represent the radioactive nuclei that have not lost a neutron, and the marked side to represent nuclei that have lost a neutron and become a different isotope of the candy "element." On average, every 100 years, one-half of the "blank" candies present will decay (lose a neutron) to become the "marked" isotope, so the half-life of the candy element is 100 years. Count out 80 candies and place them in a cup. To simulate decay, simply empty the candies out onto a flat surface. If the candy lands with the blank side up, it has not decayed, but if lands with the marked side up, it has decayed. Count all the candies that have decayed, and enter your observations in a table starting with the first simulation (the first 100 years) and record the number of blank candies and the marked candies in a data table. Try to predict the number of candies that will decay in the next 100 years (i.e., the next time you empty the cup on the table).

Simulation	Blank Candies	Marked Candies	Ratio of Blank to Marked Candies	Prediction for Next Simulation
1 (100 years)				
2 (200 years)				
3 (300 years)				
4 (400 years)				

Remove all the "marked" isotopes (you can eat them if you're hungry), and repeat the simulation using only the remaining radioactive "nuclei" (the blank candies). Continue this process until there are no radioactive isotopes left. Add more rows to the data table as needed.

Plot the data on a graph with the number of years that passed on the x-axis and the number of radioactive nuclei on the y-axis.

- Could you predict how many candies decayed after each simulation?

- What do you think would happen to your graph if you had more data?

- About how many nuclei would decay if you started with twice as many candies? What about if you started with half as many candies? What does that tell you about how the quantity of "radioactive isotopes" affects the number that decay?

- If scientists can't know *exactly* how many nuclei decayed, how do they use radiometric dating as a tool?

You can calculate the number of unstable atoms, N, left remaining from an original supply N_0, with the equation $N = N_0\, e^{-\lambda t}$, where λ is the probability of an atom decaying in a given time interval (also known as the decay constant)."

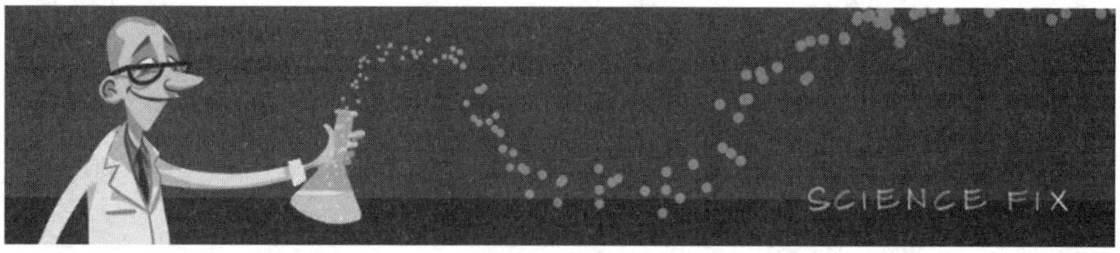

This activity is based on Radioactive M&Ms by Darren Fix at ScienceFix.com (http://www.sciencefix.com/).

You can watch this simulation unfold in a video from ScienceFix.com. The video comes with a downloadable instruction sheet for understanding radiometric dating.

http://www.sciencefix.com/home/2010/4/8/lesson-m-m-radiometric-dating.html

Radiometric Dating Isochrons

John Weber developed this exercise to introduce students to the isochron method for determining radiometric ages. Students can analyze a real data set and plot the results.

http://serc.carleton.edu/NAGTWorkshops/time/activities/60693.html

Explore!

How the Earth Was Made

From the *History Channel*, this short clip discusses Lord Kelvin's ideas about the Earth and the discovery of radiometric dating.

http://www.history.com/shows/how-the-earth-was-made/videos/the-age-of-earth#the-age-of-earth

What Are Isotopes?

Tyler DeWitt does a great job explaining exactly what isotopes are by comparing them with the different models of automobiles.

http://www.youtube.com/watch?v=EboWeWmh5Pg

Virtual Dating

This website goes into a bit more detail about radiometric dating, compete with quizzes to test yourself.

http://www.sciencecourseware.org/virtualdating/

Rareresouce.com

Shawn Mike, a graphic designer from Chicago, developed Rareresource. The website includes a wealth of information and some videos on dinosaurs.

http://www.rareresource.com/dinosaur_videos.htm

Did Dinosaurs Travel in Herds or Packs?

The American Museum of Natural History discusses how trackways can provide important clues to the behavior of birds and their dinosaur relatives.

http://www.amnh.org/exhibitions/past-exhibitions/dinosaurs-ancient-fossils-new-discoveries/trackways

Watch a quick video about the evidence for herds here:

http://www.youtube.com/watch?v=jtbpusl0Vo0.

Overcoming Misconceptions

Age of the Earth

Scientists no longer debate the overall age of the Earth—the 4.568 billion year age is widely accepted simply because of the quantity of evidence from many sources that supports that age. Scientists certainly continue to debate specific aspects of the formation of the Earth, such as how mineral grains condensed and aggregated to form the Earth.

Radiometric Dating

A lot of misinformation about radiometric dating exists, but the process is a valid scientific tool that has been tested and retested by scientists worldwide. Even though radiometric dating provides a range of ages, that doesn't mean it's inaccurate. Rocks contain many different kinds of isotopes, and scientists consistently derive the same ages of rocks using decay rates of different isotopes. Radiometric dates are not only supported by volumes of consistent results, they have been validated with independent lines of evidence, such as tree rings and varved sediments (for more on varved sediments, see Hughen and Zolitschka 2007 http://www.sciencedirect.com/science/article/pii/B0444527478000648).

Which of the following is a true statement?

a. Scientists debate about the Earth's age.

b. Scientific tools that only give results as a range of numbers are not valuable tools.

c. Scientists use independent lines of evidence to validate results.

d. When aging rocks, scientists use isotopes that will give them the results they think are best.

Go the Distance: Examine the Primary Literature

Fossilized dung? Yes, dung can fossilize, too, under the right conditions. Now, fossilized dinosaur dung, known as a coprolite, is giving scientists insight into dinosaur diets. Dolores Piperno and Hans-Dieter Sues discuss recent evidence that suggests not only that dinosaurs ate grasses, but also that grasses originated and diversified right along with dinosaurs. These two groups may be an ancient example of the evolutionary interactions between plants and the animals that eat them.

- Does this article explain the results of *the authors'* original research?

- What is the purpose of this article?

- Do the authors provide additional evidence to support their claims?

Piperno, D. R., and H.-D. Sues. 2005. Dinosaurs Dined on Grass. *Science* 310 (5751): 1126–28. doi:10.1126/science.1121020. http://www.sciencemag.org/content/310/5751/1126.

Delve Deeper

1. Why do scientists consider radiometric dating a valid way to measure the age of rocks?

2. The beginning of the Cambrian period, 542 million years ago, is often referred to as the Cambrian Explosion. The name is appropriate in some ways and inappropriate in others. Why is this a fitting name for this time? Why isn't the term *explosion* a good descriptor?

3. In the space provided, indicate whether you think each statement is a Hypothesis (H), a Prediction (P), or an Observation (O).

___ C_4 plants have higher levels of carbon-13 than both living and extinct C_3 plants.

___ The ratio of carbon isotopes in extant animals reflects the kinds of animals and plants that they eat.

___ Fossil animals will show the same relationship between the ratios of carbon isotopes and diet as living animals.

___ Hominin fossils in East Africa dating from 4.2 million years ago have relatively low ratio of ^{13}C to ^{12}C.

___ The relatively low ratio of ^{13}C to ^{12}C in 4.2 million-year-old hominin fossils indicates a diet rich in C_3 plants.

___ The relatively low ratio of ^{13}C to ^{12}C in hominin fossils is similar to the ratio found today in the teeth of chimpanzees.

___ Chimpanzees feed on fruits and leaves.

___ Plant fossils from the same sites where the teeth of 4.2 million-year-old hominin fossils were discovered will contain C_4 plants, consistent with early hominins living in grassy woodlands where they easily could have found C_4 plants.

___ 4.2 million-year-old hominin fossils were actively selecting C_3 plants for their diet.

Test Yourself

1. Why was Lord Kelvin's estimate for the age of the Earth inaccurate?
   a. Because he failed to account for plate tectonic and how that might affect heat flow
   b. Because he could not directly measure the age of any rock
   c. Because he used a model to generate a prediction for the age of the Earth
   d. Because he was trying to prove that evolution by natural selection could not occur
   e. Because he could not estimate cooling of rocks found deeper than in mines
2. Why can't scientists hope to find conventional fossils from very early in Earth's history?
   a. Because plate tectonics have destroyed almost all of the planet's original surface
   b. Because most fossils from that period are tiny zircons that do not have body forms
   c. Because the process of fossilization has changed over the course of Earth's history
   d. All of the above
   e. None of the above
3. Which of the following molecules has NOT been used as a biomarker, an organic chemical signature of once living organisms?
   a. Carbon isotopes
   b. Oxygen isotopes
   c. Sodium ions
   d. Cellular pigments
   e. All of the above have been used as organic chemical signatures.

4. How confident are scientists about the discoveries related to the earliest signs of life on Earth?
   a. Very confident. Two scientists found organic carbon produced by photosynthetic bacteria, so now all scientists accept the 3.7-billion-year-old carbon as the earliest sign of life.
   b. Fairly confident. Stromatolite formation is a process observable today, and scientists have discovered stromatolite fossils that are 3.45 billion year old.
   c. Note very confident. Scientists all have different opinions about the oldest signs of life on Earth.
   d. Not confident. Scientists constantly have to test new techniques, so they can't be sure until they find an identifiable fossil.
   e. It's all just a guess.
5. Which of the following statements about the history of life is TRUE?
   a. Because cyanobacteria, the lineage of bacteria that carries out photosynthesis, have not changed in 2.6-billion years, evolution must not be occurring.
   b. Scientists have not found any fossils from before the Cambrian period, 542 million years ago.
   c. Because scientists have not been able to resolve many of the relationships among the three domains of life, they cannot know anything about the early history of life.
   d. Living bacteria can offer clues about how the first multicellular animals evolved.
   e. None of the above is a true statement.
6. Which fossils often took the form of disks, fronds, or blobs?
   a. Prokaryotes
   b. Cambrian
   c. Stromatolites
   d. Ediacaran fauna
   e. None of the above
7. Did all the different organisms classified as Ediacaran fauna appear in the fossil record at the same time?
   a. No. The fossil record indicates that some organisms appeared as early as 575 million years ago, but other organisms did not appear until 20 million years later.
   b. Yes. The fossil record indicates that most organisms in the Ediacaran fauna coexisted, and radiometric dating is not accurate enough to distinguish when different organisms appeared.
   c. Maybe. The fossil record is incomplete, and not enough is known about the Avalon Assemblage to state that the Ediacaran fauna didn't appear at the same time.
   d. Yes. Even though the fossil record indicates that some organisms existed for long periods, and many organisms existed for relatively short periods, scientists cannot know whether these organisms appeared at the same time or not.
   e. No. Scientists are unable to determine the relationships of Ediacaran fauna because they are so diverse.

8. Which group does the following idea not apply to "many of the most diverse animal and plant species alive today belong to relatively recent radiations"?

   a. Flowering plants
   b. Insects
   c. Birds
   d. Bacteria
   e. Mammals

Contemplate

Why is understanding how organisms fossilize important to paleontology?	**How certain are evolutionary biologists about the age of the earliest life on Earth?**

4 The Tree of Life

How Biologists Use Phylogeny to Reconstruct the Deep Past

Check Your Understanding

1. How does homology relate to the theory of evolution?
   a. Homology refers to traits that only superficially look alike, like bat wings and human arms; the theory of evolution cannot explain these similarities.
   b. Homology refers to traits that look alike but have entirely different origins, like the fins of dolphins and of fish; the theory of evolution cannot explain the origins.
   c. Homology refers to traits that are structurally similar in different organisms, like bat wings and human arms, because they each were inherited from a shared common ancestor with those traits; the theory of evolution provides a mechanism for those observations.
   d. Homology refers to traits that have converged on a shared form; the theory of evolution provides a mechanism for those observations.
2. How accurate is radiometric dating?
   a. Not very accurate because scientists cannot know how much of a particular isotope was originally present in a rock, and thus, they cannot know how much of it has decayed
   b. Very accurate because scientists can determine the exact age of rocks and the fossils found within them
   c. Accurate because radiometric dating can determine estimates for the ages of rocks and fossils, often with relatively small margins of error
   d. Not very accurate because scientists use probabilities to determine decay rates for each isotope they use in radiometric dating
3. Besides radiometric dating, what other lines of evidence can be used to determine the ages of fossils?
   a. Stratigraphy
   b. Homology
   c. Complexity
   d. a and b only
   e. All of the above

Learning Objectives for Chapter 4

Add important definitions and notes next to each learning objective for this chapter to help guide your understanding.

Learning Objective	Important Definitions	Notes
Identify the different components of phylogenies and the functions of each.		
Discuss how different lines of evidence can lead to different conclusions about species' taxonomical relationships.		
Analyze the relationships of characters in a phylogeny.		
Demonstrate how scientists can determine the timing of branching events.		
Explain how phylogenies can be used to develop hypotheses about the evolution of tetrapods.		
Explain how the bones of the middle ear can be used to trace the evolution of mammals.		
Discuss how a phylogenetic approach can be used to explore the role of feathers in dinosaur evolution.		

homoplasy ×

clade = single branches in tree of life
monophyletic = group of organisms that form clade

Identify Key Terms

Match terms and definitions by filling in the blank to the right of the term with the appropriate letter.

1. **Branches** k
2. **Characters** g
3. **Clades** r a
4. **Convergent evolution** d
5. **Evolutionary reversal** e
6. **Exaptation** s
7. **Homoplasy** m n
8. **Horizontal gene transfer** q
9. **Internal nodes** i
10. **Monophyletic** a B
11. **Nodes** j
12. **Outgroups** f
13. **Paraphyletic** l
14. **Parsimony** h
15. **Phylogeny** o
16. **Polyphyletic** p t
17. **Polytomy** t p
18. **Synapomorphy** m
19. **Taxa (singular, taxon)** b
20. **Tips** c

a. Single "branches" in the tree of life; each represents an organism and all of its descendants

b. A group of organisms that form a clade

c. The terminal ends of an evolutionary tree, representing species, molecules, or populations being compared

d. The independent origin of similar traits in separate evolutionary lineages

e. The reversion of a derived character state to a form resembling its ancestral state

f. Groups of organisms (e.g., a species) that are outside of the monophyletic group being considered. In phylogenetic studies, these can be used to infer the ancestral states of characters

g. Heritable aspects of organisms that can be compared across taxa

h. A principle that guides the selection of alternative hypotheses; the alternative requiring the fewest assumptions or steps is usually (but not always) best. In cladistics, scientists search for the tree topology with the least number of character state changes.

i. Nodes that occur within a phylogeny and represent ancestral populations or species

j. Points in a phylogeny where a lineage splits (a speciation event or other branching event, such as the formation of subspecies)

k. Lineages evolving through time that connect successive speciation or other branching events

l. A group of organisms that share a common ancestor although the group does not include all the descendants of that common ancestor

m. A shared derived character (i.e., one that evolved in the immediate common ancestor of a clade and was inherited by all of its descendants)

n. A character state similarity not due to shared descent (e.g., produced by convergent evolution or evolutionary reversal)

o. The branching evolutionary history of populations, genes, and species. It is a visual representations of these relationships

p. An internal node of a phylogeny with more than two branches (i.e., the order in which the branchings occurred is not resolved)

q. The transfer of genetic material—other than from parent to offspring—to another organism, sometimes a distantly related one, without reproduction. Once this material is added to the recipient's genome, it can be inherited by descent.

r. A group of organisms that a taxonomist judges to be a cohesive unit, such as a species or order.

s. A trait that initially carries out one function and is later co-opted for a new function. The original function may or may not be retained.

t. A taxon that does not include the common ancestor of all members of the taxon

polyphletic = taxon doesn't include

branches lineages that connect success speciation or other event

doesn't include common ancesto & all members

taxon
different ancestors – polyphyletic
– paraphyletic
ancestor & all descents – orthophyletic

character
shared, derived
unique, derived
same common ancestry

Link Concepts

Fill in the bubbles with the appropriate terms.

- ancestral
- character state
- clade
- common ancestor
- homoplasy
- hypothesis
- monophyly
- phylogeny
- shared derived
- synapomorphy

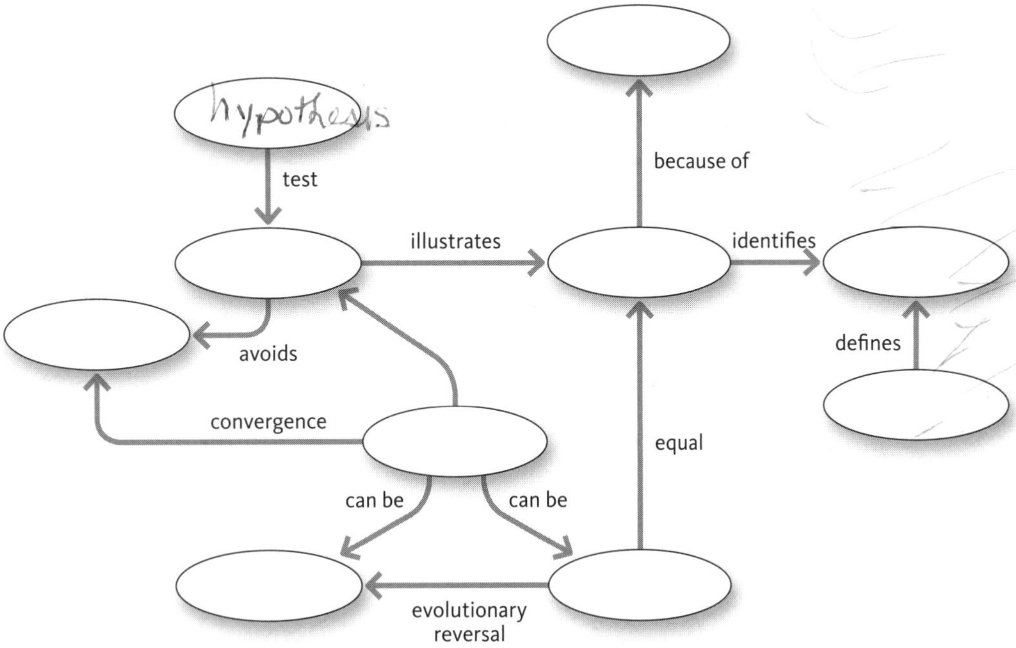

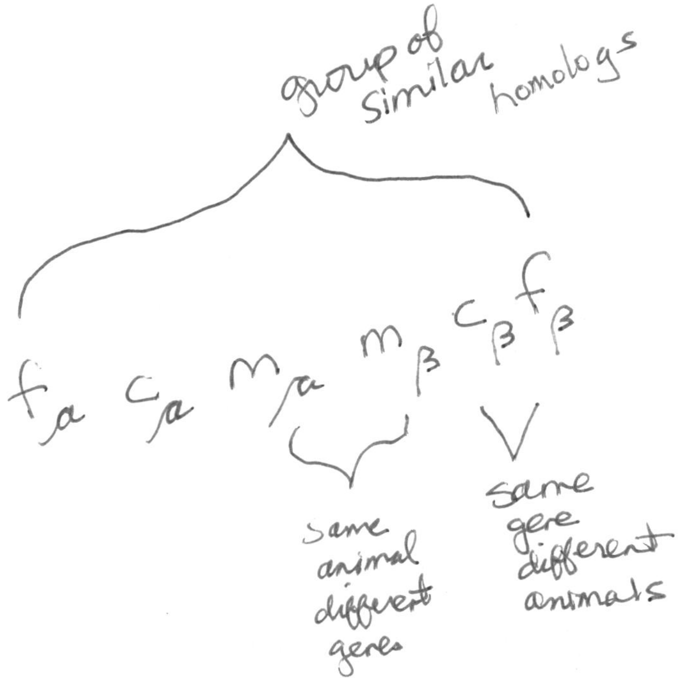

Key Concepts

Fill in the blanks for the key concepts from the chapter.

4.1 Tree Thinking

____________ represent the branching pattern of evolution over time.

4.2 Phylogeny and Taxonomy

Different conclusions can result from different lines of ____________ . Compared with the tools of scientists today, Linnaeus built his taxonomic system on a relatively basic understanding of ____________ . Evolutionary theory has added new ____________ for these relationships and new insight to the meaning of taxonomic groupings.

4.3 Reconstructing Phylogenies

Unique character states that arose in the ____________ of a clade, and are inherited by daughter species in that ____________ , are called *shared derived* characters, or ____________ .

Phylogenetic trees are most ____________ when they are constructed using shared derived characters.

Phylogenetic trees are ____________ that describe the relationships among taxa based on the best available evidence . Phylogenies help scientists identify questions that can be answered with additional evidence.

4.4 Fossils, Phylogeny, and the Timing of Evolution

Combining evidence from ____________ with morphological evidence from ____________ species can offer insight to the timing of ____________ events.

4.5 Testing Hypotheses with Phylogenies: From Sea to Land

Phylogenies illustrate relationships among ____________ species —not *all* species. Combining different lines of evidence can lead scientists to new fossil ____________ and insight to why ____________ may have evolved.

4.6 Homology as a Window into Evolutionary History

The mammalian ____________ is made up of modified parts of the ____________ and surrounding skull.

Phylogenetic trees can reveal the ____________ of small changes and adjustments to ____________ over the course of their ____________ history.

4.7 How Feathered Dinosaurs Took Flight

_____________ are an exaptation; they evolved originally for functions other than _____________ .

Interpret the Data

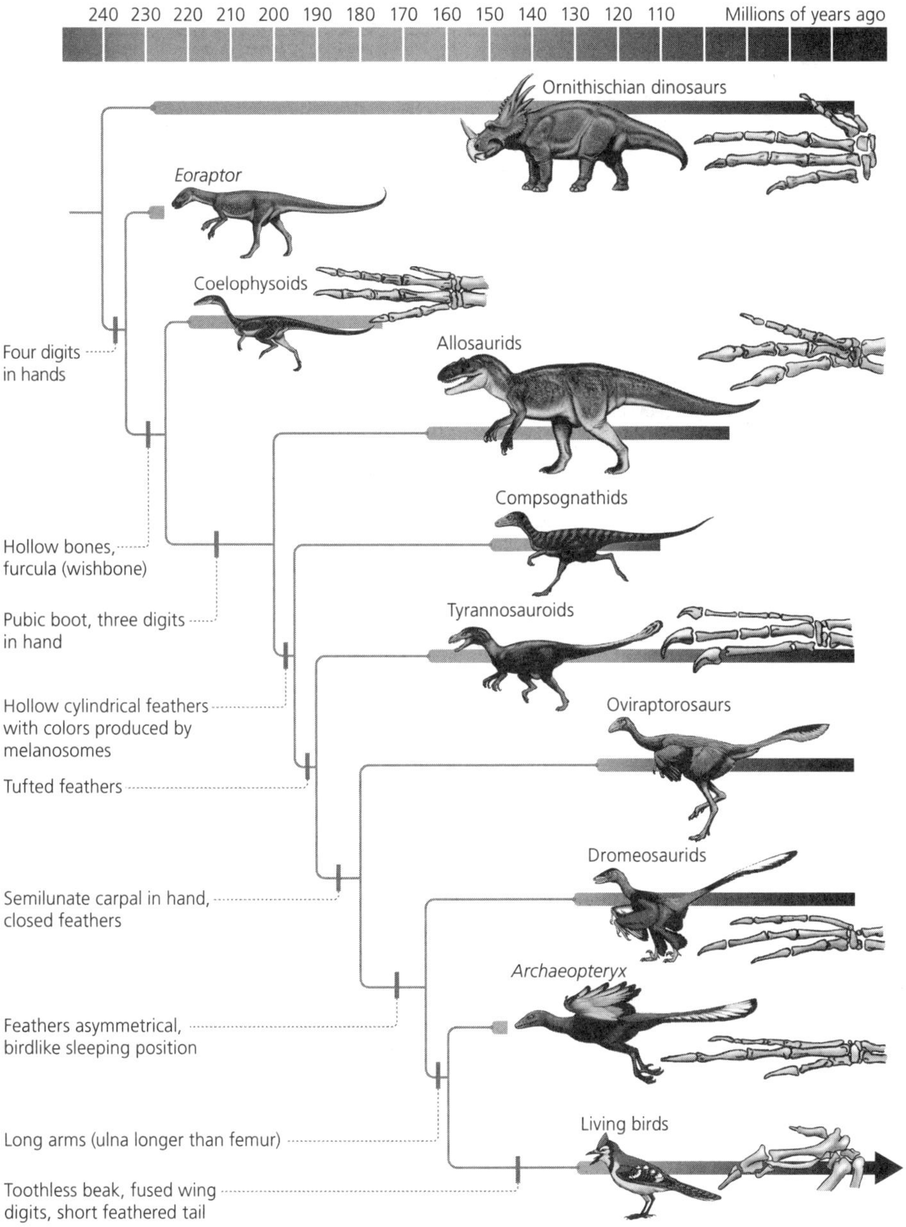

This evogram (Figure 4.15) illustrates the lineages of therapods by arranging the branches of a cladogram along a timeline (top). Some of the key character changes that distinguish each branch of theropods are listed at each node. The thick lines show the ages of known fossils, and the ages of the nodes are estimates based on the known ages of the fossils.

- Explain why living birds are considered dinosaurs.

- What trait(s) distinguish living birds from theropod dinosaurs?

- Which branches of theropods have feathers?

- Which branches may have been able to fly?

Games and Exercises

Phylogenies can be incredibly powerful tools for studying relationships in more than just biological organisms. Practically, the analysis is the same whether you want to look at the relationships among beetles or the evolution of languages; they are simply nested hierarchies—groups within groups within groups.

A cladogram groups organisms based on their shared derived characteristics. A simple example adapted from an ENSI/SENSI lesson plan called "Making Cladograms" (http://www.indiana.edu/~ensiweb/lessons/mclad.html) illustrates how this nesting works.

Characters	Shark	Bullfrog	Kangaroo	Human
Vertebrae	yes	yes	yes	yes
Two pairs of limbs		yes	yes	yes
Mammary glands			yes	yes
Placenta				yes

One way to appreciate the relationships among the organisms in the table above is to make a Venn diagram. Venn diagrams are like nesting doll toys—with the largest on the outside and the smallest on the inside. Start with the character that is shared by all the taxa, and then each successive group is smaller because they share a more limited set of characters.

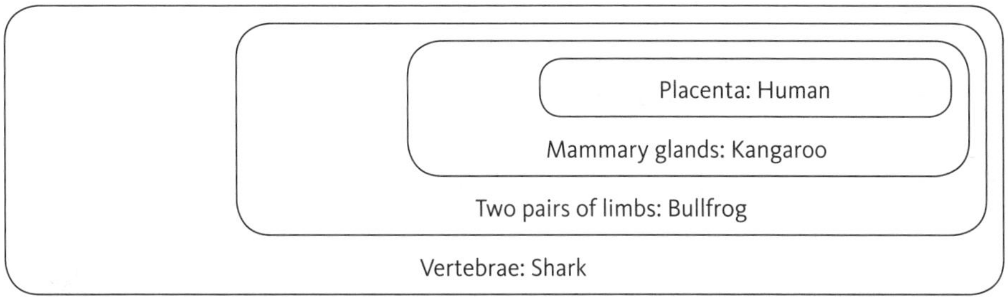

The resulting cladogram looks like this:

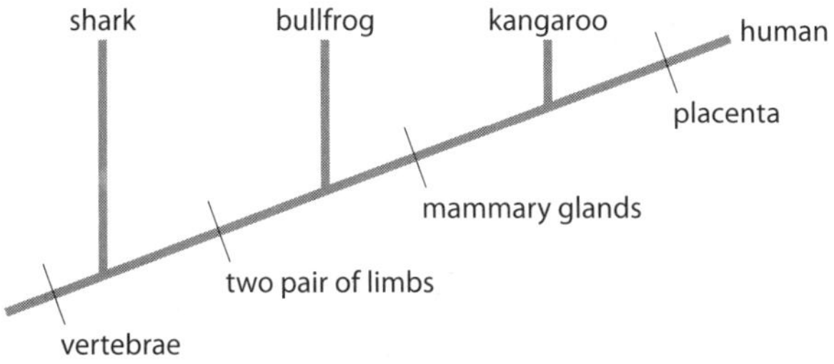

Other analyses can be really complex, but learning how to develop a phylogenetic tree just takes a little practice. Try this exercise with some evidence from stick insects using the shared differences to develop a tree.

Stick insects are an order of insects, many of which have striking adaptations in camouflage. Some species look like sticks; others look like leaves—they even "behave" like leaves, swaying as if blown by a breeze. Within the Phasmatodea, however, classification of more specific taxonomic levels is not well resolved. The Euphasmatodea is essentially a synonym for the order, and taxonomists may divide the order into two or three suborders. In fact, scientists are actively generating new hypotheses about the relationships among this interesting group, especially as new fossils are discovered that can help resolve the history of the major clades.

Here is a simplified cladogram of the extant (living species) stick insects:

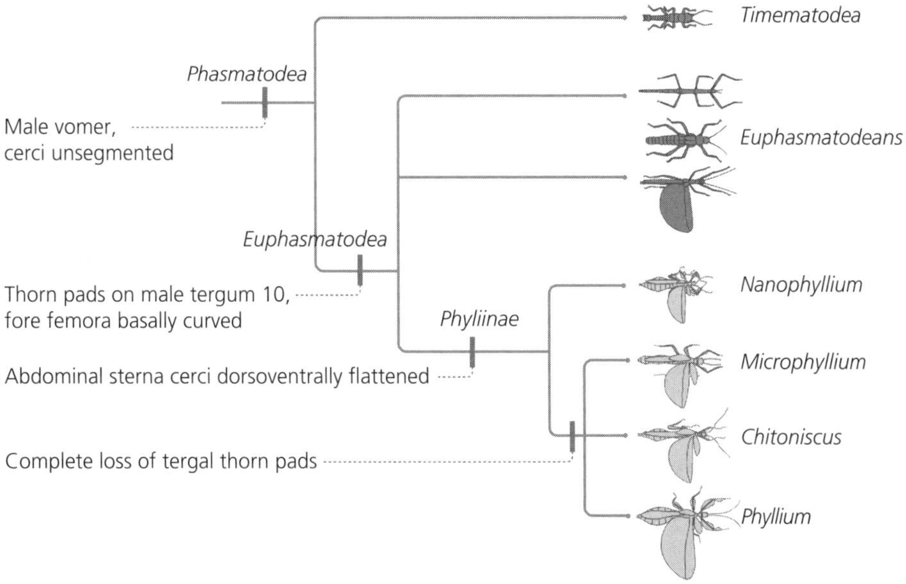

Some helpful definitions:

vomer = A hardened section of the male's 10th abdominal sternum that is used during copulation;

cerci = paired appendages extending from the anal segment of many arthropods;

tergum (plural: terga) = the dorsal part of a segment (see sternum);

femora (singular: femur) = the thigh, typically the longest segment of the insect leg; and

sternum (plural: sterna) = the ventral part of a segment (see tergum).

You can use this cladogram to read "backward" and determine the characters used to distinguish each group of stick insects. Develop a table, like the one below, that includes the character states for each taxon shown. Use 1 to represent shared derived characters and 0 to represent ancestral characters.

	Vomer Present **Cerci Unsegmented**	**Thorn Pads on Male T10** **Fore Femora Curved**	**Abdominal Sterna** **Cerci Dorsoventrally Flat**	**Loss of Tergal Thorn Pads**
Timematodea (T)				
Euphasmatodeans (E)				
Nanophyllium (N)				
Microphyllium (M)				
Chitoniscus (C)				
Phyllium (P)				

- Are the Euphasmatodea a monophyletic clade?

- Draw a Venn diagram that indicates the nested relationship among the stick insects.

Now consider some fossil evidence. *Timematodea* and *Archipseudophasmatidae* were both found in amber that dates from approximately 40 million years ago. Male *Archipseudophasmatidae* have thorn pads on their 10th terga, and the fore femora are curved—they do not have dorsoventrally flattened cerci or abdominal sterna. *Eophyllium* has abdominal sterna, but it lacks dorsoventrally flattened cerci (Wedman et al. 2007). *Eophyllium* was found in deposits that date to 47 million years ago. Where would you place these two new genera given your current understanding? Draw a new Venn diagram:

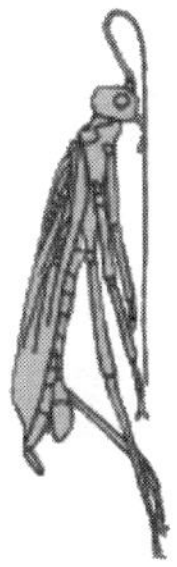

Archipseudophasmatidae

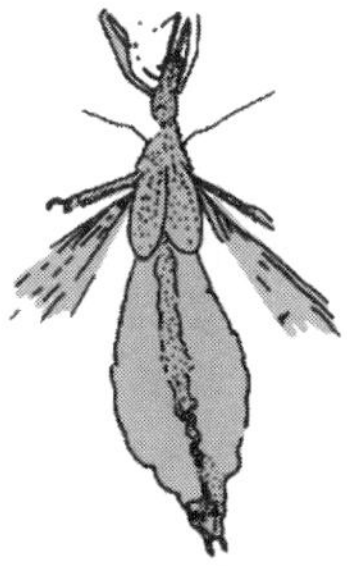

Eophyllium

- Now draw a new cladogram:

- Did the addition of *Eophyllium* clarify the resolution of the Euphasmatodea?

Carnivorans Revisited

Building phylogenies is not always straightforward. Usually, the more evidence, the more confident scientists can be about the hypothetical relationships among taxa. Figure 4.15A shows the most parsimonious phylogeny that could be developed for the carnivorans based on 12 characters. Add the character-state scoring number from the table to the phylogenetic tree below.

MORPHOLOGICAL DATA MATRIX FOR CARINVORA

Taxa	Character-State Scoring											
#	1	2	3	4	5	6	7	8	9	10	11	12
ancestral state	0	0	0	0	0	0	0	0	0	0	0	0
cat	0	1	0	1	0	0	1	1	1	0	0	0
hyena	0	1	0	1	0	0	1	0	1	0	0	0
civet	0	1	0	0	0	0	0	0	1	0	0	0
dog	1	0	0	0	1	0	0	0	0	0	0	0
raccoon	1	0	0	0	1	0	0	0	0	0	0	0
bear	1	0	0	0	1	1	0	0	0	1	0	0
otter	1	0	0	0	1	0	0	0	0	1	0	0
seal	1	0	1	0	1	1	0	0	0	1	1	1
walrus	1	0	1	0	1	1	0	0	0	1	1	1
sea lion	1	0	1	0	1	1	0	0	0	1	0	0

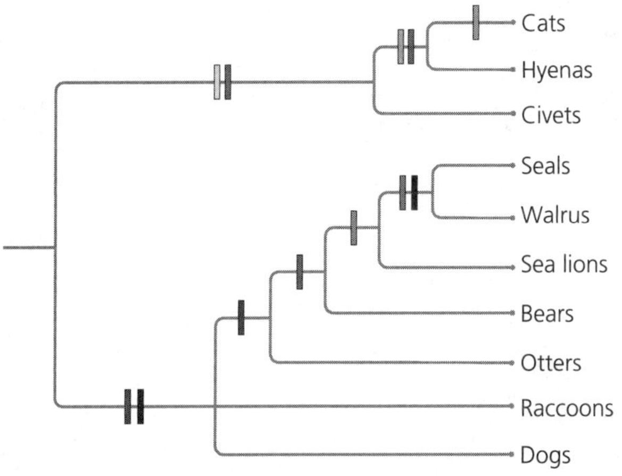

There are 34 million possible hypotheses that can be developed based on those 12 characters; the tree above represents only one possibility. So, consider how adding an additional character, such as the presence or absence of the lower premolar 1, might help scientists resolve this phylogeny. Map the character onto the phylogeny developed from the previous 12 characters.

Additional Character #13: Presence (0) or Absence (1) of Lower Premolar 1

Taxa	Character-State Scoring	Taxa	Character-State Scoring
ancestral state	0	bear	0
cat	1	otter	1
hyena	1	seal	0
civet	0	walrus	0
dog	0	sea lion	0
raccoon	0		

- Does this additional character help resolve the phylogeny? Why or why not?

- Is the presence of the lower premolar 1 a homologous trait in the carnivorans?

Tree Challenge: Develop a Phylogenetic Tree That Illustrates the Relationships among Your Shoes

Phylogenies can also be used (cautiously) to ask questions about human artifacts, like how fashion has changed over time. To keep it simple, grab one shoe from five different pairs. Identify five different characters that can be used to identify the shoes, such as presence/absence of shoe laces, sole material (natural/man-made), type of shoe (boot, sandal, tennis shoe), toe shape (round/pointed)—you will likely have to come up with your own categories depending on the shoes you choose. Develop a data matrix by listing the characters in columns and the species in rows. Then identify the differences between species pairs in another data matrix. Build the tree by plotting out the differences. It may be difficult, but that's what evolutionary biologists have to deal with every day. They make educated guesses about the relationships—the phylogenetic tree is a hypothesis that they can test with additional data.

As far as shoes go, people are notorious for borrowing other people's ideas (and ideas in general). So the observed "homology" may not really result from "descent"—it might result because one lineage (i.e., one fashion designer) took an idea from another lineage—something akin to horizontal gene transfer (see Box 4.3). Do you think "borrowing" might influence the relationships in the tree?

Explore!

Why Study the Tree of Life?

Why Study the Tree of Life is a 4-minute video produced by the Yale Peabody Museum of Natural History that highlights the role phylogenies play in understanding and responding to challenges we face on this planet, from increasing yields of our food crops to protecting endangered species to curing disease. The site also has a variety of information and other links.

http://archive.peabody.yale.edu/exhibits/treeoflife/film_study.html

What Did T. rex Taste Like?

Jennifer Johnson Collins, Judy Scotchmoor, and Caroline Strombergfrom the University of California Museum of Paleontology created this extensive look into how cladograms can be used to ask interesting and bizarre questions. Going through the exploration takes a bit of time, but wouldn't it be cool to tell your friends you know what *Tyrannosaurus rex* tasted like?

http://www.ucmp.berkeley.edu/education/explorations/tours/Trex/index.html

Cladograms

In this video, Paul Andersen shows how to construct a simple cladogram from a group of organisms using shared characteristics.

https://www.youtube.com/watch?v=ouZ9zEkxGWg

ScienceShot: Stubby-Tailed Dinosaurs Shook Their Thing

This site shows new research about the function of feathers in early theropod dinosaurs.

http://news.sciencemag.org/2013/01/
scienceshot-stubby-tailed-dinosaurs-shook-their-thing

Tree Thinking Challenge

Also from *Science*, David A. Baum, Stacey DeWitt Smith, and Samuel S. Donovan developed two quizzes designed to help you understand the phylogenetic relationships.

http://www.sciencemag.org/content/suppl/2005/11/07/310.5750.979.DC1/
Baum.SOM.pdf

Overcoming Misconceptions

Reading phylogenetic trees can be difficult because they are not necessarily intuitive. Misunderstanding is easy. Following are several common misconceptions related to reading trees.

The Flow of Time

Trees can be depicted a number of different ways. In all of these figures, A–D represent the species or groups of interest, and the nodes and branches show their historical relationships.

In this style, time is represented from bottom to top—the extant (living) species or groups are listed at the top, and each successive node down represents an earlier and earlier common ancestor.

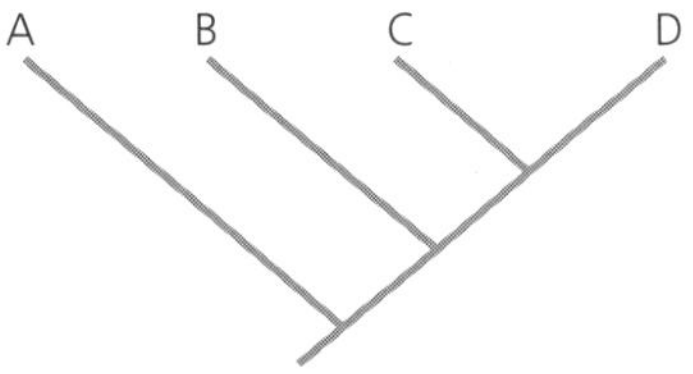

In this style, time is just the opposite—the extant (living) species or groups are listed at the bottom, so each successive node up represents an earlier and earlier common ancestor.

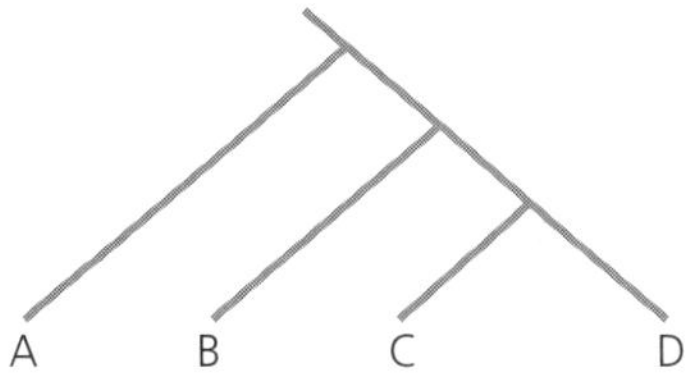

In this style, the extant (living) species or groups are listed at the right, so each successive node to the left represents an earlier and earlier common ancestor.

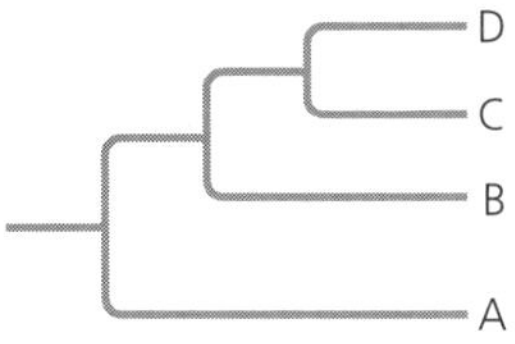

And in this style, time is depicted as concentric circles. Extant species are on the outside, and earlier and earlier common ancestors can be found closer and closer to the center.

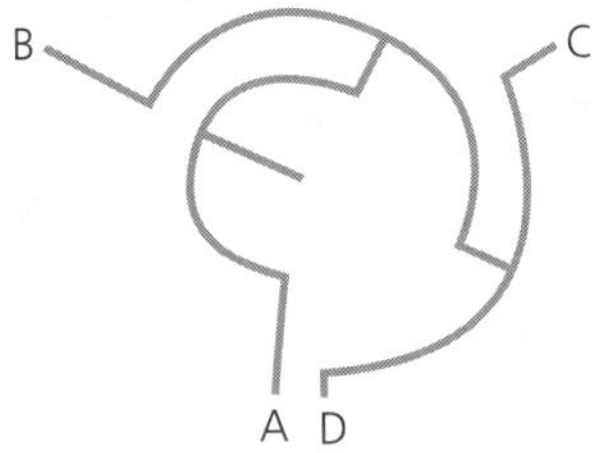

Tips and Relatedness

The tips of phylogenetic trees aren't locked in to where they appear on the page. Relatedness comes from the shared common ancestors (i.e., the nodes). Think of the tree as a mobile, with each group of branches spinning freely at the nodes. Do these two trees illustrate the same relationship or not?

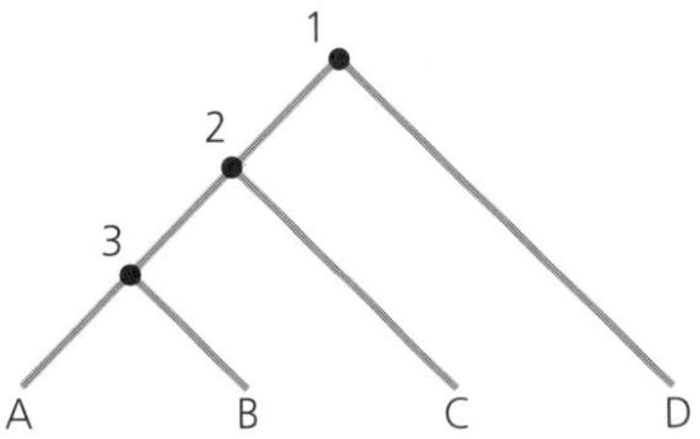

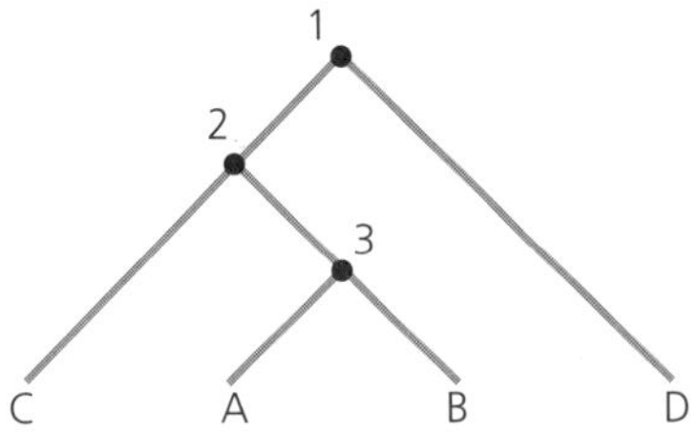

Is C more closely related to D or to A?

Node 1 represents the hypothetical common ancestor of all four groups (A-D). Node 2 represents the hypothetical common ancestor of groups A–C, and node 3 represents the hypothetical common ancestor of A and B. C shares a more recent common ancestor with A (node 2) than it does with D (node 1), so C is more closely related to A.

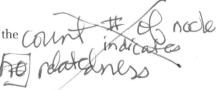

Counting the number of nodes does not indicate relatedness either. C is not more closely related to A because only two nodes separate them. Again, think of the nodes as points that can swing freely.

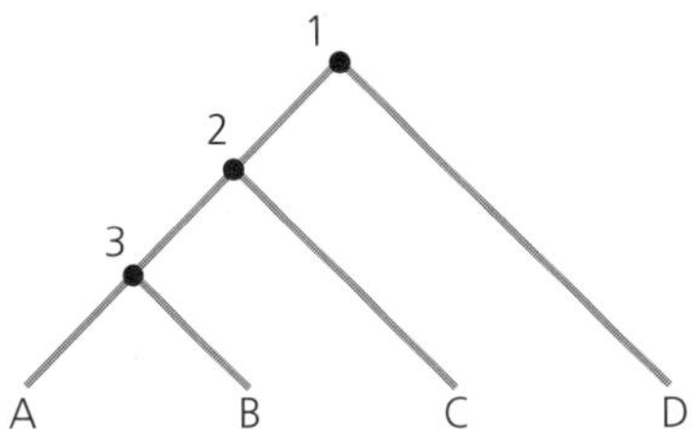

Here, only two nodes separate C from A and C from D.
And here,

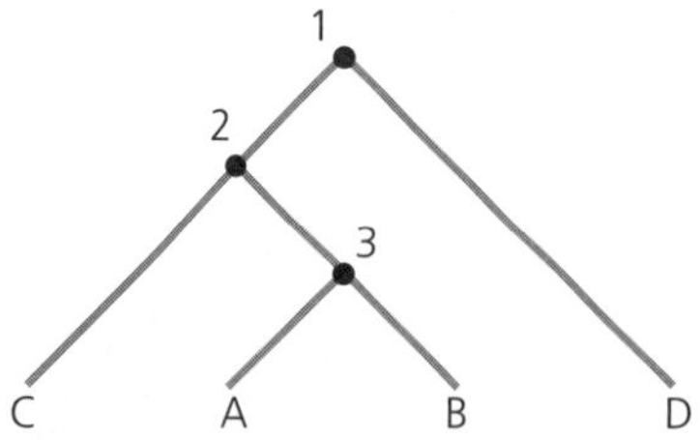

two nodes separate C from A. But C is more closely related to A because it shares a more recent common ancestor with A than it does with D.

Straight Lines and Change

A straight line doesn't mean that a species has not changed—the lines are an artifact of the graphic representation. Take this tree:

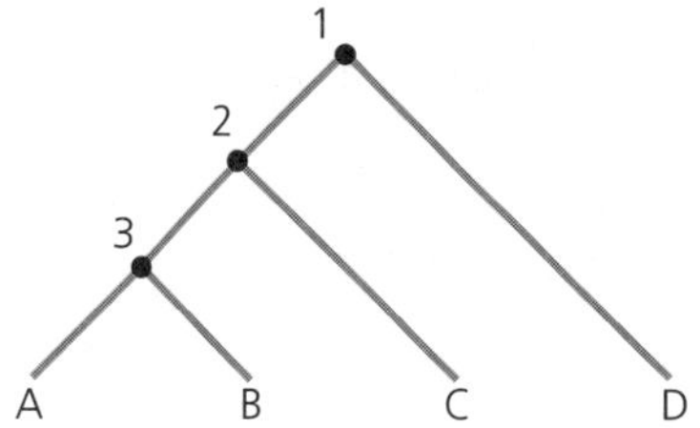

Just because the line to A is the straight line with other lines branching off it, doesn't mean that A didn't change and B–D did. Because the branches can spin freely, A doesn't have to be depicted as a straight line:

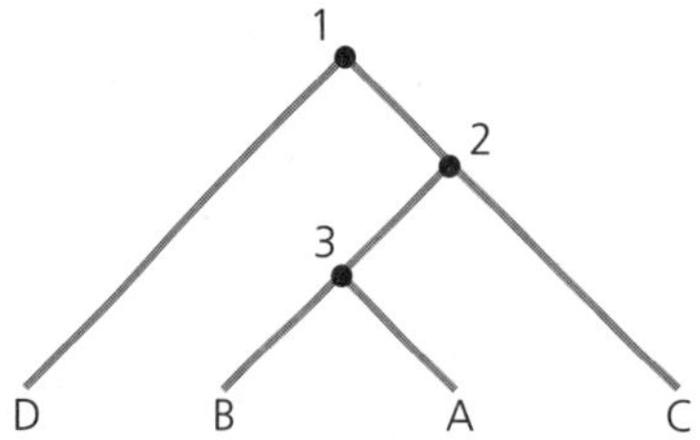

The important thing to understand is that nodes represent the hypothetical common ancestors of the branches, and common ancestors are found earlier and earlier in time.

"Lower" and "Primitive"

Lineages that branch off earlier in the tree are not "lower" or "primitive" forms (see Box 4.1). So D is not more primitive (or higher or advanced) than A. D simply shares an earlier common ancestor with A than B or C.

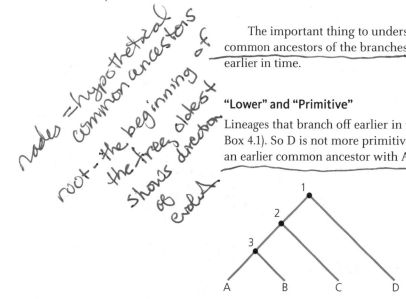

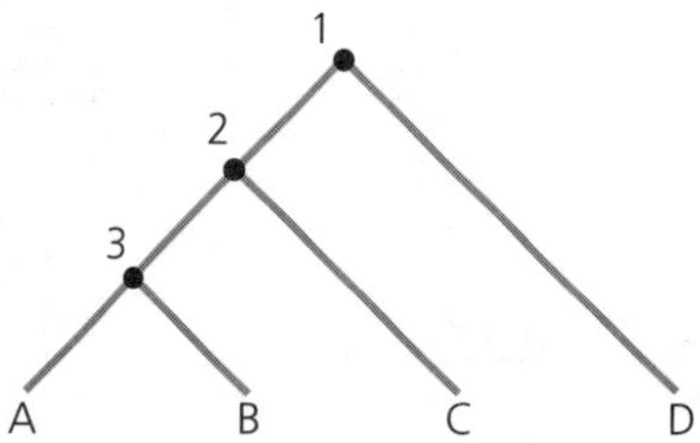

Missing Links

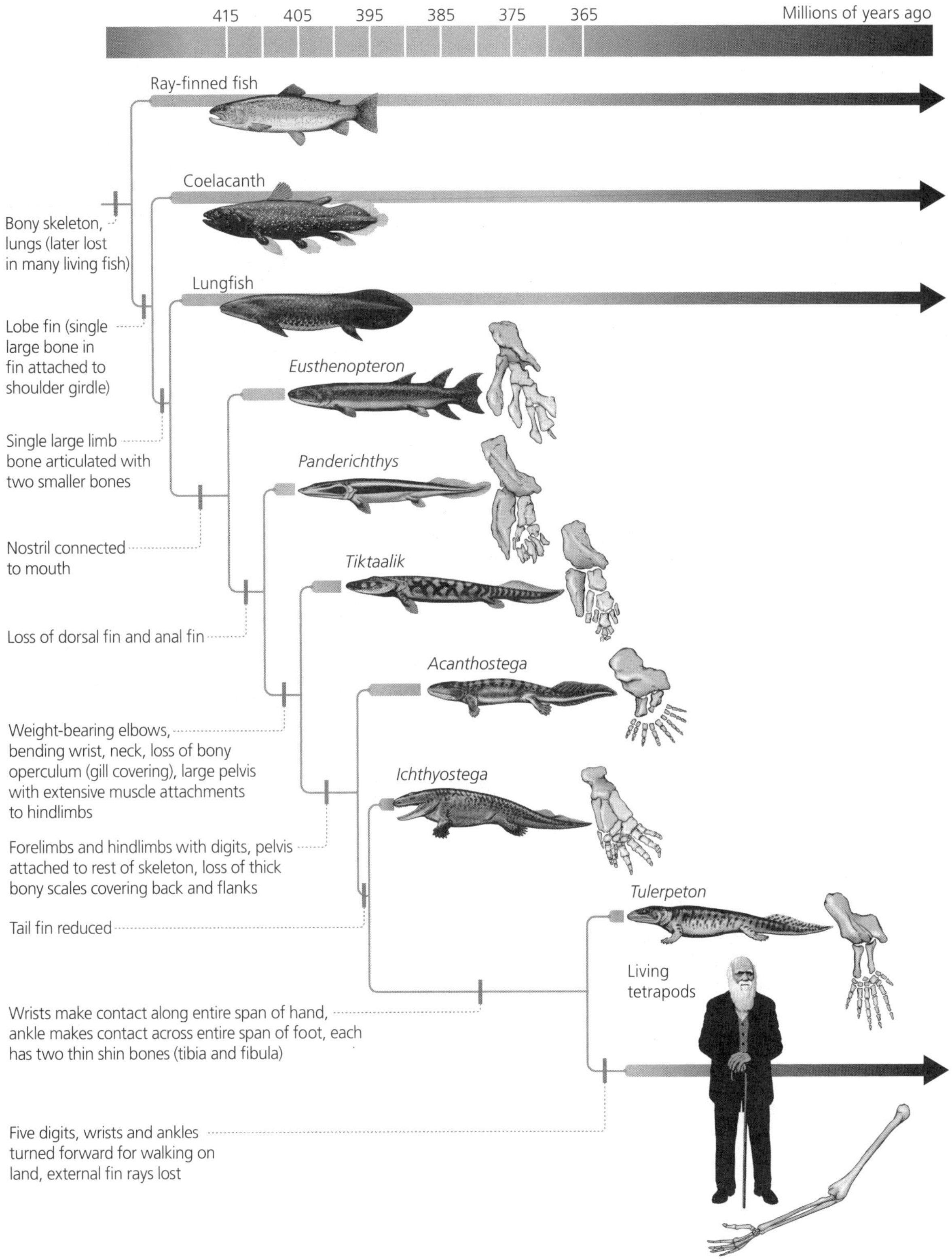

This tree shows the relationship of lobe-fins to tetrapods and how new tetrapod traits evolved over time. The tetrapod "body plan" evolved gradually, over perhaps 40 million years. The earliest tetrapods probably still lived mainly underwater. This tree includes only a few representative species; paleontologists have discovered many others that provide even more detail about this transition from sea to land.

In the phylogeny of tetrapods shown in Figure 4.22, the branches represent hypothesized lineages. Yet all of the fossils that paleontologists have placed in this phylogeny are located at the tips of branches, rather than along the branches themselves. This may be puzzling at first. If we can trace our ancestry back to Devonian lobe-fins, then why can't paleontologists identify any of our direct ancestors? Why can't they find the so-called missing link?

The answer has to do with the nature of paleontology. There are over 50,000 known species of vertebrates alive today. Some 360 million years ago, there was likely a comparable diversity. But since living tetrapods are a monophyletic group, they must have descended from a single ancestral species. Paleontologists have managed to find the fossils of only a small fraction of those species. The odds that they will find the direct ancestor of living tetrapods are extremely small.

When Neil Shubin and other paleontologists examine the fossils of early tetrapods, such as *Tiktaalik*, they find many synapomorphies uniting them in a clade with living tetrapods. But they also find other synapomorphies that are not found in other tetrapods. The result is a phylogeny that represents both of those relationships—the characters shared with all tetrapods, represented by the common ancestor, and the evolution of those distinct lineages, represented by each of the branches.

Newspaper headline writers are fond of the term *missing link,* but the fact is that paleontologists do not search for missing links in the sense of a direct line of evolutionary history. Instead, they look for fossils that can help resolve uncertainties in phylogenies and that can help resolve the history by which characters evolved in major clades. In this sense, paleontologists have found an abundance of links that have added to our understanding of the history of life.

Which of the following is a true statement?

a. The fewer the nodes between species, the more related to each other they are.

b. Straight lines in phylogenies indicate species that have not evolved.

c. Lineages that branch off later are more advanced than lineages that branch off earlier.

d. Scientists are not looking for missing links.

Go the Distance: Examine the Primary Literature

Bradley Livezey and Richard Zusi developed a phylogeny of modern birds to help resolve some of the conflicting evidence for how birds are related to each other—a heated topic among birders around the world. Many questions still remain, but for now, the birding industry has some restructuring to do in bird books.

- Why did Livezey and Zusi argue that avian systematics needed to be examined?

- Scientists use "outgroups" to determine what characters are "primitive" in the analysis; the organisms being classified are considered the "ingroup." Why did Livezey and Zusi use the Theropoda as an outgroup?

Livezey, B. C., and R. L. Zusi. 2007. Higher-Order Phylogeny of Modern Birds (Theropoda, Aves: Neornithes) Based on Comparative Anatomy. II. Analysis and Discussion. *Zoological Journal of the Linnean Society* 149 (1): 1–95. doi:10.1111/j.1096-3642.2006.00293.x. http://www.ncbi.nlm.nih.gov/pubmed/18784798.

Delve Deeper

1. Why are phylogenies based on shared derived character? What does including an outgroup do for understanding those relationships?

2. Do you agree with how time is portrayed in this phylogeny? Why or why not?

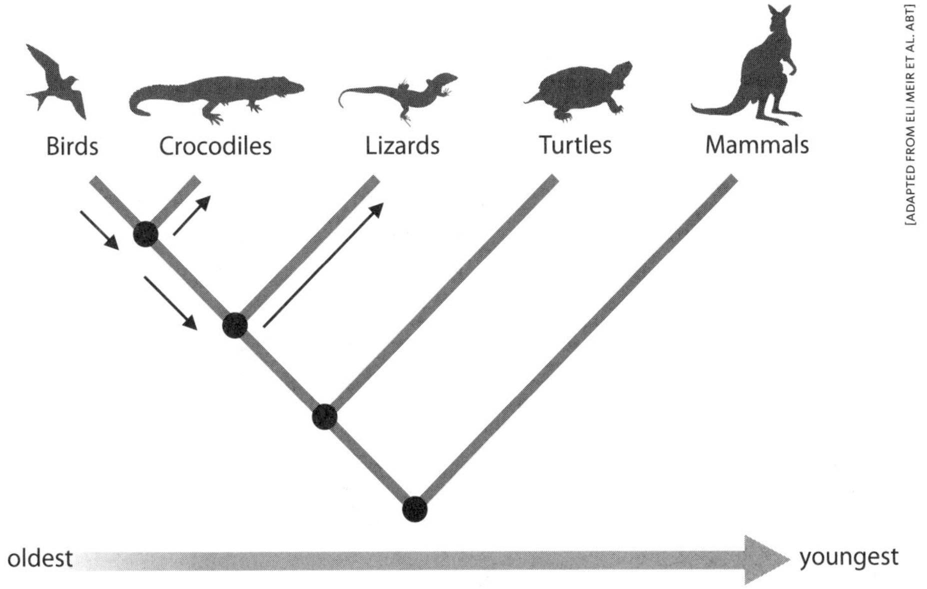

3. Scientists have been examining the correlation between brain size and basal metabolic rate. Some scientists have suggested that basal metabolic rate influences brain size because the brain tissue is metabolically active and costly to run. Maternal investment, either through gestation or lactation, can influence the transfer of metabolic energy during growth and development of offspring. Therians provide a unique opportunity to test hypotheses about basal metabolic rate because eutherians invest in gestation (by providing a placenta to feed developing offspring), and marsupials invest in lactation. How might tests of these hypotheses be influenced by phylogenetic relationships? What might be the effect on the relationship between basal metabolic rate and brain size if one group, the primates (eutherians), have inordinately large brain sizes?

   Read Weisbecker, V. and A. Goswami. 2012. Brain size, life history, and metabolism at the marsupial/placental dichotomy. Proceedings of the National Academy of Sciences (http://www.pnas.org/content/early/2010/08/31/0906486107.abstract). What did the authors find?

Test Yourself

1. Which of the following is NOT a monophyletic group?
    a. A clade
    b. Reptilia
    c. Aves
    d. Amphibia
    e. All of the above are monophyletic groups.
2. Based on your understanding of Carl Linnaeus' classification system, how would you treat his conclusions?
    a. His concept of nested hierarchies accurately characterized and organized species relationships, so his conclusions are valid.
    b. His concept of nested hierarchies doesn't reflect modern understanding of species relationships, so Linnaean classification has no place in modern science.
    c. His classification scheme is based on evidence that has been refuted with modern techniques and should be ignored.
    d. His classification scheme provides a useful convention for naming species and thinking about species relationships, so his conclusions could be used as a starting point for examining taxonomic relationships.
    e. His classification scheme provides a useful convention for naming species, so the taxonomic units he named should be maintained.

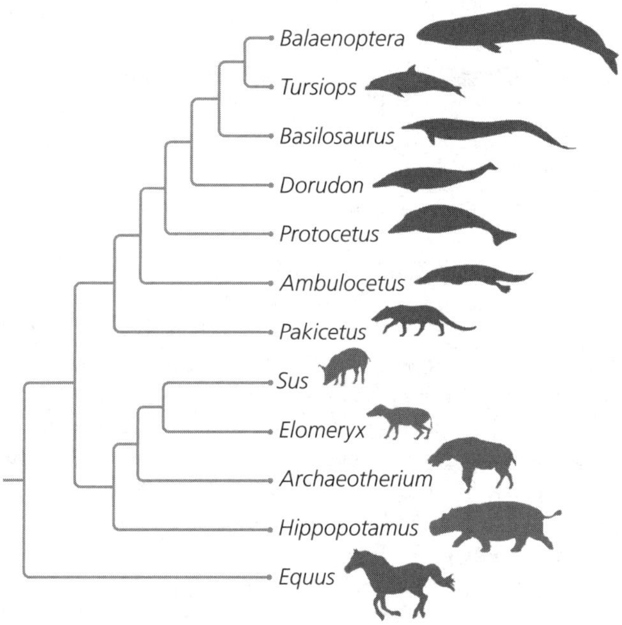

3. In the past, morphological evidence alone was used to examine the relationships of the cetaceans to other groups. According to this phylogeny:
   a. The extinct genus *Ambulocetus* is more distantly related to the genus *Balaenoptera* than to the extinct genus *Protocetus*.
   b. The genus *Equus* (horses) is more closely related to the genus *Hippopotamus* (hippos) than to the extinct genus *Pakicetus*.
   c. More changes have occurred in the whale lineage than in the horse lineage.
   d. Horses (genus *Equus*) were the ancestor of whales and hippos.
   e. None of the above are correct statements.
   f. All of the above are correct statements.
4. How does including fossils in phylogenies of extant taxa affect the conclusions scientists can draw?
   a. Including fossils can change the hypothesis generated by the phylogeny.
   b. Including fossils can define the timing of branching events.
   c. Including fossils can affect understanding of common ancestors.
   d. Including fossils can generate new questions about clades.
   e. All of the above
5. Which group of theropod dinosaurs did NOT have feathers according to the evogram in Figure 4.29?
   a. Allosaurids
   b. Compsognathids
   c. Tyrannosauroids
   d. Oviraptorosaurs
   e. None of the theropod dinosaurs had feathers

Contemplate

What questions might you ask about fashion using a phylogenetic analysis?	**Can technology, such as the evolution of video games, be examined using phylogenies?**

5 Raw Material
Heritable Variation among Individuals

Check Your Understanding

1. Who incorporated the idea that the theory of evolution required the capacity for one generation to pass on its traits to the next?
   a. Charles Darwin
   b. Jean-Baptiste Lamarck
   c. Alfred Russel Wallace
   d. All of the above
   e. None of the above
2. Why is heritable variation among individuals an important factor for natural selection?
   a. Because variation has to be heritable for a species to survive
   b. Because variation has to be heritable for individuals to pass down their beneficial mutations
   c. Because when individuals respond to the environment, they can pass the traits they acquired on to offspring
   d. Because natural selection cannot act when all individuals are absolutely identical.
   e. Both b and d
3. How can the phylogeny on the left be used to understand the phylogeny on the right?

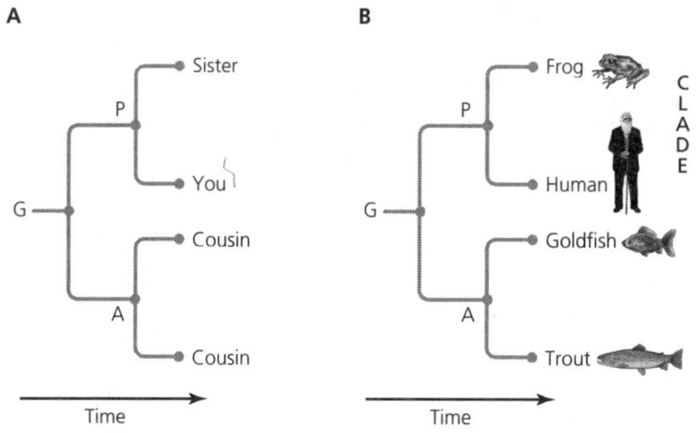

   a. The phylogeny on the left shows that you and your sister are more closely related to each other than you are to your cousins, just as humans and frogs are more closely related to each other than they are to goldfish or trout.
   b. The phylogeny on the left shows that the relationship between humans and frogs cannot be compared with the relationship between siblings.
   c. The phylogeny on the left shows that your cousins must be more closely related to each other than you and your sister because trout and goldfish are more closely related to each other than frogs and humans.
   d. The phylogeny on the right shows that humans are more closely related to goldfish than to trout.

Learning Objectives for Chapter 5

Add important definitions and notes next to each learning objective for this chapter to help guide your understanding.

Learning Objective	Important Definitions	Notes
Describe the structure of proteins.		
Describe the organization of DNA within a human cell.		
Compare and contrast the events that occur in transcription and translation.		
Describe three mechanisms that influence gene expression.		
Explain the difference between coding and noncoding segments of DNA.		
Differentiate between somatic mutations and germ-line mutations and their roles in variation within a population		
Explain why independent assortment and genetic recombination are important to evolution.		
Discuss the complex relationship between genotypes and phenotypes.		
Analyze the role of the environment in gene expression.		

Identify Key Terms

Match terms and definitions by filling in the blank to the right of the term with the appropriate letter.

1. **Allele** m
2. **Alternative splicing** o
3. **Amino acids** e
4. **Autosomes** ag
5. **Base** c
6. **Cis-acting elements** z
7. **Dominant allele** al
8. **Enhancers** j
9. **Epigenetic** w
10. **Gene** as
11. **Gene control region** aj
12. **Gene expression** k
13. **Genetic polymorphism** am
14. **Genetic recombination** f
15. **Genome** q
16. **Genotype** x
17. **Germ-line mutations** an
18. **Hormones** af
19. **Independent assortment** n
20. **Meiosis** i
21. **Messenger RNA (mRNA)** y
22. **MicroRNA** b
23. **Mobile genetic elements** v
24. **Morphogen** ____
25. **Mutation** ak
26. **Nucleotides** l
27. **Phenotype** t
28. **Phenotypic plasticity** ae
29. **Plasmids** g
30. **Ploidy** ab
31. **Polyphenism** aa
32. **Pseudogenes** ao
33. **Quantitative traits** s
34. **Recessive allele** ad
35. **Repressors** aq

a. Sequences of DNA located away from the focal gene (e.g., on another chromosome). These stretches of DNA generally code for a protein, microRNA, or other diffusible molecule that then influences expression of the focal gene.

b. One group of RNAs that act as post-transcriptional regulators of gene expression. This type of RNA binds to complementary sequences on specific mRNAs and can enhance or silence the translation of genes. The human genome encodes more than 1000 of these tiny RNAs.

c. One of four nitrogen-containing molecules in DNA: adenine (A), cytosine (C), guanine (G), and thymine (T). In RNA, uracil (U) replaces T.

d. The process of modifying RNA after transcription but before translation, during which introns are removed and exons are joined together into a contiguous strand.

e. The structural units that, among other functions, link together to form proteins.

f. The exchange of genetic material between paired chromosomes during meiosis. This process can form new combinations of alleles and is an important source of heritable variation.

g. Molecules of DNA, found most often in bacteria, that can replicate independently of chromosomal DNA

h. The process that takes place when a strand of mRNA is decoded by a ribosome to produce a strand of amino acids.

i. A form of cell division that occurs only in eukaryotes, in which the number of chromosomes is cut in half. This process gives rise to gametes or spores and is essential for sexual reproduction.

j. Short sequences of DNA within the gene control region where activator proteins bind to initiate gene expression.

k. The process by which information from a gene is transformed into a product.

l. The structural units that link together to form DNA (and RNA). Each one includes a base.

m. One of any number of alternative forms of the DNA sequence of the same locus.

n. The random mixing of maternal and paternal copies of each chromosome during meiosis, resulting in the production of genetically unique gametes.

o. The process of combining different subsets of exons together, yielding different mRNA transcripts from a single gene.

p. Mutations that affect cells in the body ("soma") of an organism. These mutations affect all the daughter cells produced by the affected cell and can affect the phenotype of the individual. In animals, these mutations are not passed down to offspring. In plants, they can be passed down during vegetative reproduction.

q. All the hereditary information of an organism. The genome comprises the totality of the DNA, including the coding and noncoding regions.

r. Proteins that bind to specific DNA sequences and act, in essence, like a light switch by turning all the sequences on or off simultaneously.

s. Measurable phenotypes that vary among individuals over a given range to produce a continuous distribution of phenotypes. Quantitative traits are sometimes called complex traits; they're also sometimes called polygenic traits because their variation can be attributed to polygenic effects (i.e., the cumulative action of many genes).

t. An observable, measurable characteristic of an organism. It may be a morphological structure (e.g., antlers, muscles), a developmental process (e.g., learning), a physiological process or performance trait (e.g., running speed), or a behavior (e.g., mating display). They can even be the molecules produced by genes (e.g., hemoglobin).

u. The process that takes place when RNA polymerase reads a coding sequence of DNA and produces a complementary strand of RNA, called messenger RNA (mRNA)

v. Types of DNA that can move around in the genome. Common examples include transposons ("jumping genes") and plasmids

36. **RNA (ribonucleic acid)** ar
37. **RNA polymerase** ap
38. **RNA splicing** d
39. **Sex chromosomes** ac
40. **Somatic mutations** ____
41. **Trans-acting elements** ____
42. **Transcription** u
43. **Transcription factors** ____
44. **Transfer RNA (tRNA)** ai
45. **Translation** h
46. **Vertical gene transfer** ah

w. The functional modifications to DNA that don't involve changes to the sequences of nucleotides. Epigenetics is the study of the heritability of these modifications.

x. The genetic makeup of an individual. Although it includes all the alleles of all the genes in that individual, the term is often used to refer to the specific alleles carried by an individual for any particular gene.

y. Molecules of RNA that carry genetic information from DNA to the ribosome, where it can be translated into protein

z. Stretches of DNA located near a gene—either immediately upstream (adjacent to the promoter region), downstream, or inside an intron—that influence the expression of that gene. These regions often code for binding sites for one or more transposable factors.

aa. A trait for which multiple, discrete phenotypes can arise from a single genotype depending on environmental circumstances

ab. The number of copies of unique chromosomes in a cell (n). Normal human somatic cells are diploid (2n); they have two copies of 23 chromosomes.

ac. Chromosomes that pair during meiosis but differ in copy number between males and females. For organisms such as humans with XY sex determination, X and y are these chromosomes. Females are the homogametic sex (XX) and males are the heterogametic sex (XY).

ad. An allele that produces its characteristic phenotype only when it is paired with an identical allele (i.e., in homozygous states)

ae. Changes in the phenotype produced by a single genotype in different environments

af. Molecular signals that flow from cells in one part of the body to cells in other parts of the body. They act directly or indirectly to alter expression of target genes.

ag. Chromosomes that do not differ between sexes

ah. The process of receiving genetic material from an ancestor

ai. A short piece of DNA that physically transfers a particular amino acid to the ribosome

aj. An upstream section of DNA that includes the promoter region as well as other regulatory sequences that influence the transcription of DNA cis/trans

ak. Any change to the genomic sequence of an organism

al. An allele that produces the same phenotype whether it is paired with an identical allele or a different allele (i.e., a heterozygotic state)

am. The simultaneous occurrence of two or more discrete phenotypes within a population. In the simplest case, each phenotype results from a different allele or combination of alleles of a single gene. In more complex cases, the phenotypes result from complex interactions between many different genes and the environment.

an. Mutations that affect the gametes (eggs, sperm) of an individual and can be transmitted from parents to offspring. Because they can be passed on, these mutations create the heritable genetic variation that is relevant to evolution.

ao. DNA sequences that resemble functional genes but have lost their protein-coding ability or are no longer expressed. They often form after a gene has been duplicated, when one or more of the redundant copies subsequently lose their function.

ap. The enzyme that builds the single-stranded RNA molecule from the DNA template during transcription

aq. Proteins that bind to a sequence of DNA or RNA and inhibit the expression of one or more genes

ar. An essential macromolecule for all known forms of life (along with DNA and proteins). It differs structurally from DNA in having the sugar ribose instead of deoxyribose and in having the base uracil (U) instead of thymine (T).

as. A segment of DNA whose nucleotide sequences code for proteins, or RNA, or regulate the expression of other genes

at. A signaling molecule that flows between nearby cells and acts directly to alter expression of target genes

Link Concepts

Fill in the bubbles with the appropriate terms.

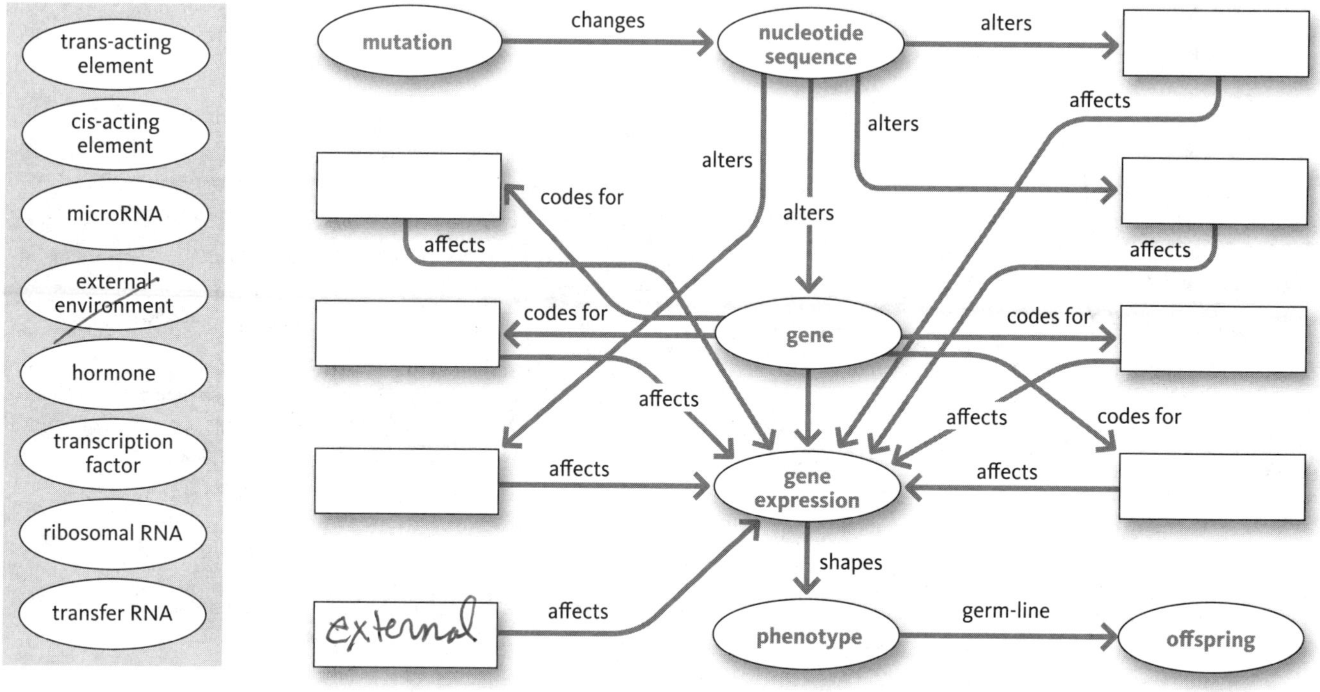

Key Concepts

Fill in the blanks for the key concepts from the chapter.

5.1 Evolution's Molecules: Proteins, DNA, and RNA

____________ serve a variety of functions within an organism, and ____________ their ____________ can affect cell structure, the ability to catalyze chemical reactions, the ability to carry information from cell to cell, or even the ability to respond to another signal molecule.

Because of its structural ____________ , DNA functions as the basis of the system that encodes and replicates information necessary to build life. Although the ____________ process is astonishingly faithful, ____________ occasionally lead to variation among individuals.

The mRNA carries genetic information used to encode ____________ , so any mistake introduced during copying can ____________ the proteins that are produced.

Mutations affecting ____________ forms of RNA, such as ribosomal RNA, transfer RNA, and microRNA, can affect the ____________ and ____________ of genes.

The genomes of most organisms are ____________ just protein-coding sequences; they often contain a diversity of ____________ elements, including pseudogenes and mobile genetic elements, that reflect the ____________ of the organism.

5.2 Mutations: Creating Variation

Mutations (in germ-line cells) are not ____________ , especially mutations that affect large parts of the genome, but they can accumulate ____________ over time. Within populations mutations comprise the ultimate source of ____________ , the raw material that is essential for evolution.

Because information in human chromosomes is present in ____________ , a deleterious mutation affecting gene expression or activity in one chromosome may be ____________ by the presence of a ____________ copy of the gene in the other chromosome. (This is not true for haploid organisms like bacteria, which have only a single copy of their genome.)

Genetic changes in ____________ arise when mutations outside of the protein-coding regions affect ____________ much a gene is transcribed.

Changes in ____________ of gene expression can have profound consequences for ____________ by adding another component to heritable genetic variation.

5.3 Heredity

Because of ______________ and the ______________ of chromosomes, meiosis can generate extraordinary genetic __Variation__ among gametes in sexually reproducing organisms.

The fusion of __egg__ and __sperm__ results in great genetic diversity among offspring, even from the same parents.

5.4 The Complex Link between Most Phenotypes and Genotypes

__Phenotype__ are a result of both an organism's genotype and the environment.

______________ often result from developmental threshold mechanisms, where organisms respond to a critical level of some ______________ cue (e.g., photoperiod or temperature) by switching from production of a default phenotype to that of an __alternative__ phenotype.

______________ varying traits are called quantitative traits, and the study of their inheritance and evolution is called __evolutionary__ genetics.

Evolutionary biologists study __genes__ in the expression of phenotypic traits.

5.5 How Do Genes Respond to the Environment?

The expression of __genes__ is often influenced by signals from the ______________. This can cause gene activity to be ______________ with particular developmental or ecological contexts.

Interpret the Data

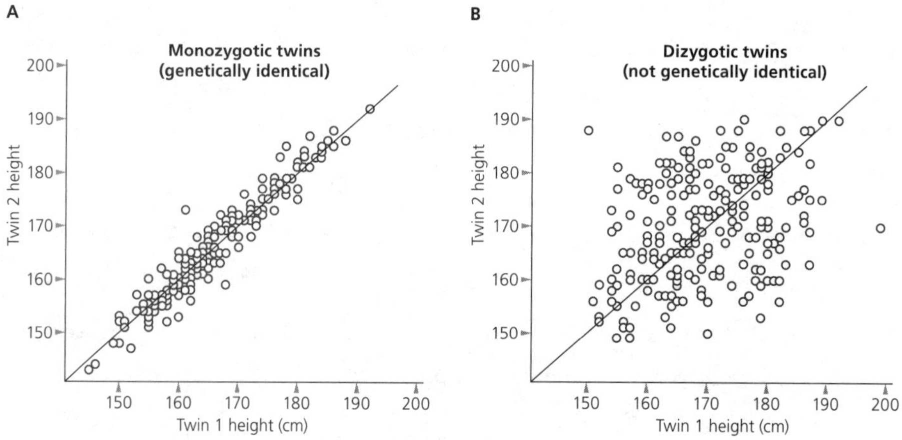

Figures 5.21 A and B show the relationship between the heights of identical and fraternal twins. One twin's height is marked along the x-axis, and his or her twin sibling's height is marked on the y-axis. (Data courtesy of David Duffy.)

- Can you predict the height of Twin 2 when twins are identical? For example, what would you expect the height of Twin 2 to be if the height of Twin 1 is 160 cm?

- Is there variation in the heights of identical twins?

- Why don't the heights of identical twins match exactly?

- Do you see any correlation in height between fraternal twins?

- What might be contributing to the variation between fraternal twins?

Games and Exercises

Turning the information coded in DNA into proteins is an amazing process. A messenger RNA (mRNA) molecule is transcribed from a gene, and a ribosome uses that mRNA to assemble a protein. The ribosome reads three bases (a codon) in the mRNA at a time, each matching one of 20 amino acids. Use the chart below to transcribe the two strands of DNA to mRNA and add the amino acids by codon.

	2nd base in Codon				
1st base in Codon	U	C	A	G	3rd base in Codon
U	Phe	Ser	Tyr	Cys	U
	Phe	Ser	Tyr	Cys	C
	Leu	Ser	**STOP**	**STOP**	A
	Leu	Ser	**STOP**	Trp	G
C	Leu	Pro	His	Arg	U
	Leu	Pro	His	Arg	C
	Leu	Pro	Gln	Arg	A
	Leu	Pro	Gln	Arg	G
A	Ile	Thr	Asn	Ser	U
	Ile	Thr	Asn	Ser	C
	Ile	Thr	Lys	Arg	A
	Met	Thr	Lys	Arg	G
G	Val	Ala	Asp	Gly	U
	Val	Ala	Asp	Gly	C
	Val	Ala	Glu	Gly	A
	Val	Ala	Glu	Gly	G

Strand 1: CACGTGGACTGAGGACTC

Strand 2: CACGTGGACTGAGGACAC

- What's the difference between the two strands?
- What's the difference between the amino acid sequences that result?
- If the mutation in Strand 2 was at the third base in that codon instead of the second, would the hemoglobin protein form normally?

HBB is the gene associated with sickle-cell anemia. Scientists have pinpointed the location of this gene on Chromosome 11 in humans. They have determined that 444 base pairs within the mRNA code for the amino acid sequence of the gene's protein product, hemoglobin, and that protein itself is 146 amino acids long (the mRNA includes 2 extra base pairs). Several hundred variations of the HBB gene are known, but sickle-cell anemia is most commonly caused by the hemoglobin variant Hb S. This variant made in the bodies of people with sickle-cell disease (Hb S) differs from normal hemoglobin (Hb A) in just one amino acid of the HBB polypeptide chain. In normal hemoglobin, this amino acid is glutamic acid. In sickle-cell hemoglobin, it is a valine.

ACAGATATA is a board game that simulates the relationship between randomness and biodiversity. You can learn about gene expression and molecular evolution with a competitive edge! (See Eterovic and Santos 2013; http://www.evolution-outreach.com/content/6/1/22.)

Smiley Face Trait Inheritance

Welcome to The Science Spot

Here's a fun way to think about traits and inheritance adapted from T. Tomm (2003) and http://sciencespot.net/.

You'll need a nickel and a dime—the nickel will represent the female parent and the dime the male parent. Both parents are heterozygous for all the Smiley Face traits. Flip both coins for each trait in the table below. If the coin lands with heads up, that parent contributes a dominant allele for that trait. If the coin lands tails up, that parent contributes a recessive allele. Circle the allele the offspring receives from each parent for each trait by circling the corresponding letter. Use the results and the Smiley Face Traits chart on the next page to fill in the genotype and phenotype for each trait. Then, create a sketch of your smiley face.

Trait	Female		Male		Genotype	Phenotype
Face shape	C	c	C	c		
Eye shape	E	e	E	e		
Hair style	S	s	S	s		
Smile	T	t	T	t		
Ear style	V	v	V	v		
Nose style	D	d	D	d		
Face color	Y	y	Y	y		
Eye color	B	b	B	b		
Hair length	L	l	L	l		
Freckles	F	f	F	f		
Nose color	R	Y	R	Y		
Ear color	P	T	P	T		

Face Shape

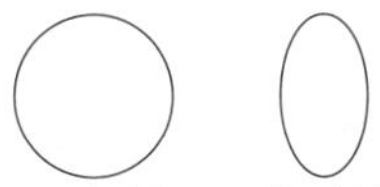

Circle (C) Oval (c)

Nose Style

Down (D) Up (d)

Eye Shape

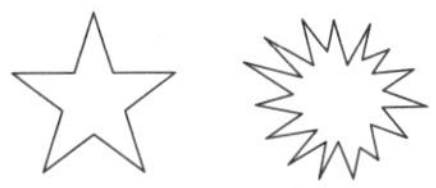

Star (E) Blast (e)

Face Color

Yellow (Y)
Green (y)

Face Color

Blue (B)
Red (b)

Hair Style

Straight (S) Curly (s)

Hair Length

Long (L)
Short (l)

Hair Length

Blue (B)
Red (b)

Smile

Thick (T) Thin (t)

Nose Color

Red (RR)
Orange (RY)
Yellow (YY)

Ear Color

Hot Pink (PP)
Purple (PT)
Teal (TT)

Ear Style

Curved (V) Pointed (v)

Sex

To determine the sex, flip the coin for the male parent. Heads equals X and tails equals Y.

XX – Female – Add pink bow in hair
XY – Male – Add blue bow in hair

Extract Your Own DNA

Anna Rothschild produced this short video for *NOVA* showing how you can extract your own DNA using bottled water, clear dish soap, food coloring, table salt, and 70 percent isopropyl alcohol.

http://www.pbs.org/wgbh/nova/body/extract-your-dna.html

Explore!

WEHI.TV

The Walter and Eliza Hall Institute produces awesome animations that illustrate scientific concepts that are difficult to observe. The following links illustrate transcription and translation.

http://www.wehi.edu.au/education/wehitv/dna_central_dogma_part_1_-_transcription/

http://www.wehi.edu.au/education/wehitv/dna_central_dogma_part_2_-_translation/

The entire 8-minute video by Drew Berry explaining DNA transcription using amino acids to turn genes into proteins can be found here on YouTube.

http://www.youtube.com/watch?v=TSv-Rq5C3K8

Learn.Genetics

Produced by the University of Utah, Learn.Genetics is an incredible resource for learning about genetics and its role in evolution. Below are two modules that are particularly relevant.

"Tour of the Basics" is a module that provides a narrated overview of the basics: What is DNA, what is a gene, what is a chromosome, what is a protein, what is heredity, and what is a trait?

http://learn.genetics.utah.edu/content/begin/tour/

"Transcribe and Translate a Gene" is a cool interactive module that lets you transcribe a gene and build a protein.

http://learn.genetics.utah.edu/content/begin/dna/transcribe/

Gene Expression

From genomicseducation, this YouTube video is an animation that illustrates gene expression within a cell.

http://www.youtube.com/watch?v=OEWOZS_JTgk

Signal Transmission and Gene Expression

Paul Anderson, of bozemanbiology, takes the animation one step further and shows what extreme base jumping can teach about gene expression and the production of glucose within the cells of the liver.

http://www.youtube.com/watch?v=D-usAds_-lU

The Gene School

This interactive page includes a number of different experiments and games to play that all have to do with genetics.

http://library.thinkquest.org/19037/teach_links.html

The Genetic Basis for Bacterial Mercury Methylation

Jerry M. Parks and colleagues identify two genes involved in the mercury methylation by bacteria in a recent issue of *Science* magazine.

http://www.sciencemag.org/content/339/6125/1332.abstract

Patients Should Get DNA Information, Report Recommends

In *Science Insider*, the American College of Medical Genetics and Genomics recently argued that patients whose genomes are sequenced for medical reasons should be given information about genes that may affect their health, including those that put them at risk of certain cancers and potentially fatal heart conditions, even if those risks are unrelated to the reason for sequencing in the first place.

http://news.sciencemag.org/health/2013/03/
patients-should-get-dna-information-report-recommends?ref=em

Overcoming Misconceptions

Our understanding of genetics has changed significantly since Mendel's time, and today it continues to richen and deepen. For example, we know that a nucleotide consists of three components: a five-carbon sugar, one or more phosphate groups, and a base. The five-carbon sugars are known as ribose in RNA and 2-deoxyribose in DNA. A, G, C, and T are the DNA bases; A, G, C, and U are RNA bases. Strung together, these bases represent sequences that can influence development. Mutations can occur within the sequences (whether they be harmful, beneficial, or simply neutral) resulting in different alleles. But this is a very simplified view. The models we use to understand the complexities of genetics can sometimes lead to misconceptions.

Rarely Are Traits Influenced by a Single Locus

Mendel discovered that some phenotypes were predictable, and some diseases, such as Huntington's, can be tied to a single mutation. The idea that a single gene "codes for" a specific phenotypic effect is often an over- simplification, however. The majority of traits are the result of polygenic effects (many genes).

Genes Are Not Analogous to Words in a Sentence

Thinking of genes as words in a sentence is another over-simplification. Genes are actually quite complex concepts, and the more scientists learn, the less acceptable this over-simplification becomes. For example, DNA sequences that encode proteins also include introns and exons, both of which may carry instructions that effect the coding of that gene or other genes and in turn the combinations of exons that are spliced together. Protein encoding regions may not even have borders, so no punctuation.

A Locus Can Have More than Two Alleles

Simple models of loci (plural of locus) often represent the alleles in upper- and lowercase letters, such as AA, Aa, and aa, implying that the locus only has two alleles. But within a population of individuals, the same locus can have many alleles—diploid individuals will have two of those alleles. A more appropriate way of representing alleles incorporates subscripts, such as A_1, A_2, A_3, A_4, reflecting the diverse number of alleles that may be present for any locus in the entire population.

Phenotypes Are Influenced by More than Just Genotypes

Genes are clearly an important influence on how an organism turns out, but environmental conditions are also important. When scientists refer to environmental conditions, they consider more than the external environment (e.g., light conditions). A host of internal environmental factors influence gene expression, including other gene products, such as the timing and quantity of the release of the hormone adrenalin from the adrenal gland. This hormone circulates around the body and attaches to the surface of muscles and other types of cells, switching on other genes inside. From the perspective of a gene inside a cell, everything from transcription factors and hormones, to nutrients, toxins, and temperature, can be thought of as environmental factors influencing whether or not the gene is expressed.

Which of the following is a true statement?

a. Scientists expect to find a gene that codes for skin color.

b. Decoding the genome is a straightforward process.

c. Because offspring receive one chromosome from each parent, any given locus can only have two alleles in a population.

d. A low concentration of transcription factors can influence the expression of a gene.

Go the Distance: Examine the Primary Literature

As Chapter 5 explains, height is a classic polygenic trait that is highly heritable. Hana Lango Allen and colleagues examined the genetic variation of 183,727 different people and found that at least 180 loci influence adult height, but they were only able to explain 10 percent of the variation. Nevertheless, their methods offer insight to the complexities of the genetic architecture of these kinds of traits.

- Why do they argue their approach is so valuable?

- How might this approach help our understanding of polygenic diseases?

Allen, H. L., K. Estrada, G. Lettre, S. I. Berndt, M. N. Weedon, et al. 2010. Hundreds of Variants Clustered in Genomic Loci and Biological Pathways Affect Human Height. *Nature* 467: 832–38. doi:10.1038/nature09410.

Delve Deeper

1. Why aren't the mutations that occur in skin cells or in other organs, such as the heart or brain, heritable?

2. Why don't all phenotypic traits occur as discrete, alternative states like Mendel's peas?

3. What sources of genotypic variation among individuals are random?

Test Yourself

1. What is alternative splicing?
   a. The process of splicing different subsets of exons from the same gene to produce different combinations and therefore different proteins
   b. The process of alternating introns and exons within the mRNA transcript
   c. The process of creating messenger RNA from various genes on the DNA template
   d. The process of switching a variety of genes on or off at the same time that results from transcription factors that can bind to identical regulatory regions near hundreds of different genes
   e. None of the above
2. When an organism incorporates genetic material from another organism that is not its parent, this is known as:
   a. Vertical gene transfer.
   b. Diagonal gene transfer.
   c. Horizontal gene transfer.
   d. None of the above

3. Can a mutation that occurs within a cell during mitosis in a diploid cell be passed on to offspring?
   a. No. Only somatic cells undergo mitosis, so the mutation cannot be passed on to offspring.
   b. No. Mutations cannot occur during mitosis.
   c. Maybe. It depends on whether the cell is part of the germ line or not.
   d. Yes. Mitosis is the process of cell division, so mutations will be passed on to offspring because as cells divide they will reproduce the mutation and pass it on.
   e. Yes. Because all mutations, beneficial and deleterious, are passed on to offspring.
4. What is an allele?
   a. One of several alternative forms of the DNA sequence of the same locus
   b. One of several alternative forms of a gene that occur at the same place on paired chromosomes
   c. One of several alternative forms of a gene that occur at the same locus in different individuals
   d. All of the above
   e. None of the above
5. What is/are the most important factor/s generating genetic diversity among individuals in a population ofeukaryotes?
   a. Mutation
   b. Independent assortment
   c. Genetic recombination
   d. b and c only
   e. All of the above are important for generating genetic diversity in eukaryotes.
6. Complete dominance occurs when:
   a. The phenotype of a heterozygote is identical to the homozygote.
   b. The phenotype of a heterozygote is not identical to the homozygote.
   c. The genotype of a heterozygote is identical to the homozygote.
   d. The genotype of a heterozygote is not identical to the homozygote.
   e. None of the above
7. How would you describe the development of horns in dung beetles?
   a. Horn development is a phenotypically plastic trait.
   b. Horn development is a result of dominant alleles.
   c. Horn development is polyphenic.
   d. All of the above
   e. a and c only
8. According to Figure 5.19, do male dung beetles produce horns if their body size is 5mm?
   a. Yes. All male dung beetles produce horns because horn length is genetically controlled.
   b. Yes. A few male dung beetles with 5mm body sizes may produce horns, but their horns are relatively small.
   c. Yes. Male dung beetles that are 5mm and produce horns are the most fit for their body size.
   d. No. Because a 5mm body size is the critical threshold, only males larger than 5mm produce horns.
   e. No. Males with body size of 5mm never produce horns.

9. Why don't all phenotypic traits occur as discrete, alternative states like Mendel's peas?
   a. Because variation in the environment can lead to variation in the phenotypes that arise from a single genotype
   b. Because the variation in some traits can be attributed to the cumulative action of many genes
   c. Because phenotypes result from complex interactions between many different genes and the environment
   d. a and c only
   e. All of the above
10. How does sexual reproduction contribute to the quantity of genetic diversity among individuals of a population?
   a. By increasing the likelihood that an individual will have multiple offspring
   b. By increasing the ploidy of individuals within the population
   c. By separating, combining, and mixing alleles as a result of independent assortment and genetic recombination
   d. By incorporating somatic mutations into the gametes or germ-line cells
   e. By reducing the ability of an autosome to mask mutations affecting gene expression by the presence of a functional copy of the gene in the other autosome

Contemplate

Why are mutations important to evolution by natural selection?	**If alcoholism is heritable, what factors can influence whether the children of an alcoholic will be alcoholics?**

6 The Ways of Change

Drift and Selection

Check Your Understanding

1. According to the text, what is an organism's phenotype?
   a. The interaction of an organism's genes with the environment to produce characteristics such as how the amount of light a plant is exposed to influences its height
   b. Characteristics of an organism that can be classified into discrete categories, such as gender or eye color
   c. Any aspect of an organism that can be measured, such as how it looks, how it behaves, how it's structured
   d. The genetic makeup of an individual
   e. Both b and d
2. How many alleles can a genetic locus in a diploid individual have?
   a. One
   b. Two
   c. More than two
   d. It depends on the locus.
3. Why is the variation of phenotypic traits often continuous, distributed around a mean in a bell-shaped curve?
   a. Because phenotypic traits are a result of dominance
   b. Because phenotypic traits are not related to genotypes
   c. Because phenotypic traits are only influenced by the environment
   d. Because phenotypic traits are often polygenic

Learning Objectives for Chapter 6

Add important definitions and notes next to each learning objective for this chapter to help guide your understanding.

Learning Objective	Important Definitions	Notes
Define population genetics.		
Calculate allele frequencies and determine whether a population is in Hardy–Weinberg equilibrium.		
Explain why the Hardy–Weinberg equilibrium is a "null model" for evolution.		
Discuss the effects of genetic drift on large and small populations.		
Predict the effect of a bottleneck or founder event on allelic diversity.		
Compare and contrast measures of fitness of a phenotype and fitness of an allele within a population.		
Describe how slight differences in fitness can change the frequencies of alleles within a population over time.		
Explain how pleiotropy affects the response to selection acting on alleles.		
Discuss how scientists use laboratory studies to gain insight into natural selection.		
Explain why natural selection cannot drive dominant alleles to fixation within a population.		
Explain how selection can act to either remove or maintain allelic diversity.		
Discuss the effects of inbreeding on an individual's fitness.		
Analyze the influence of drift and inbreeding on the genetics of populations within a landscape.		

Identify Key Terms

Match terms and definitions by filling in the blank to the right of the term with the appropriate letter.

1. **Additive allele** ____
2. **Antagonistic pleiotropy** ____
3. **Average excess of fitness (of an allele)** ____
4. **Balancing selection** ____
5. **Epistasis** ____
6. **Fitness** ____
7. **Fixed allele** ____
8. **Founder effect** ____
9. **F_{ST}** ____
10. **Gene flow** ____
11. **Genetic bottleneck** ____
12. **Genetic distance** ____
13. **Genetic locus (plural, loci)** ____
14. **Heterozygote advantage** ____
15. **Inbreeding coefficient (*F*)** ____
16. **Inbreeding depression** ____
17. **Landscape genetics** ____
18. **Negative frequency-dependent selection** ____
19. **Negative selection** ____
20. **Null hypothesis** ____
21. **Pleiotropy** ____
22. **Population genetics** ____
23. **Population structure** ____
24. **Positive selection** ____

a. Occurs when selection favors heterozygote individuals over either the dominant homozygote or the recessive homozygote

b. Selection that decreases the frequency of alleles within a population. Negative selection occurs whenever the average excess for fitness of an allele is less than zero

c. The success of the genotype at producing new individuals (its fitness) standardized by the success of other genotypes in the population (e.g., divided by the average fitness of the population)

d. An allele that yields twice the phenotypic effect when two copies are present at a given locus than occurs when only one copy is present. Additive alleles are not influenced by the presence of other alleles (e.g., there is no dominance)

e. The type of selection that increases allele frequency in a population. Positive selection occurs whenever the average excess for fitness of an allele is greater than zero.

f. The type of selection that favors more than one allele. This process acts to maintain genetic diversity in a population by keeping alleles at frequencies higher than would be expected by chance or mutation alone.

g. Mathematical statements that have been proven based on previously established theorems and axioms. They use deductive reasoning and show that a statement necessarily follows from a series of statements or hypotheses—the proof. They are not the same as theories. Theories are explanations supported by substantial empirical evidence—the explanations are necessarily tentative but weighted by the quantity of evidence that supports them.

h. An allele that remains in a population when all of the alternative alleles have disappeared. No genetic variation exists at a fixed locus within a population, because all individuals are genetically identical at that locus.

i. A reduction in the average fitness of inbred individuals relative to that of outbred individuals. It arises because rare, recessive alleles become expressed in a homozygous state where they can detrimentally affect the performance of individuals.

j. A measure of genetic distance between subpopulations

k. A type of genetic drift. It describes the loss of allelic variation that accompanies founding of a new population from a very small number of individuals (a small sample of a much larger source population). This effect can cause the new population to differ considerably from the source population.

l. The probability that the two alleles at any locus in an individual will be identical because of common descent. *F* can be estimated for an individual, $F_{pedigree}$, by measuring the reduction in heterozygosity across loci within the genome of that individual attributable to inbreeding, or it can be estimated for a population, by measuring the reduction in heterozygosity at one or a few loci sampled for many different individuals within the population.

25. **Relative fitness (of a genotype)** ___	m. The study of the distribution of alleles within populations and the mechanisms that can cause allele frequencies to change over time
26. **Theorems** g	n. Occurs when rare genotypes have higher fitness than common genotypes. This process can maintain genetic variation within populations.
	o. The specific location of a gene or piece of DNA sequence on a chromosome. When mutations modify the sequence at a locus, they generate new alleles—variants of a particular gene or DNA region. Alleles are mutually exclusive alternative states for a genetic locus.
	p. The occurrence of populations that are subdivided by geography, behavior, or other influences that prevent individuals from mixing completely. Population subdivision leads to deviations from Hardy–Weinberg predictions.
	q. The success of an organism at surviving and reproducing and thus contributing offspring to future generations
	r. Occurs when the effects of an allele at one genetic locus are modified by alleles at one or more other loci
	s. The condition when a mutation in a single gene affects the expression of more than one different phenotypic trait pleiotropy
	t. A measure of how different populations are from each other genetically. This information can inform population geneticists about levels of inbreeding within a population or about the historic relationships between populations or species.
	u. The difference between the average fitness of individuals bearing the allele and the average fitness of the population as a whole.
	v. A relatively new field of research that combines population genetics, landscape ecology, and spatial statistics
	w. The movement, or migration, of alleles from one population to another
	x. Occurs when a mutation with beneficial effects for one trait also causes detrimental effects on other traits
	y. An event in which the number of individuals in a population is reduced drastically. Even if this dip in numbers is temporary, it can have lasting effects on the genetic variation of a population.
	z. A default hypothesis that there is no relationship between two measured phenomena. By rejecting this hypothesis, scientists can provide evidence that such a relationship may exist.

$$\frac{H_e - H_o}{H_e} = F \quad \frac{\frac{\#\,exp}{total\ pop} - \frac{obs}{total\ pop}}{\frac{exp}{total\ pop}}$$

Link Concepts

Fill in the bubbles with the appropriate terms.

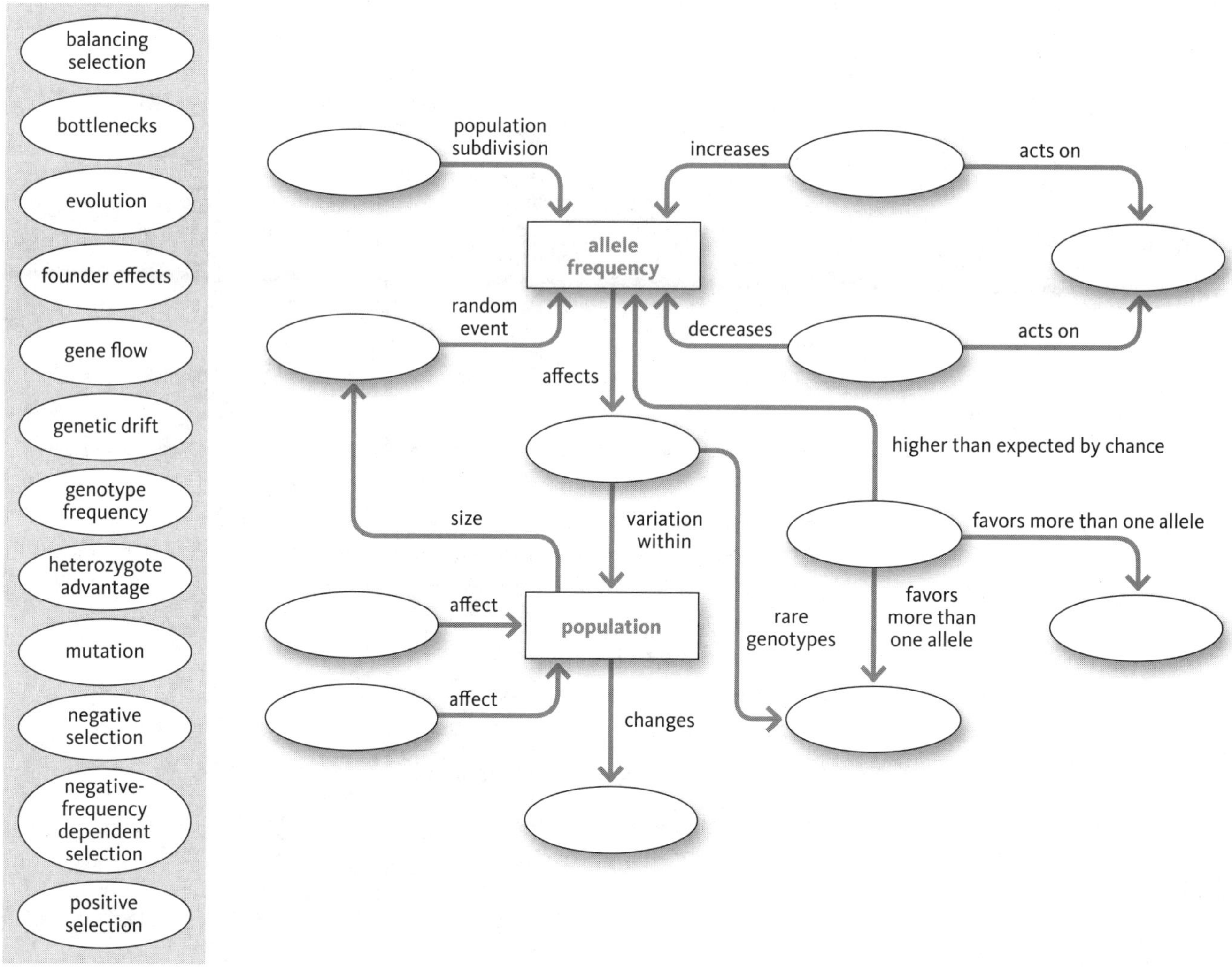

Key Concepts

Fill in the blanks for the key concepts from the chapter.

6.1 The Genetics of Populations

Because diploid organisms carry two copies of each autosomal chromosome, they can have up to two alleles for each gene or locus. Individuals carrying two copies of the same allele are homozygous at that locus, whereas individuals carrying two different alleles are het. for the locus.

Populations contain mixtures of individuals, each with a unique ______ reflecting the alleles that they carry at all of their genetic loci. At any time, some alleles will be more common than others.

Population geneticists study how and why the ______ of allelic diversity changes over time (i.e., evolution). △ in allele frequency

6.2 Change over Time—or Not

The Hardy–Weinberg theorem proves that in the absence of drift, mutation, immigration emigration, and ______, allele frequencies at a genetic locus will not change from one generation to the next.

Mechanisms of evolution are processes that can change allele frequencies in a population from one generation to the next.

For any measured set of allele frequencies, the Hardy–Weinberg theorem ______ the genotype frequencies ______ for a population that is not evolving.

6.3 Evolution's "Null Hypothesis"

Because it describes the conditions in which evolution will not occur, the Hardy–Weinberg theorem serves as the fundamental theorem of population genetics.

6.4 A Random Sample

______ is the random, nonrepresentative sampling of alleles from a population during breeding. It is a mechanism of evolution because it causes the genetic composition of a population to change from generation to generation.

Alleles are lost due to genetic drift much more rapidly in new populations than in ~~populations~~ older ~~founded~~ populations.

6.5 Bottlenecks and Founder Effects

Even brief __catastrophic__ events can lead to drastic reductions in the amount of __alleles__ within a population, and this loss of allelic diversity can persist for many generations after the event.

6.6 Selection: Winning and Losing

Selection occurs whenever genotypes differ in their ______ fitness.

The outcome of selection depends on the ______ of an allele as well as its effects on ______ .

Both drift and selection are potent ______ of evolution, and their relative importance depends critically on population size. When populations are very ______ , the effects of drift can enhance the action of selection (e.g., by removing harmful alleles that have been driven to low frequency by selection), or oppose it (e.g., by removing alleles that are beneficial). When populations are ______ , the effects of drift are minimal, and selection is the more important force.

Alleles often affect the ______ of an organism in more than one way. When these fitness effects oppose each other, the ______ between them will determine the net direction of ______ acting on the allele. This balance may tip one way in one ______ and a different way in others.

Laboratory studies of experimental evolution help reveal how new alleles can ______ and ______ through a population in response to selection.

Rare alleles are almost always carried in a __het.__ state.

When recessive alleles exist in heterozygous individuals, they are __immune__ to the action of __selection__ .

______ cannot drive dominant ______ alleles all the way to __zeros__ , because once the alternative (recessive) alleles become rare, they can hide indefinitely in a heterozygous state.

The mutation rate for any specific ______ may be extremely low, but it is much higher when considering an entire ______ or ______ . The gradual accumulation of mutations within populations is the ultimate source of ______ genetic variation.

Balancing selection actively maintains multiple ______ within a population. Two mechanisms are negative frequency-dependent selection and heterozygote advantage. ______ occurs if the fitness of an allele is higher when that allele is rare than when it is common. ______ occurs when the heterozygotes for the alleles in question have higher fitness than either of the homozygotes.

Alleles are selectively ______ if they have no effect on the ______ of their bearers. This phenomenon often occurs when genetic variation at a locus does not affect the ______ of an individual.

Selection acts on whole pop. ______ of individuals.

6.7 Inbreeding: The Collapse of a Dynasty

Inbreeding increases the ______ of genetic loci that are ______ for alleles.

Inbreeding changes allele ______ frequencies but not genetic ______ frequencies and therefore is not a direct mechanism of evolution ______. Inbreeding can, however, set the stage for strong selection on recessive ______ alleles that typically would be masked in het ______ individuals.

Genetic bottlenecks often go hand in hand with inbreeding and selection if low ______ numbers of individuals establish a new population. These "founding events" can be important episodes of rapid evolution because the effects of genetic drift are pronounced, and the increased ______ arising due to inbreeding exposes ______ alleles to positive and negative selection. If the new population survives this bottleneck, it may be very ______ from its parent population.

6.8 Landscape Genetics

Population ______ enhances the effects of genetic drift, eroding genetic ______ from within local subpopulations and causing allele frequencies to ______ from place to place.

______ counteracts the effects of population subdivision, increasing genetic ______ within subpopulations and homogenizing allele frequencies across the landscape.

Because extreme population subdivision can lead to inbreeding, two metrics of heterozygosity are important to population geneticists: ______ is a measure of how inbred any particular individual is, and ______ is a measure of the genetic distance between populations.

Population subdivision can reduce the frequency of ______ .

Two metrics of ______ important to population geneticists are $F_{pedigree}$, a measure of how ______ any particular individual is, and F_{ST}, a measure of the ______ between populations.

Interpret the Data

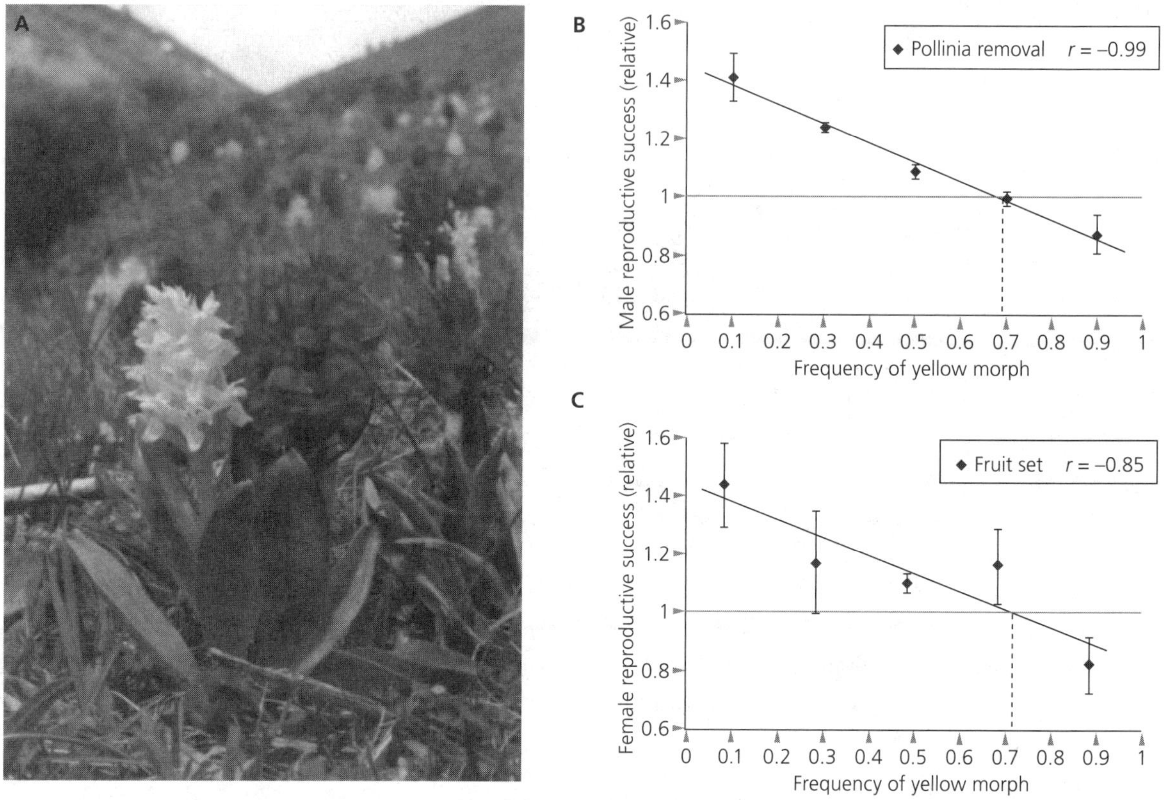

Figure 6.18 A: Elderflower orchids in Europe have a polymorphism for yellow and purple flowers. Unlike most other flowers, these orchids do not produce nectar for pollinating bumblebees to eat. Naïve bumblebees looking for a meal end up pollinating the flower without reward. Luc Gigord and his colleagues set up an experiment to determine the effects of the color of the orchid flower on its fitness. They examined the fitness of the yellow genotype relative to its frequency in the population B: Male fitness was measured as the number of pollen packets attached to unsuspecting bees. C: Female fitness was measured as the number of fertilized seeds. (Adapted from Gigord et al. 2001.)

- At the rarest frequency of the yellow morph, what was the relative reproductive success of males? Of females?

- At the most common frequency of the yellow morph, what was the relative reproductive success of males? Of females?

- Why might the foraging experiences of bumblebees drive this relationship?

Games and Exercises

When Can You Say That What You Observed Doesn't Match What You Expected to Find?

X^2 (pronounced chi squared) is a statistical test used to determine whether the difference between what is observed in populations and the frequencies expected based on models like Hardy–Weinberg are is statistically significant. X^2 is the likelihood that the difference may be due to chance. There are several statistical tests based on the chi-squared distribution. The Pearson's chi-squared test is calculated by summing of each observed genotype frequency minus its expected frequency, squaring that sum, and dividing it by the expected frequency. The sum is compared to a known distribution to determine significance.

Cavalli-Sforza and colleagues measured the frequency of the *A* allele as 0.877, so the expected frequency of the *AA* genotype is p^2 = 0.877 x 0.877 = 0.769. They measured the frequency of the *S* allele as $q = 1 - p$, or q = 1 – 0.877 = 0.123, so the expected frequency of the *SS* genotype in the population is q^2 = 0.123 x 0.123 = 0.015. Heterozygotes should occur with a frequency of $2pq$ = 2 x 0.877 x 0.123 = 0.216. They examined 12,387 individuals, so they expected to find:

AA = 12,387 x 0.769 = 9,527.2

AS = 12,387 x 0.216 = 2,672.4

SS = 12,387 x 0.015 = 187.4

Genotype	Expected by HW	Observed
AA	9527.2 (76.9%)	9365 (75.6%)
AS	2672.4 (21.6%)	2993 (24.2%)
SS	187.4 (1.5%)	29 (0.2%)
Total	12,387	12,387

What they found was 9,365 homozygotes for the A allele, 2,993 heterozygotes, and 29 homozygotes for the S allele. So, calculating chi square:

$(\text{observed}_{AA} - \text{expected}_{AA})^2/\text{expected}_{AA}$ = (9,365 – 9,527.2) / 9,527.2 = 2.76

$(\text{observed}_{AS} - \text{expected}_{AS})^2/\text{expected}_{AS}$ = (2,993 – 2,672.4) / 2,672.4 = 38.46

$(\text{observed}_{SS} - \text{expected}_{SS})^2/\text{expected}_{SS}$ = (29 – 187.4) / 187.4 = 133.89

X^2 = 2.76 + 38.46 + 133.89 = 175.11

But what does that mean? If the observed values are pretty close to the expected values, then the value of the X^2 statistic should be relatively small. If, on the other hand, the observed values deviate from the expected by a lot, then the value of the X^2 statistic will be large. The larger the value, the more evidence refuting the null hypothesis that the observed match the expected. The critical value (also known as the *p*-value) is the area under the curve to the right of the calculated chi-square value. It represents the probability of finding that chi-square or larger by chance, given the null hypothesis.

Determining the value of chi squared depends on the "degrees of freedom." The degrees of freedom (df) are crucial to defining the shape of the chi-square curve. The shape of the curve is different for every degree of freedom.

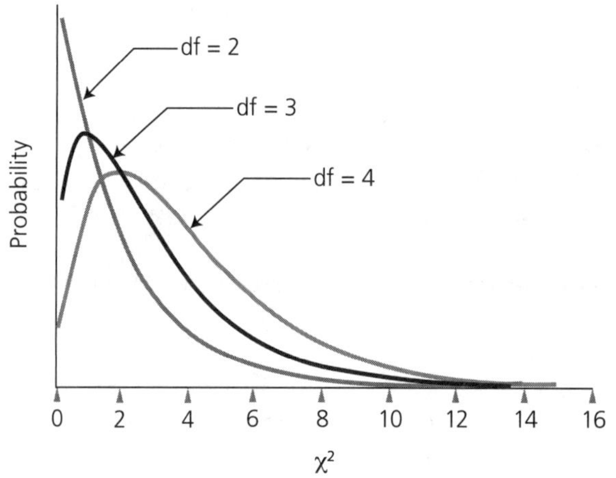

In general, df = (the number of rows in the table – 1) x (the number of columns in the table – 1). For counting data, like the numbers of individuals with different genotypes, the df = the number of classes – 1. We have 3 classes of genotypes (AA, AS, or SS), so df = 2.

Now you're ready to see how significant the evidence that Cavalli-Sforza and colleagues found actually was. The critical values for curves for 1 to 100 degrees of freedom are readily available on the web. The table below gives the critical values for df 1 to 4. The numbers across the top of the table are the area under the curve to the right of the critical value—the probability of calculating a specific chi square or larger.

Start with the degrees of freedom (df = 2), then move along that row to the right until you find a critical value near 175.11. You can't! The largest value in the table for df = 2 is 15.20—not even close! So, the probability of finding that chi-square or larger by chance has to be far less than 0.0005. Cavalli-Sforza and colleagues found a statistically significant deviation in the observed versus the expected frequencies of individuals with AA, AS, and SS genotypes. What likely caused that deviation?

A Chi-Square Table

df	**Critical Value**											
	0.25	0.20	0.15	0.10	0.05	0.025	0.02	0.01	0.005	0.0025	0.001	0.0005
1	1.32	1.64	2.07	2.71	3.84	5.02	5.41	6.63	7.88	9.14	10.83	12.12
2	2.77	3.22	3.79	4.61	5.99	7.38	7.82	9.21	10.60	11.98	13.82	15.20
3	4.11	4.64	5.32	6.25	7.81	9.35	9.84	11.34	12.84	14.32	16.27	17.73
4	5.39	5.59	6.74	7.78	9.49	11.14	11.67	13.23	14.86	16.42	18.47	20.00

JBstatistics does a really great job of explaining the *p*-value in chi-square tests in this YouTube video:

http://www.youtube.com/watch?v=HwD7ekD5l0g

White Bears and Hardy–Weinberg

Kermode bears (*Ursus americanus kermodeii*) are white bears that can be found in the rainforest on Princess Royal and Gribbell islands along the coast of British Columbia. This subspecies of black bears is known as the Spirit Bear. Its white fur is the result of a single nucleotide replacement in the melanocortin-1 (*Mc1r*) receptor portion of that gene that leads to the production of adenine instead of guanine. The white fur allele is a recessive allele unique to this subspecies. Dr. Kermit Ritland and his colleagues found that approximately 20 percent of the bears on Princess Royal and Gribbell islands are white; the rest are black. The percentage of white bears on the mainland drops off drastically with distance from those islands.

Assuming complete dominance, calculate the expected genotype frequencies.

- What does q^2 equal?

- What does p^2 equal?

- What does $2pq$ equal?

Thus, for a randomly mating population at equilibrium, the Hardy–Weinberg theorem gives us the genotype frequencies expected for Kermode bears.

BB genotypes should occur at a frequency of ____________________

BW genotypes should occur at a frequency of ____________________

and *WW* genotypes should occur at a frequency of ____________________

- If you had a population of 37 bears on the two islands, how many would you expect to be black?

- How many would you expect to be white?

Ritland and his colleagues also found that the occurrence of white bears dropped off dramatically on the mainland. Say they counted 112 bears 100 kilometers from the coast and found the frequency of the *W* allele to be 0.08. Calculate the expected genotype frequencies for this population.

- How many bears would you expect to have black fur? How many Spirit Bears?

To find out more about Spirit Bears, see Ritland et al. and the NA Bear Center and NatGeo:

Ritland, K., C. Newton, and H. D. Marshall. 2001. Inheritance and Population Structure of the White-Phased "Kermode" Black Bear. *Current Biology* 11(18): 1469-1472.

http://www.bear.org/website/bear-pages/black-bear/basic-bear-facts/101-what-is-a-spirit-bear.html

National Geographic offers this incredible view into the British Columbia rainforest on their trek to capture Spirit Bears on film.

http://video.nationalgeographic.com/video/magazine/ngm-kermode-bear

White Bears and Selection

The distribution of Spirit Bears is highly intriguing. Spotting a Spirit Bear is a rare event—and breathtaking. This intrigue spurred Ritland and his colleagues to investigate further. They found that the frequency of the G nucleotide allele was $q = 0.369$ for the population of Kermode bears on three islands off the coast of British Columbia: Gribbell, Princess Royal, and Roderick. The population on these three islands consisted of 42 bears that were *BB*, 24 that were *BW*, and that 21 were *WW*. Do those observations match what they should have expected if the population was at Hardy–Weinberg equilibrium?

Complete this table:

- What is *p*?

Hints:

$\text{expected}_{BB} = p^2 \times 87$

$\text{expected}_{BW} = 2pq \times 87$

$\text{expected}_{WW} = q^2 \times 87$

	Observed	**Expected**	**(Observed – Expected)²**
BB (black)	42		
BW (black)	24		
WW (white)	21		
Total			

- What are the degrees of freedom?

- What is the chi square value?

- How would you interpret this result?

- Which genotypes are overrepresented in the population? Which are underrepresented?

White fur color may actually provide a selective advantage while fishing for salmon by functioning as camouflage against a cloudy sky. A white fur advantage shouldn't be limited to the rainforest coast, however, because bears hunt salmon inland as well. So, Philip Hedrick and Kermit Ritland examined factors leading to the prevalence of white bears on the northwest coast of British Columbia. They looked for evidence of natural selection, along with drift, and mating preferences (see Hedrick and Ritland 2011). But for now, consider how natural selection alone may be driving changes in genotype and allele frequencies.

One way to determine the effects of natural selection is by using selection coefficients to determine changes in frequencies after selection. The selection coefficient (s) = 1 – the relative fitness of a phenotype. It is a measure of *disadvantage.* You can also calculate the relative fitness of the disadvantaged phenotype as $1 - s$. If the selection coefficient for black bears is 0.2, that means their relative fitness is 0.8 (so they generally survive and reproduce at 80 percent of white bears). The *relative* fitness of white bears will be 1.0 (or 100 percent) by definition.

If black bears are at a disadvantage with a selection coefficient of $s = 0.2$, then what should the genotype frequencies be after selection in time_{t+1}?

To calculate the genotype frequencies after selection (time_{t+1}), first multiply the frequency of each genotype by its relative fitness to represent the offspring produced:

Genotype:	*BB*	*BW*	*WW*
f_{t+1}	$p^2 \text{ x } w_{11}$	$2pq \text{ x } w_{12}$	$q^2 \text{ x } w_{22}$

Determine the average fitness of the population, $\overline{w}$. $\overline{w}$ is the sum of the fitnesses of each genotype multiplied by (i.e., weighted by) the frequencies at which they occur:

$\overline{w} = p^2 \text{ x } w_{11} + 2pq \text{ x } w_{12} + q^2 \text{ x } w_{22} =$

Now, determine the relative frequencies of individuals with each genotype after selection using the average fitness of the population.

Genotype:	BB	BW	WW
f_{t+1}	$(p^2 \text{ x } w_{11}) / \overline{w}$	$(2pq \text{ x } w_{12}) / \overline{w}$	$(q^2 \text{ x } w_{22}) / \overline{w}$

And then calculate each *allele* frequency in this new generation by considering the frequency of homozygote individuals plus half the frequency of heterozygote individuals:

$p_{t+1} = [(p^2 \text{ x } w_{11}) / \overline{w}] + (pq \text{ x } w_{12}) / \overline{w})$

$= (p^2 \text{ x } w_{11} + pq \text{ x } w_{12}) / \overline{w}$

=

=

and

$q_{t+1} = [(q^2 \text{ x } w_{22}) / \overline{w}] + (pq \text{ x } w_{12}) / \overline{w})$

$= (q^2 \text{ x } w_{22} + pq \text{ x } w_{12}) / \overline{w}$

=

=

- Is there another way to calculate q_{t+1} if p_{t+1} is known?

- What would be the proportion of each genotype found in time_{t+1}?

p^2 = ____________ = ____________ or ____________ % homozygous black bears

$2pq$ = ____________ = ____________ or ____________ % heterozygous black bears

q^2 = ____________ = ____________ or ____________ % white bears

- What does the selection coefficient mean in terms of persistence of the *B* allele within the population?

So, for a trait with complete dominance, both the selection coefficient and the average excess of fitness offer fairly straightforward approaches to determining allele frequencies and genotype frequencies. In diploids (and other polyploid organisms), complete dominance is rare. More importantly, selection acts to change allele frequencies through the phenotype. Where the selection coefficient focuses on genotype frequencies, the average excess of fitness focuses on allele frequencies. It examines the overall fitness effect of having a specific allele within a population. If the selection coefficient or the average excess of fitness stays the same, you can calculate allele frequencies in any number of generations.

- Determine the change in allele frequencies resulting from selection in time$_{t+2}$ as a function of the average excess of fitness. Assume the fitnesses have stayed the same.

  $\Delta p = (p_{t+1} / \overline{w}) \text{ x } a_B$

  where a_B is the average excess of fitness of the *B* allele. Remember that the average excess of fitness for the *B* allele is $[p \text{ x } (w_{11} - \overline{w})] + [q \text{ x } (w_{12} - \overline{w})]$.

Genotype:	BB	BW	WW
f_{t+1}	$p_{t+1}{}^2 \text{ x } w_{11}$	$2p_{t+1}q_{t+1} \text{ x } w_{12}$	$q_{t+1}{}^2 \text{ x } w_{22}$

  $\overline{w}$ =

  a_B =

  Δp =

- What does the average excess of fitness of the *B* allele indicate?

- Calculate the average excess of fitness for the W allele:

  $a_W = [p \text{ x } (w_{12} - \overline{w})] + [q \text{ x } (w_{22} - \overline{w})]$

  and the predicted change in frequency of the *W* allele as a result of selection:

  $\Delta q = (q / \overline{w}) \text{ x } a_W$

- Record the allele frequencies in a table:

	p	q
t		
$t+1$		
$t+2$		

- What would be the proportion of each genotype in time$_{t+2}$?

Think back to the original population of 87 bears:

- 42 (48 percent) were homozygous black bears
- 24 (28 percent) were heterozygous black bears
- 21 (24 percent) were white Spirit Bears
- Based on your understanding of Hardy–Weinberg and the distribution of observed phenotypes, what can you say about this population?

Hedrick and Ritland developed models to explain the population genetics of Kermode bears. Their results can be found here:

Hedrick, P. W., and K. Ritland. 2011. Population Genetics of the White-Phased "Spirit" Black Bear of British Columbia. *Evolution* 66(2): 305-313.

Punnett Squares

The Khan Academy has a series of videos that explain heredity and genetics, including how to use Punnett squares in a variety of situations.

https://www.khanacademy.org/science/biology/heredity-and-genetics/v/introduction-to-heredity

Drifting Along

Alan R. Lemmon developed a simulation of the Hardy–Weinberg theorem that shows what happens to genotype frequencies when assumptions are violated. The simulation also allows you to see the outcomes in different formats.

http://www.evotutor.org/EvoGen/EG1A.html

You can also see simulations of the effects of directional, disruptive, and stabilizing selection.

http://www.evotutor.org/Selection/Sl5A.html

Kent Holsinger, professor of ecology and evolutionary biology at the University of Connecticut, put together a variety of simple simulations to help students understand principles of population genetics. Here are links to three of the simulations:

- Genetic drift: This simulation allows you to select different allele frequencies to start, different population sizes, and different numbers of generations for simulations.

  http://darwin.eeb.uconn.edu/simulations/drift.html

- Natural selection: This simulation allows you to select different fitness levels for each of three genotypes and view the outcome of natural selection after 100 generations.

  http://darwin.eeb.uconn.edu/simulations/selection.html

- Natural selection and genetic drift: This simple simulation allows you to examine the effects of both natural selection and drift by selecting from different starting allele frequencies, different population sizes, and different numbers of generations given a specific fitness, representing selection for an initially rare allele.

  http://darwin.eeb.uconn.edu/simulations/selection-drift.html

If you're interested in seeing more, go to:

http://darwin.eeb.uconn.edu/simulations/simulations.html

[Note: You will have to add these sites to the exceptions in Java setup for these sites to run.]

Explore!

Natural Selection: Crash Course Biology #14

In his "crash course" YouTube video, Hank Green discusses natural selection, adaptation, and fitness.

http://www.youtube.com/watch?v=aTftyFboC_M

Genetic Drift: Random Evolutionary Change

Paul Andersen shows how genetic drift can be a mechanism for evolutionary change in this YouTube video. He also discusses bottlenecks in northern elephant seals and the high incidence of total colorblindness that resulted from a typhoon that hit the small island of Pingelap.

http://www.youtube.com/watch?v=mjQ_yN5znyk

Lactase and Me

From Genome BC, this YouTube video is an animation (complete with sound effects) that explains the genetics behind lactose intolerance.

http://www.youtube.com/watch?v=U4w-0qkYnjg

Galápagos: The Finches

Open University introduces the Galápagos finches in this YouTube video, including one species that uses tools to access food.

http://www.youtube.com/watch?v=l25MBq8T77w

Multiple Instances of Ancient Balancing Selection Shared between Humans and Chimpanzees

Ellen M. Leffler and her colleagues examined balancing selection in humans and chimpanzees by conducting genome-wide scans for sequence variations that differed by a single nucleotide between humans and chimpanzees (SNPs).

http://www.sciencemag.org/content/339/6127/1578
.abstract?sid=9c531c35-0af9-4a0c-9e52-2cfda69e9322

Evolution in Our Own Time

In "Evolution via Roadkill," *Science Now* reports on recent research that shows that cliff swallow wings are shorter than they were historically, largely because of the selective action of cars on swallow fitness.

http://news.sciencemag.org/sciencenow/2013/03/evolution-via-roadkill.html?ref=em

See the original report here:

http://www.cell.com/current-biology/retrieve/pii/S0960982213001942

Fluttering from the Ashes?

This News & Analysis from *Science* discusses how scientists are trying to identify genes that distinctly identify some extinct species, with the idea of bringing them back to life.

http://www.sciencemag.org/content/340/6128/19.summary

Overcoming Misconceptions

Dominance Doesn't Necessarily Mean Dominance

Alleles are considered dominant when only a single copy is enough to produce a trait, but that doesn't mean that the dominant allele of a trait will always have the highest frequency in a population and the recessive allele will always have the lowest frequency. For example, Huntington's disease is a rare disease caused by an uncommon dominant allele, and blood type O is the most common blood type in humans but results from recessive alleles. Mendel's simple model crossing two heterozygous (Aa) parents yields a 3:1 ratio of A to a alleles in the offspring, so more of the dominant allele. But that cross is based on heterozygous individuals. If most individuals in a population are homozygous for the recessive allele (aa), then most offspring will be aa. Ultimately, allele frequencies are determined by natural selection and genetic drift.

Allele and Gene Cannot Be Used Interchangeably

An allele is one of any number of alternative forms of a gene or genetic locus, whereas a gene can be defined as segments of DNA whose nucleotide sequences code for proteins, or RNA, or regulate the expression of other genes. Although a gene is a complex concept (and the more scientists learn about DNA and gene expression the more difficult defining the concept becomes), an allele is a very specific difference in DNA sequence that scientists are able to identify relative to other alleles.

Genetic Drift Occurs in Large Populations as Well as Small

Genetic drift is not just a phenomenon affecting small populations; all populations experience random changes in allele frequency from one generation to the next. The difference is that large populations simply have more individuals and likely more copies of any particular allele, so all the alleles are more likely to be represented in the next generation, and the rates of allele frequency change due to drift will be lower.

Natural Selection and Genetic Drift Are Both Mechanisms for Evolution

Scientists debate over the role of natural selection and genetic drift in the evolution of organisms, but they agree that both are mechanisms for evolutionary change. Genetic drift is a nonselective process—it is random. But like natural selection, this random process does influence the frequency of alleles in the next generation.

Which of the following is a true statement?

a. A dominant allele is an allele that produces the same phenotype whether it is homozygous or heterozygous.

b. The dominant allele of a trait will always have the highest frequency in a population.

c. Genetic drift does not occur in large populations.

d. Alleles are different forms of genes.

Go the Distance: Examine the Primary Literature

Warren Johnson and his colleagues review the effects of an historic conservation decision: to move eight female pumas (*Puma concolor stanleyana*) from Texas to Florida to increase genetic diversity in a remnant population of Florida panthers (*Puma concolor coryi*). Fifteen years after the event, the authors found that population size, genetic heterozygosity, survival, and fitness had increased, and measures of inbreeding had declined.

- What led to the levels of heterozygosity found in the Florida panther population prior to the management decision?
- Why did moving eight females affect the current levels of heterozygosity in the Florida panther population?
- Texas pumas are a different subspecies than Florida panthers. What impact, if any, would you argue this move had on those taxonomic designations?

Johnson, W. E., Onorato, D. P., Roelke, M. E., Land, E. D., Cunningham, M., Belden, R. C, et al. 2010. Genetic restoration of the Florida panther. *Science*, 329:1641-1645. DOI: 10.1126/science.1192891. http://www.sciencemag.org/content/329/5999/1641.full

Delve Deeper

1. If a mutation that produces a new allele that is deleterious arises in a population, what will most likely happen to the frequency of that allele?

2. Why has the evolution of resistance to insecticides in mosquitoes been so rapid?

Test Yourself

1. What are genetic loci?
   a. Variants of a particular gene or DNA region
   b. The plural form of genetic locus
   c. The specific locations of genes or base pairs
   d. Both a and b
   e. Both b and c
   f. All of the above
2. In a tetraploid organism, how many copies of the same allele will it carry at a locus if it is heterozygous at that locus?
   a. one
   b. two
   c. three
   d. four
   e. a, b, or c
3. How can population genetics help us understand the evolution of mosquito resistance?
   a. By examining how the applications of pesticides cause mortality of mosquitoes
   b. By examining the distribution of alleles within populations and the mechanisms that cause allele frequencies to change over time
   c. By examining the mating and reproductive tactics of mosquitoes across populations that are resistant and non-resistant
   d. By examining the phylogenetic relationships among species of mosquitoes and their common ancestors
   e. Populations genetics examines all of the above.
4. What are population geneticists referring to when they say that sexually reproducing organisms are mating at random?
   a. That, for a focal genetic locus, the probability of fertilization of one gamete by another will not vary depending on which allele is carried by the gametes
   b. That mate choice is not an important factor for sexually reproducing organisms
   c. That some gametes are not any more likely to encounter other gametes
   d. That models of population genetics have to make unrealistic assumptions about sexually reproducing organisms to be useful
   e. That sexually reproducing organisms choose mates randomly within the population
5. In a population of 100 offspring, how many individuals would you predict will be heterozygous at a particular locus with two alleles if the frequency of one of the alleles in the parent generation is 0.4?
   a. 16
   b. 36
   c. 48
   d. 52
   e. 100

6. Which of these statements about balancing selection is TRUE?
   a. It is possible only in sexually reproducing species.
   b. It is responsible for maintaining the S allele for sickle-cell anemia within humans.
   c. It is not possible when heterozygotes have a higher fitness.
   d. It is why sickle-cell anemia is selected against in areas with high levels of malaria.
   e. None. All are false statements.
7. How does drift affect the frequencies of alleles within a population?
   a. Some individuals are more likely to breed with other individuals, and so only their alleles will appear in the next generation.
   b. Random mating does not equal uniform mating, and as a result of this imperfect sampling, some alleles do not get represented in the next generation.
   c. In large populations, the likelihood that all individuals will be able to mate is low, so the likelihood that all alleles being represented in the next generation is also low.
   d. Drift results in a variety of genotypes over many generations because the heterozygotes mate randomly leading to some homozygotes of each allele and some heterozygotes, changing the frequency of the alleles.
   e. Random mating within a population mixes alleles at a particular locus into many different combinations, and when this happens, frequencies of alleles change across generations.
8. According to Figure 6.8, what is the difference between the probabilities that a rare allele and a common allele will be lost in a population bottleneck of 20 individuals?
   a. About 65 percent
   b. 100 percent
   c. 5 percent
   d. 0.9 percent
   e. It depends on the allele.
9. What is the difference between bottlenecks and founder effects?
   a. Bottlenecks reduce genetic variation because of drift; founder effects reduce genetic variation by shrinking the population size.
   b. Bottlenecks reduce genetic variation by shrinking the population size; founder effects reduce genetic variation because of drift.
   c. Bottlenecks can happen to any population; founder effects only happen when small numbers of individuals start a new population.
   d. Bottlenecks are events that reduce the number of individuals; founder effects describe the loss of genetic variation that accompanies events like bottlenecks.
   e. Bottlenecks only affect the diversity of alleles when they severely cut down the population; founder effects always severely cut down the population.
10. Which type of natural selection favors rare genotypes?
    a. Balancing selection
    b. Negative-frequency dependent selection
    c. Negative selection
    d. Positive selection
    e. None of the above favor rare genotypes.

11. What does an inbreeding coefficient of 0.25 signify?
    a. 0.25 is the inbreeding coefficient assigned to Charles II.
    b. 0.25 is the average probability that alleles at two loci in an individual are identical by descent.
    c. 0.25 signifies the level of relatedness of most populations.
    d. 0.25 is the inbreeding coefficient of most royal families.
    e. 0.25 is the inbreeding coefficient of a child produced by a brother and sister mating.
12. Why would conservation biologists be concerned about inbreeding depression?
    a. Because endangered species often have small populations prone to inbreeding
    b. Because rare, recessive alleles can become expressed in a homozygous state where they can reduce the fitness of individuals
    c. Because inbreeding depression can affect reproductive rates of endangered species
    d. Because inbreeding depression can reduce the genetic variation within a population
    e. All of the above

Contemplate

Why can't scientists conduct experiments like Richard Lenski's with organisms such as birds or mammals?	**What might happen to a flower adapted to mountain habitats over time? Would climate change affect your predictions?**

7 Beyond Alleles

Quantitative Genetics and the Evolution of Phenotypes

Check Your Understanding

1. What is the difference between polyphenic traits and polygenic traits?
   a. Polyphenic traits are the different traits that arise because different alleles lead to different phenotypes; polygenic traits are traits influenced by many genes leading to a continuous distribution of phenotypes over a given range.
   b. Polyphenic traits are the multiple, discrete phenotypes that can arise from different alleles within a population; polygenic traits vary continuously within a population because of heritable variation.
   c. Polyphenic traits are the multiple, discrete phenotypes that can arise from a single genotype depending on environmental circumstances; polygenic traits are traits influenced by many genes leading to a continuous distribution of phenotypic variation over a given range.
   d. Polyphenic traits are traits that result when natural selection favors rare genotypes leading to multiple phenotypes within a population; polygenic traits are traits that arise because a single gene affects the expression of many different phenotypic traits.
2. Which of the following were important facts in Charles Darwin's development of theory of evolution by natural selection?
   a. No two individuals are exactly the same; rather, every population displays enormous variability.
   b. Much of the variation among individuals within a population is heritable.
   c. Organisms can inherit characters that were acquired during their parents' lifetime.
   d. Both a and b
   e. Both b and c
3. Can genes that respond to environmental stimuli be passed on to offspring?
   a. Yes. Individuals that learn how to respond to the environment can pass that information to their offspring.
   b. Yes. Individuals can inherit the mechanisms that respond to the environment.
   c. No. The only "environmental" influences on gene expression that can be inherited come from other gene products, such as hormones, transcription factors, and cis- and trans-acting elements. Any effects of external environmental factors are not encoded by the genome, so they cannot be heritable.
   d. No. Environmental factors, such as the amount of food available to an individual or the temperature an egg is exposed to, are not heritable.

Learning Objectives for Chapter 7

Add important definitions and notes next to each learning objective for this chapter to help guide your understanding.

Learning Objective	Important Definitions	Notes
Explain how continuous variation in phenotypes arises.		
Analyze the differences between broad sense and narrow sense heritability.		
Demonstrate the value of quantitative trait loci for examining the genetics of phenotypic traits.		
Compare and contrast plastic and evolutionary changes within a phenotype.		

Identify Key Terms

Match terms and definitions by filling in the blank to the right of the term with the appropriate letter.

1. **Broad sense heritability (H^2)** a
2. **Genome-wide association mapping (GWA)** e
3. **Narrow sense heritability (h^2)** b
4. **Quantitative genetics** g
5. **Quantitative trait loci (QTLs)** f
6. **Reaction norm** c
7. **Selection differential (S)** h
8. **Variance** d

a. The proportion of the total phenotypic variance of a trait that is attributable to genetic variance, where genetic variance is represented in its entirety as a single value (i.e., genetic variance is not broken down into different components).

b. The proportion of the total phenotypic variance of a trait attributable to *the additive effects of alleles* (the additive genetic variance). This is the component of variance that causes offspring to resemble their parents, and it causes populations to evolve predictably in response to selection.

c. The pattern of phenotypic expression of a single genotype across a range of environments. In a sense, reaction norms depict how development maps the genotype into the phenotype as a function of the environment.

d. A statistical measure of the dispersion of trait values about their mean.

e. This process involves of scanning through the genomes of many different individuals, some with, and others without, a focal trait of interest, to search for markers associated with expression of the trait.

f. Stretches of DNA that are correlated with variation in a phenotypic trait. These regions contain genes, or are linked to genes, that contribute to population differences in a phenotype.

g. The study of continuous phenotypic traits and their underlying evolutionary mechanisms.

h. A measure of the strength of phenotypic selection. The selection differential describes the difference between the mean of the reproducing members of the population who contribute offspring to the next generation and the mean of all members of a population.

Link Concepts

Fill in the bubbles with the appropriate terms.

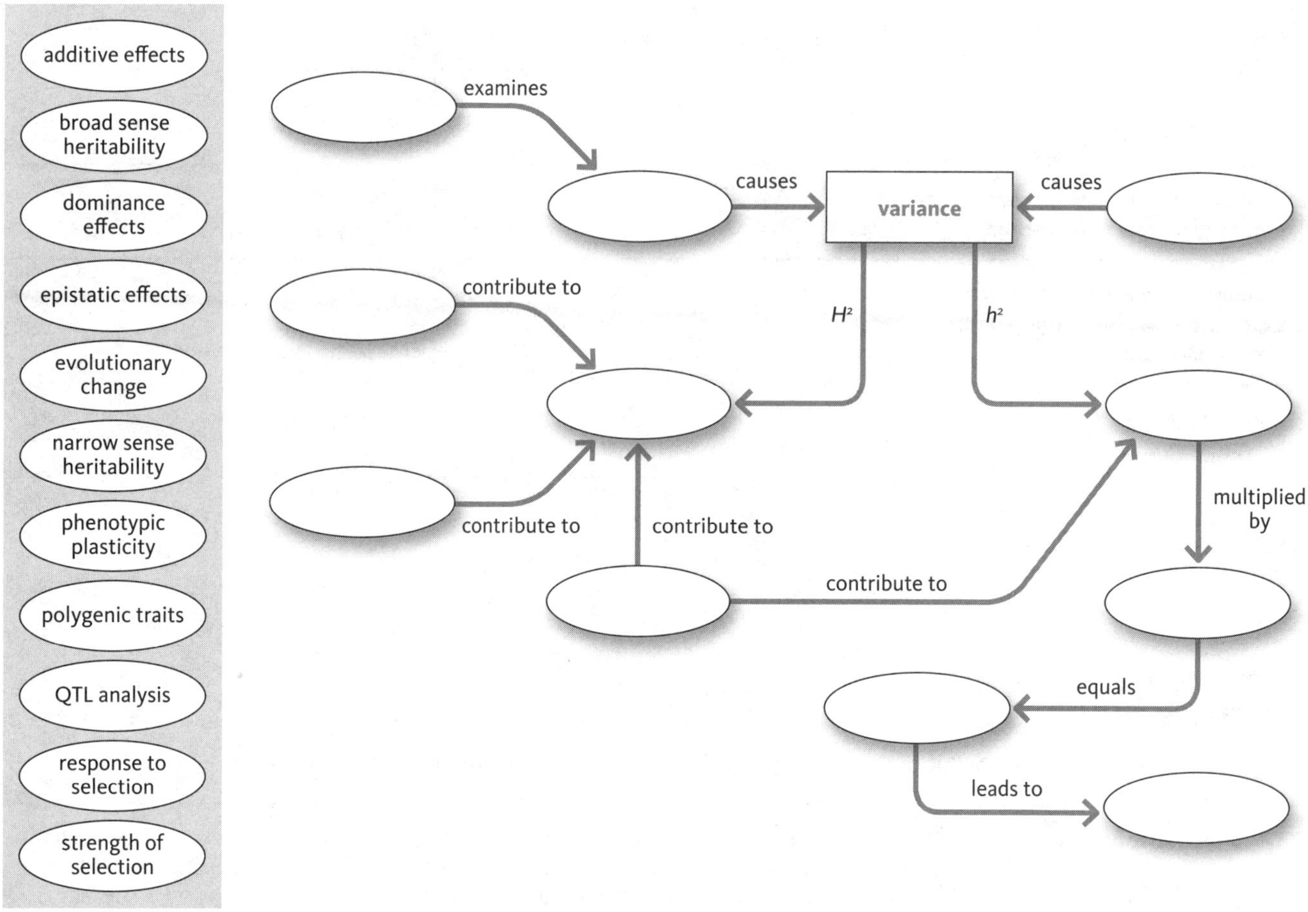

Key Concepts

Fill in the blanks for the key concepts from the chapter.

7.1 Genetics of Quantitative Traits

When the components of variation act independently, their effects are ____________, so that variation attributable to genes and variation attributable to the environment sum to yield the total ____________ variance of the sample. This allows biologists to estimate the relative contributions of different sources of ____________ to the phenotypic distribution observed.

The ____________ of a trait is the proportion of ____________ variance that is due to ____________ differences among individuals.

Broad sense heritability reflects all of the genetic contributions to a trait's phenotypic variance including ____________, ____________, and ____________ gene effects. It also includes influences of the parent phenotype on the environment of offspring that can cause ____________ to resemble each other.

Narrow sense heritability more accurately reflects the contributions of specific, ____________ components of genetic ____________ to offspring.

7.2 The Evolutionary Response to Selection

Selection can shape populations by favoring individuals with trait values at one end of a distribution (____________), by favoring individuals with trait values near the middle of a distribution (____________), or by favoring individuals with trait values at both ends of a distribution (____________).

____________ and ____________ are *not* the same thing. Populations can experience selection even if they cannot evolve in response to it.

The speed of evolution is a product of the ____________ and the extent to which offspring resemble their parents for that trait (the ____________).

7.3 Dissecting Complex Traits: Quantitative Trait Locus Analysis

QTL mapping studies permit quantitative geneticists to identify regions of the ____________ responsible for genetic variation in ____________ traits. This method can serve as a first step toward elucidating the genes responsible for phenotypic ____________ .

7.4 The Evolution of Phenotypic Plasticity

Organisms may differ in how they react to environmental situations, and these differences may be ____________. When this occurs, the ____________ themselves can evolve, leading to the evolution of adaptive ____________. (Think of this as genetic changes to the underlying physiological and developmental response mechanisms.)

Phenotypic changes occurring within the ____________ of an individual are not evolution. Evolution occurs only when the ____________ of alleles within a population change from one generation to the next.

Phenotypically plastic traits are especially confusing because the plastic change in phenotype is *not* evolution, yet a ____________ change in the amount or nature of plasticity *is* evolution. The crucial distinction is whether or not the change in question arises due to shifts in the ____________ of alleles within the population.

Interpret the Data

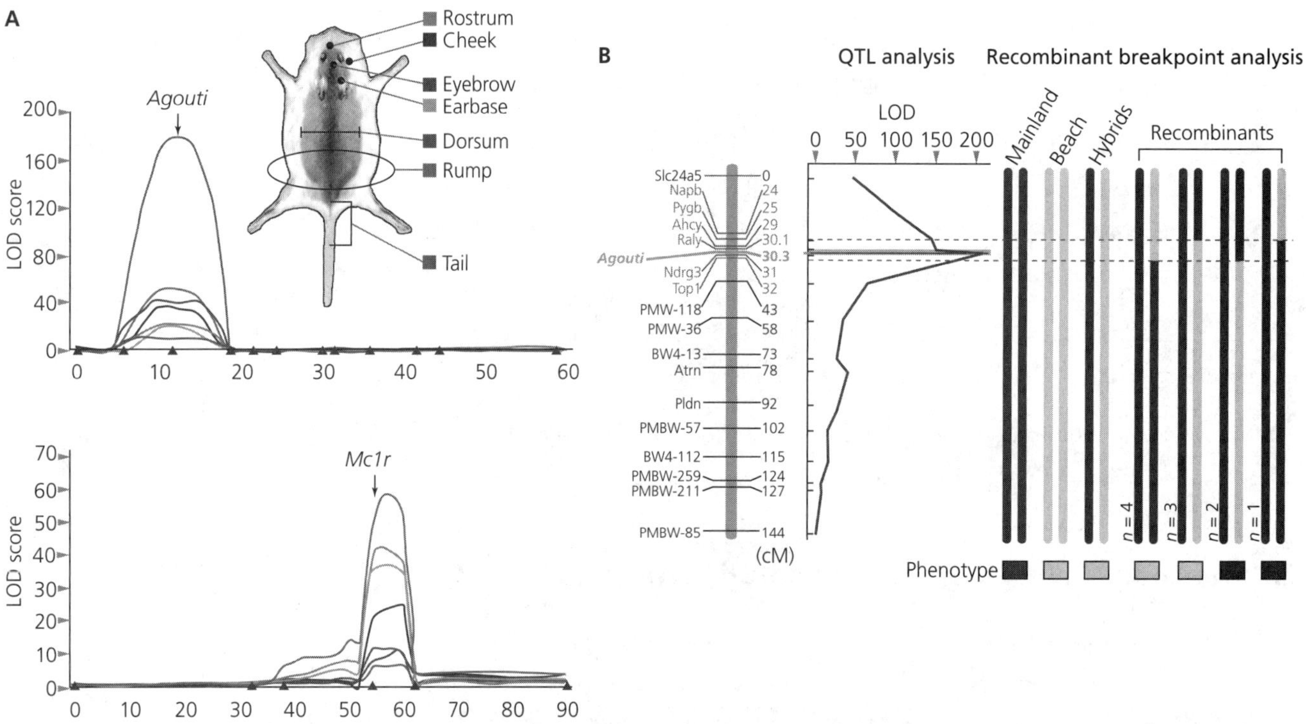

Hopi Hoekstra and her colleagues conducted an experiment to examine the coloration of oldfield mice by crossing beach mice with mainland mice. Beach mice are mostly white, and mainland mice are dark on their dorsal surface. Because hybrids inherit one copy of each chromosome from their parents, the offspring had intermediate coat colors. When the scientists bred the hybrids, they produced a second F_2 mouse generation. They used QTL analysis on the F_2 generation to look for correlations among alleles and seven locations on the coat. In A, LOD refers to the "logarithm of the odds" score, a statistical estimate of whether two loci (e.g., a marker and a gene influencing a focal trait) are likely to lie next to each other on a chromosome. cM = centimorgans, units of distance along a chromosome. B is a closer look at the genetic region containing *Agouti.* (Figure adapted from Steiner et al. 2007 and Manceau et al. 2011.)

- Which region had the highest LOD for *Agouti*? For *Mc1r*?

- Based on this evidence, why might the researchers looked more closely at Agouti?

- How far along on the chromosome does the *Agouti* sequence lie?

- What does the diagram in B indicate about the phenotype associated with heterozygotes at this locus? Why is this important?

Games and Exercises

What Does a *Mean* Mean?

Scientists and statisticians use the terms mean and variance to describe the distribution of a continuous variable—the way the data is spread out. The "normal" distribution is that classic bell curve where most of the values are in the center of the distribution, and as you move away from the center, values get more and more rare. In its simplest form, the x-axis lists the value of the variable, and the y-axis lists the number of occurrences of those values. If you tally the number of occurrences of each value, you can plot them as a histogram, which should match the curve of a bell fairly closely—if the data are normally distributed.

Not every measure is normally distributed, however, and it's important to know when that assumption is being violated. If you were to plot the heights of all adults, you'd probably see a bimodal distribution—that is two peaks—one for males and one for females. You'd have to plot the sexes separately, but the distributions of adult male heights and adult female heights should be shaped like a bell. Wages, on the other hand, definitely aren't normally distributed. (Would you *expect* them to be?)

So a mean is the average of a set of measurements. It's a measure of the center of the distribution. The *population* mean (μ) is an expected outcome—if you could record every value ever, you would expect the mean to be the most common value. The *sample* mean ($\bar{x}$), is the average of the values you actually have. The distinction is important because you don't really know if the values you actually have truly reflect every value ever. So, the sample mean is an *estimate* of the population mean.

To calculate the sample mean ($\bar{x}$), simply add all the values you've got and divide by how many values you added together. That equation looks like this:

$$\bar{x} = \frac{1}{n} \sum_{i} x_1$$

where n is the number of values and x_i are just each value (x_1, x_2, x_3, etc.). Most spreadsheets have an "average" function that does this for you.

Ask some of your friends how tall they are and record their heights (for now, stick with all boys or all girls). Calculate their mean height. Here's an example:

Friend #	Height (cm)
1	152.4
2	162.56
3	165.1
4	157.48
5	152.4
Sum	789.94

$\bar{x} = 789.94 / 5 = 157.99$

The variance and the standard deviation basically describe the spread of the distribution: how wide the bell is. Like the population mean, the population variance (σ^2) and population standard deviation (σ) are expected outcomes: if you could record every value ever. But since you can't record every value ever, the sample variance serves as an estimate of the population variance. It describes how far each individual measurement is from the mean of your sample.

$$s^2 = \frac{1}{n-1} \sum_{i} (x_1 - \bar{x})$$

So, for each value you record, subtract the sample mean and square the remainder. Add all those squares together and divide by the total number of values you recorded minus one. (You subtract the one because you are using the sample mean as an estimate of the population mean when you subtract the individual measurements.) Take the square root of the sample variance, and you've got the sample standard deviation.

Friend #	Height (cm)	$x_i - x$	$(x_i - \bar{x})^2$
1	152.40	152.40 − 157.99 = -5.59	31.226
2	162.56		
3	165.10		
4	157.48		
5	152.40		
Sum	789.94		

Calculate s^2 and s for this example:

Alone these measurements probably don't mean that much to you, but as you look at different variables and different sample sizes, you'll start to get an idea of what the sample variance and standard deviation are telling you about the width of the bell, especially when combined with your understanding of the mean. For "normal" samples, most of the values—95 percent—will lie within two standard deviations of the mean.

Check out the *Math Is Fun* website (http://www.mathsisfun.com/data/standard-normal-distribution.html) for more information about using standard deviations in the real-world, some practice questions, and even some games (some that may even help you do better with math).

The Dating Game

James Jones from Richland Community College developed a way for you to find your "perfect mate" using standard deviations. The idea is that if your names have similar standard deviations, you must be made for each other! (Not really, but who knows?)

Start by converting your name to a number by counting off the letters of the alphabet A=1, B=2, all the way to Z=26. Here's a conversion table to make it easy:

A	1	N	14
B	2	O	15
C	3	P	16
D	4	Q	17
E	5	R	18
F	6	S	19
G	7	T	20
H	8	U	21
I	9	V	22
J	10	W	23
K	11	X	24
L	12	Y	25
M	13	Z	26

For James, he first needed to find "his" standard deviation. "James" converts to:

		$x_i - x$	$(x_i - \bar{x})^2$
J	10	0.4	0.16
A	1	-8.6	73.96
M	13	3.4	11.56
E	5	-4.6	21.16
S	19	9.4	88.36
	48		195.2

$\bar{x} = 48 / 5 = 9.6$ and $s = 6.99$.

With a bit of rounding, you can see that the distribution of the name "James" is centered around the mean "J" with a standard deviation of about 7 letters.

So using this tool, who is better for James, Sandi or Brenda?

		$x_i - x$	$(x_i - \bar{x})^2$
S	19		
A	1		
N	14		
D	4		
I	9		

		$x_i - x$	$(x_i - \bar{x})^2$
B	2		
R	18		
E	5		
N	14		
D	4		
A	1		

- Based on standard deviations, who's a closer match for James?
- Go back and look at the distributions of letters in each of the names. Could you have made an educated guess about whose names had more similar distributions?
- What about name length? What should that do to the standard deviation?

- What's your standard deviation?

Find more from James Jones here: https://people.richland.edu/james/ictcm/2001/

Agouti Horses!

Jennifer Hoffman developed this cool interactive exploration of coat color in horses. You can play with the interaction of genes to create different coats.

http://www.jenniferhoffman.net/horse/index.html (This work is licensed to Jennifer Hoffman under a Creative Commons Attribution-Noncommercial-Share Alike 3.0 Unported License. See www.jenniferhoffman.net.)

Explore!

Phenotypic Plasticity in Development and Evolution

A whole issue of the *Philosophical Transactions of the Royal Society* devoted to phenotypic plasticity! This introduction reviews some of the basic concepts about phenotypic plasticity, including the types and components of variation and the evolution of plastic traits.

Fusco G., and A. Minelli. 2010. Phenotypic plasticity in development and evolution: facts and concepts. *Philosophical Transactions of the Royal Society B* 365(1540): 547-556. doi: 10.1098/rstb.2009.0267

http://rstb.royalsocietypublishing.org/content/365/1540/547.full

Genotype-Environment Interaction and Phenotypic Plasticity

This short video introduces phenotypic plasticity and multifactorial traits.

http://education-portal.com/academy/lesson/genotype-environment-interaction-and-phenotypic-plasticity.html#lesson

Siamese Cat Genetics: Your Cat Is a Heat-Map

The dark "points" in Siamese cats are a great example of phenotypic plasticity. Siamese have a mutation that affects the temperature sensitivity of melanin production in their coats.

http://www.brighthub.com/science/genetics/articles/40501.aspx

Imes, D. L., L. A. Geary, R. A. Grahn, and L. A. Lyons. 2006. Albinism in the domestic cat (*Felis catus*) is associated with a tyrosinase (TYR) mutation. *Animal Genetics* 37(2): 175-178. doi: 10.1111/j.1365-2052.2005.01409.x

http://www.ncbi.nlm.nih.gov/pmc/articles/PMC1464423/

How the Tree Frog Has Redefined Our View of Biology

From *Smithsonian Magazine,* the tree frog is an amazing story of survival and adaptation and surprising phenotypic plasticity.

http://www.smithsonianmag.com/science-nature/how-the-tree-frog-has-redefined-our-view-of-biology-165716397/#BUCJ3xqXbOAqj2mi.99

Urban Versus Forest Dwelling in Birds

This video by the International Max Planck Research School (IMPRS) for Organismal Biology follows Catarina Miranda, a PhD student at the Max Planck Institute for Ornithology, as she looks to see if there are differences between urban and forest birds in personality and physiology, and whether those differences are due to microevolutionary changes or just a result of phenotypic plasticity.

https://www.youtube.com/watch?v=S4F_X81sYVs

EvoTutor

Alan Lemmon offers this Java-based online tool that allows students to teach the concepts of evolution, including simulations, and you can explore the influence of directional, stabilizing, and disruptive selection on populations.

http://www.evotutor.org/Selection/Sl4A.html

Overcoming Misconceptions

Peter Visscher, William Hill, and Naomi Wray review several common misconceptions about heritability in their 2008 article in *Nature Reviews Genetics.* These misconceptions aren't just common to students, either. Really understanding heritability takes some work. Here are a couple important concepts to keep straight:

- High Heritability Does NOT Imply Genetic Determination

  Genetic determination is the idea of the inevitable consequence of genes—a gene determines a trait. The environment can play a large role in the development of a phenotype, even if variation in a trait has high heritability. Even though we know the genotype of an individual, we don't *necessarily* know the phenotype. Just because height has a heritability of 0.8 doesn't mean that the offspring of tall parents will all be tall. When there is high heritability of a trait, however, the phenotype is good *predictor* of an individual's phenotype because much of the variation observed in the population is caused by variation in genotypes (see Visscher et al. 2008).

- Heritability Is NOT the Proportion of a Phenotype That Is Passed on to the Next Generation

  Don't get confounded: The phenotype is not passed down to offspring, the genotype is. Narrow-sense heritability examines the additive genetic effects on variation passed down from the parents—half from each—but which half is passed down is unique to each offspring (see Visscher et al. 2008).

  Peter M. Visscher, William G. Hill, and Naomi R. Wray. 2008. Heritability in the genomics era — concepts and misconceptions. *Nature Reviews Genetics* 9 (April): 255–266.

Phenotypic Plasticity Is a Property of the Genotype

Phenotypic plasticity is the capacity of a genotype to produce different phenotypes in different environments. It is a property of the genotype. This capacity also can be adaptive (consider the visibility of being a brown hare on brown background in summer and a white hare on a snowy background in winter). Phenotypic plasticity is not evolution, but it can evolve, again, because it is a property of the genotype.

Go the Distance: Examine the Primary Literature

L. Scott Mills and his colleagues examined the seasonal coat color change in snowshoe hares. Snowshoe hares are typically brown in the summer and white in the winter. They argue that a mismatch between coat color and the environment can be big problem for these important prey animals.

- Which components of snowshoe hare coat color change were plastic in the study? Which were not?
- Why did Mills and his colleagues examine climate projections?
- Will snowshoe hares be able to alter their seasonal coat color change if the climate changes?

Mills, L. S., M. Zimova, J. Oyler, S. Running, J. T. Abatzoglou, and P. M. Lukacs, 2013. Camouflage mismatch in seasonal coat color due to decreased snow duration. *Proceedings of the National Academy of Sciences* 110: 7360–7365. http://www.pnas.org/content/early/2013/04/10/1222724110

Delve Deeper

1. Do populations that experience selection always evolve? Why or why not?

Test Yourself

1. What do the terms in the equation $H2 = \frac{V_G}{V_P} = \frac{V_G}{V_G + V_E}$ mean?

a. Broad-sense heritability equals the ratio of variation due to genotype to the variation due to phenotype.

b. Phenotypic variation results from variation due to the genotype plus variation due to the environment.

c. Heritability is the proportion of the total phenotypic variance that is attributable to genetic variation among individuals.

d. The variation due to genotype can be calculated by multiplying the phenotypic variance by broad-sense heritability.

e. All of the above

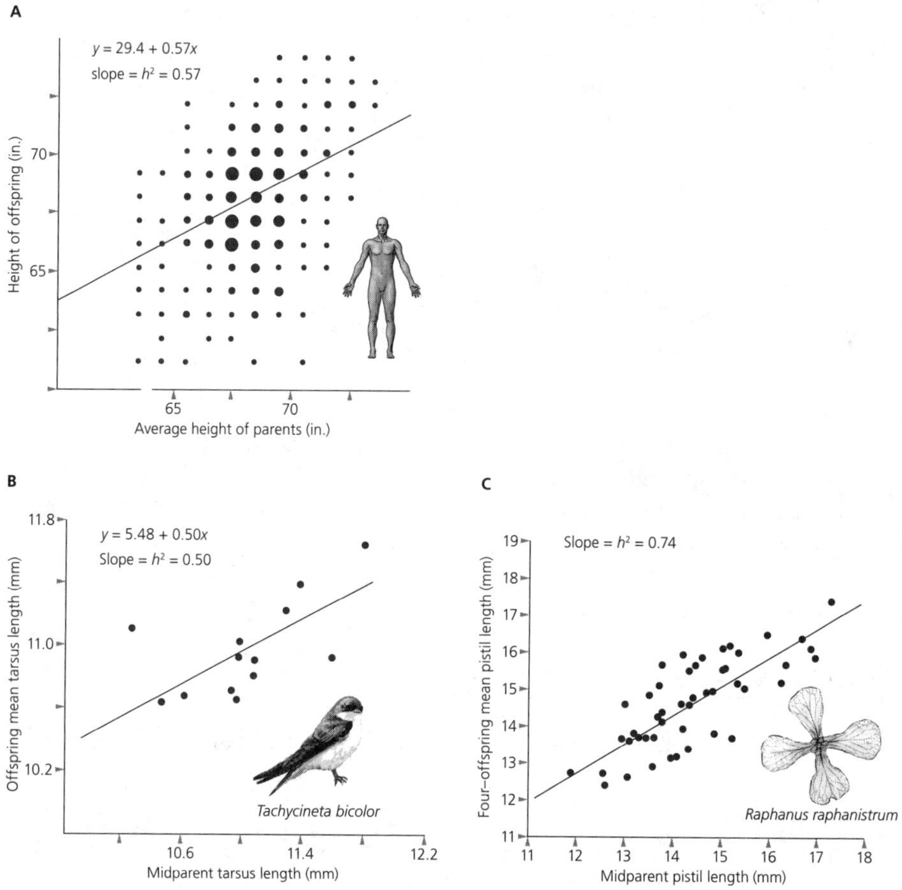

2. Which of the populations in the figure above (Figure 7.5) would you predict will evolve most in response to selection?
   a. Humans
   b. Tree swallows
   c. Wild radish
   d. Both a and b should evolve most.
   e. None of the populations should evolve.

3. According to Figure 7.13, how did Hoekstra and colleagues test whether the *Agouti* QTL they had discovered was coding for coat color?
   a. They conducted a second hybridization study and found that coat color phenotypes were most closely correlated with the *Agouti* QTL.
   b. They went back to their original study and mapped genes paying particular attention to the *Agouti* QTL.
   c. They determined the "logarithm of the odds" score to estimate whether the marker and the gene influencing coat color were likely to lie next to each other on the chromosome.
   d. They examined the color of each mouse's coat at seven locations and determined which had the greatest influence on coat color overall.
   e. They conducted a second hybridization study and found that coat color was influenced by more than one gene.
4. Is coat color in oldfield mice a polygenic trait?
   a. No. QTL regions are components of DNA produced by hybridization; they do not represent the actual DNA of oldfield mice.
   b. No. *Agouti* and *Mc1r* are QTL regions that encode for receptors and repressors; they are not actually genes, so coat color cannot be polygenic.
   c. Maybe. Scientists don't really understand the genes behind coat color: *Agouti* and *Mc1r* are just regions they can identify.
   d. Yes. Two genes are primarily involved in coat color: *Agouti* encoding a repressor that shuts down the *Mc1r* receptor; *Mc1r* encoding a receptor that triggers the production of pigment.
   e. Yes. All animal traits are polygenic.
5. According to Figure 7.21, which trait(s) of *Caenorhabditis elegans* is/are phenotypically plastic?
   a. Age at maturity
   b. Fertility
   c. Chromosome IV
   d. Both a and b
   e. None of the above

Contemplate

Is animal behavior heritable? Is it plastic?	What kind of phenotypic plasticity might you find in plants and why would you expect them to have plastic traits?

8 Natural Selection
Empirical Studies in the Wild

Check Your Understanding

1. From which sources does variation among individuals ultimately arise?
   a. Genetic recombination
   b. Mutation
   c. Independent assortment
   d. All of the above
   e. None of the above
2. Within a population, allele frequencies can change as a result of genetic drift, natural selection, migration, and mutation. Which mechanism would you argue is the most sensitive to variation among individuals?
   a. Genetic drift because it can only occur when individuals are very different from each other
   b. Natural selection because variation among individuals is the foundation for relative success of individuals, and that relative success affects the frequencies of alleles within the population
   c. Migration because individuals with more allelic variation will be less likely to migrate than individuals with less variation
   d. Mutations because they damage individuals and remove them and their genes from the population
   e. Allele frequencies within a population do not change.
3. What is the selection differential (S)?
   a. Negative selection
   b. The difference between the trait mean of reproducing individuals and the trait mean of the general population
   c. A measure of evolution
   d. The difference between variation in a phenotypic trait caused by the environment and variation caused by the underlying genetic architecture.
   e. The difference between narrow sense heritability and response to selection

Learning Objectives for Chapter 8

Add important definitions and notes next to each learning objective for this chapter to help guide your understanding.

Learning Objective	Important Definitions	Notes
Compare and contrast the factors leading to directional and stabilizing selection and the outcomes of each.		
Demonstrate how predators can act as agents of selection.		
Explain how selection can vary across a species' range.		
Explain how natural selection can act on an extended phenotype.		
Analyze the role of natural experiments in our understanding of evolutionary change in response to selection.		
Explain how selective sweeps can be detected within genomes.		
Evaluate the evidence for the role of humans as selective agents in the evolution of plants and animals.		

Identify Key Terms

Match terms and definitions by filling in the blank to the right of the term with the appropriate letter.

1. **Aposematism** ____	a. The transfer of alleles from one population to another. It occurs when organisms or their gametes move from one location to another.
2. **Artificial selection** ____	b. The situation in which strong selection can "sweep" a favorable allele to fixation within a population so fast that there is little opportunity for recombination. In the absence of recombination, alleles in large stretches of DNA flanking the favorable allele will also reach high frequency.
3. **Ecological character displacement** ____	c. Structures constructed by organisms that can influence their performance or success. Although they are not part of the organism itself, their properties nevertheless reflect the genotype of each individual. Animal examples include the nests constructed by birds and the galls of flies.
4. **Extended phenotypes** ____	d. Similar to natural selection, except that it results from human activity. When breeders nonrandomly choose individuals with economically favorable traits to use as breeding stock, they impose strong artificial selection on those traits.
5. **Gene flow** ____	e. The physical proximity of alleles at different loci. Genetic loci that are physically close to one another on the same chromosome are less likely to be separated by recombination during meiosis. Thus they are said to be genetically linked.
6. **Genetic linkage** ____	f. An antipredator strategy used by a potential prey item to signal danger or a lack of palatability. The most commonly known form is warning coloration, in which the bright coloration of prey that are potentially dangerous can act as a deterrent to potential predators.
7. **Selective sweep** ____	g. Evolution driven by competition between species for a shared resource (e.g., food). Traits evolve in opposing directions, minimizing overlap between the species.

Link Concepts

Fill in the bubbles with the appropriate terms.

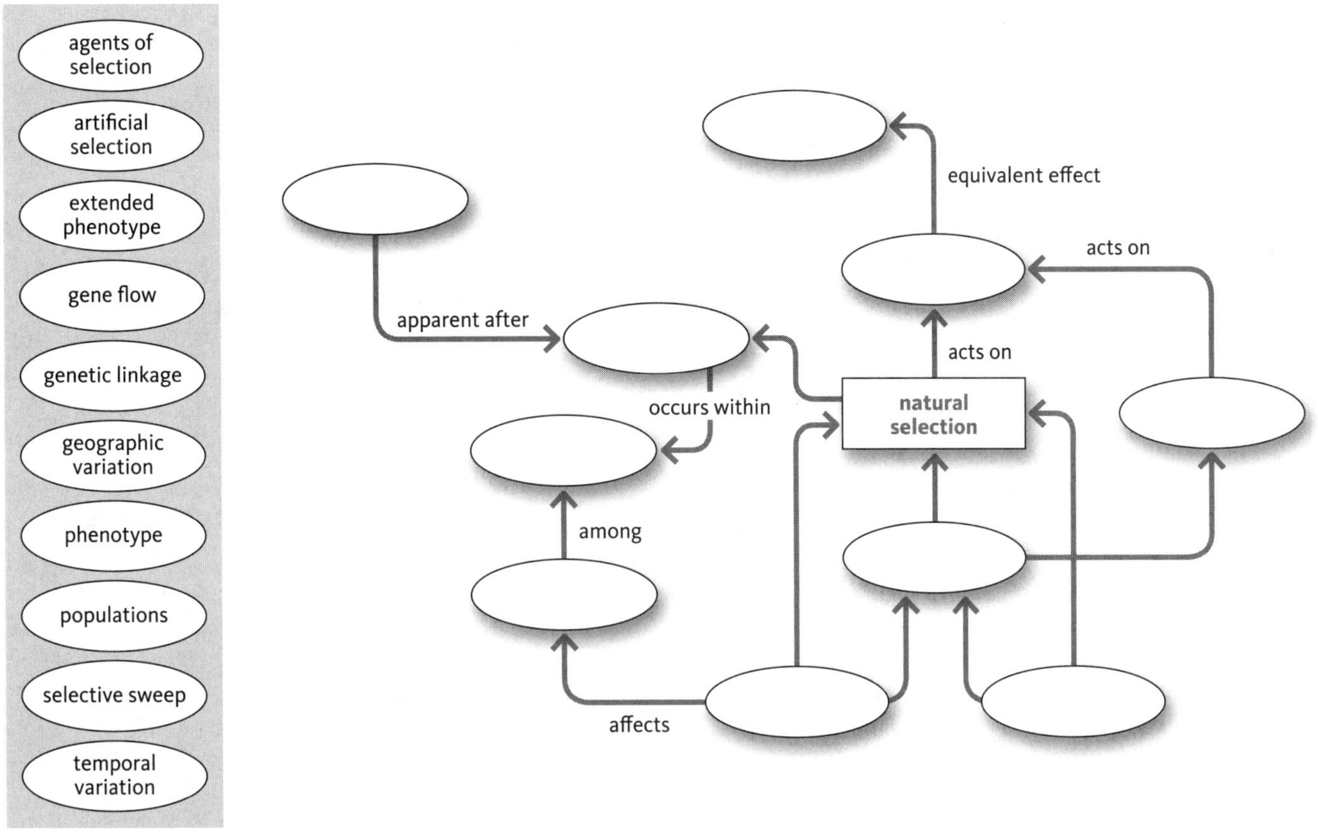

Key Concepts

Fill in the blanks for the key concepts from the chapter.

8.1 Evolution in a Bird's Beak

During a severe drought on Daphne Major, subtle differences in beak thickness among medium ground finches affected who lived and who died. Because beak depth is highly ____________, natural selection could lead to ____________ evolution of beak size.

Long-term studies of natural selection often show ____________ in the direction and magnitude of selection.

8.2 Mice in Black and White

When oldfield mice moved into new coastal habitats the nature of ____________ they experienced ____________. Cryptic individuals still survived ____________, but the colors conferring the best crypsis ____________ from brown to white.

Atlantic and Gulf coast mouse ____________ each evolved white fur, but the specific ____________ responsible are different.

8.3 The Geography of Fitness

Natural selection can lead to ____________ in space—across habitats or environments—just as dramatically as it can lead to ____________ in a single habitat over time.

8.4 Predators versus Parasitoids

When agents of selection act in ____________, the net effect can be a ____________: stabilizing selection for an ____________ trait value.

8.5 Replicated Natural Experiments

Sometimes multiple populations ____________ experience the same change in their selection environment. These populations are ideal for evolutionary studies because they act like ____________ natural experiments. The nature of the evolutionary response can be observed for each population and ____________ across the different populations.

8.6 Drinking Milk: A Fingerprint of Natural Selection

Some mutations can increase the ____________ of their carriers so greatly that the response to natural selection is ____________.

Scientists can use different lines of evidence, such as ____________ in DNA, to look for ____________ of natural selection in wild populations.

8.7 Humans as Agents of Selection

The speed of evolution is a product of the amount of available ____________ and the ____________. Weed and pest populations can be highly variable, and herbicides and pesticides can impose extremely strong ____________. The result: ____________ evolution of resistance.

An understanding of evolutionary biology can lead to novel management practices, which slow the ____________ in pest populations, or minimize undesirable evolutionary ____________ of harvesting.

Interpret the Data

Figure 8.16 is a map of the estimated frequency of the lactase persistence phenotype in the world (adapted from Itan et al. 2010). Individuals with the lactase persistence phenotype retain the ability to digest milk after weaning.

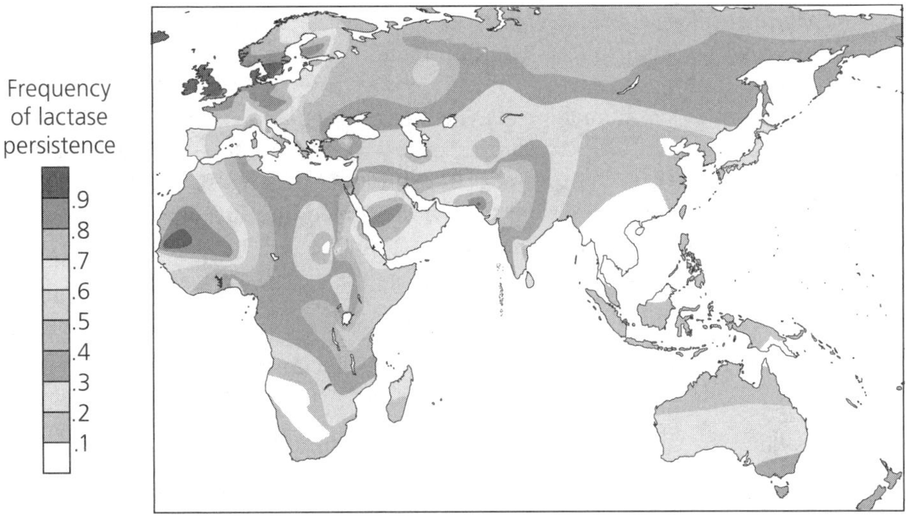

- Label the two sites where cattle were domesticated (East Africa and Northern Europe).
- At what frequency do scientists expect to find the lactase persistence phenotype in Australia?
- Why might the frequency of the lactase persistence phenotype be high in Pakistan?
- What explains the pattern of frequencies around the world?

Games and Exercises

The Evolution Experience

Does a trait, such as color, lead to differential survival? Thanks to the Natural History Museum in London you can test your skills as a predator and watch what happens to populations of blue and green beetles. You have to survive and reproduce, too, so make sure you get enough to eat.

http://www.nhm.ac.uk/nature-online/evolution/what-is-evolution/natural-selection-game/the-evolution-experience.html

Natural Selection

Bunnies breed like, well, bunnies. In this interactive simulation of natural selection from PhET at the University of Colorado—Boulder, you can develop your own ecological hypotheses that link the genetic control of coat color, and other adaptations, to the environment.

Go here and open the simulation:
https://phet.colorado.edu/en/simulation/natural-selection

You start with a single bunny. On the left side of the screen are the mutations you can add (including some control over dominance of alleles). On the right side are selective agents. The simulation also includes a population graph; you can zoom in and out of it to determine population size exactly. Be sure to note the generation bar and the play and pause buttons near the bottom of the screen.

To start, add a friend and just let the bunnies breed for three or four generations. Press pause and examine the population graph.

- Record the number of bunnies in the population:
- How many ticks along the x-axis represent one generation?

Now add wolves as a selective factor and let the simulation run.

- How many generations did it take before the white bunnies went extinct?
- Why might the wolves be successful at finding the bunnies?

Use the reset all button to return to a single white bunny and add a friend. Add "Brown fur" as a mutation, and change brown fur to a recessive trait. Start the simulation and let it run for two complete generations. Add wolves and let the simulation run for two more generations.

- Record the number of bunnies in the population:
- What proportion of the population is brown?
- What will happen to the population of bunnies if you let the simulation continue?

Let the simulation run and see if your prediction was correct.

Use the reset all button to return to a single white bunny and add a friend. Add "Brown fur" as a mutation, and change brown fur to a recessive trait. Start the simulation and let it run for two complete generations. Add wolves and let the simulation run for two more generations (stop the experiment before the white bunnies disappear completely from the population).

Now suppose it's winter. Change the habitat to "winter" by clicking on "arctic" as an environmental selective factor (don't reset). Let the simulation run for three generations.

- Record the number of bunnies in the population:
- What proportion of the population is brown?

- What will happen to the population of bunnies if you let the simulation continue?

Let the simulation run and see if your prediction was correct.

What would happen if you ran the same set of experiments, but the mutation to brown fur was dominant? Record your prediction, then run the simulations and see if you were correct.

- Based on your understanding of coat color genes, what mechanism might you propose for brown and white coat colors in snowshoe hares?

Here's some research scientists are doing on coat color in snowshoes:

Fontanesi, L., L. Forestier, D. Allain, E. Scotti, F. Beretti, S. Deretz-Picoulet, E. Pecchioli, C. Vernesi, T. J. Robinson, J. L. Malaney, V. Russo, and A. Oulmouden. 2010. Characterization of the rabbit *agouti signaling protein* (*ASIP*) gene: Transcripts and phylogenetic analyses and identification of the causative mutation of the nonagouti black coat colour. *Genomics* 95(3):166-75. doi: 10.1016/j.ygeno.2009.11.003.

- Individual snowshoe hares change colors over the seasons, however. How might that mechanism be modified to change seasonally? Can you come up with an experiment to test your predictions?

PhET Interactive Science Simulations from the University of Colorado—Boulder
https://phet.colorado.edu/

Explore!

The Story of Sticklebacks

Check out this video about the evolution of sticklebacks from Howard Hughes Medical Institute BioInteractives.

http://www.hhmi.org/biointeractive/making-fittest-evolving-switches-evolving-bodies

Sticklebacks Animated

The Genetic Science Learning Center has a great animation that tells the story of armor in sticklebacks in Loberg Lake. You can also watch lectures on the differences between freshwater and ocean-dwelling sticklebacks, as well as how scientists are uncovering important genes in armor evolution.

http://learn.genetics.utah.edu/content/selection/stickleback/

Recent News from AAAS.org and ScienceNOW: "Supergene" Is a Lifesaver for Colorful Butterfly

A linked group of genes may allow the Common Mormon swallowtail to mimic the look of a toxic species.

http://news.sciencemag.org/biology/2014/03/supergene-lifesaver-colorful-butterfly

ScienceShot: Elephants Shaped Their Own Evolution

Dietary preference impacted the structure of teeth, and ultimately their survival.

http://news.sciencemag.org/evolution/2013/06/sciencesshot-elephants-shaped-their-own-evolution

ScienceShot: How Some Goldfish Got Two Tails

Chinese breeders apparently selected for a rare mutation hundreds of years ago.

http://news.sciencemag.org/biology/2014/02/scienceshot-how-some-goldfish-got-two-tails

Getting Crops to "Talk" to Insects

Eco-friendly pest control with genetically modified plants that produce bug pheromones.

http://news.sciencemag.org/plants-animals/2014/02/getting-crops-talk-insects

How Farming Reshaped Our Genomes

Analysis of 8000-year-old DNA suggests that the cultivation of crops and livestock humans.

http://news.sciencemag.org/archaeology/2014/01/how-farming-reshaped-our-genomes

ScienceShot: What Did Corn's Ancestor Really Look Like?

Corn's ancestral form was more like a spindly grass.

http://news.sciencemag.org/plants-animals/2014/02/scienceshot-what-did-corns-ancestor-really-look

Extreme Diets Can Quickly Alter Gut Bacteria

We are much more than what we eat.

http://news.sciencemag.org/biology/2013/12/extreme-diets-can-quickly-alter-gut-bacteria

Overcoming Misconceptions

Evolution Does NOT Provides Species with the Adaptations They "Need"

Adaptations do not evolve because an organism "needs" them. The match between species and their environments can be striking, but as the Galápagos finch data shows, some individual organisms don't match their environments. The *population* of finches generally has beaks that match the seed size available, however, and *that* is remarkable. Evolution is based on processes such as random mutations and the change of gene frequencies from one generation to the next. It cannot identify "needs." Natural selection is a mechanism of evolution that causes gene frequencies to change because some individuals with some traits survive and reproduce better than other individuals with other traits.

Evolution Does NOT Always Go from Simple to Complex

Complex traits have indeed evolved from simpler precursors, but they also frequently evolve in the "reverse" direction. Indeed, studies of natural selection show that evolution can change directions in time and space. Some complex traits, like aposematic coloration, can evolve in populations under some conditions, but they can be selected against under other circumstances. This complicated process has shaped the evolution of life on Earth, leading to some traits that are highly complex and other traits that are exquisite in their simplicity. For example, we'll look at the evidence for the evolution of our eyes from light-sensitive cells in Chapter 9. And in Chapter 13, we'll see how bacteria lost most of the genes essential for surviving as free-living bacteria, giving rise to the energy-producing structures in eukaryote cells called mitochondria.

Mechanisms of Evolution Involve More than Natural Selection

Chapter 8 summarizes some of the evidence showing the crucial role natural selection plays in evolutionary theory. Depending on the environment, the traits natural selection favors within a population can shift or natural selection can favor different traits in different subsets of a population. Selection can be swift, sweeping traits that increase survival or reproduction through a population. However, evolutionary theory is far more than the mechanism of natural selection. Other important processes, such as genetic drift (Chapter 6) and sexual selection (Chapter 10), are integral to the theory.

Which of the following is a true statement?

a. Scarlet kingsnakes need to look like coral snakes to survive.
b. Natural selection always leads to more complex traits.
c. Natural selection and genetic drift are both important mechanisms of evolution.
d. Gall flies need galls that are safe from both predators and parasitoid wasps.

Go the Distance: Examine the Primary Literature

Joshua M. Akey and his colleagues provide a genome-wide analysis of selection in dogs using single nucleotide polymorphisms (SNPs). They were able to develop a fine-scale map that can be used to trace the genetic basis of phenotypes.

- What was their research question?

- Why do the authors argue that artificial selection likely affected regulatory regions rather than protein producing regions?

- Why might this technique might be important to understanding selection in humans?

Akey, J. M., A. L. Ruhe, D. T. Akey, A. K. Wong, C. F. Connelly, J. Madeoy, T. J. Nicholas, and M. W. Neff. 2010. Tracking footprints of artificial selection in the dog genome. *Proceedings of the National Academy of Sciences* January 11, 2010. doi:10.1073/pnas.0909918107. http://www.pnas.org/content/early/2010/01/06/0909918107.full.pdf+html

Delve Deeper

1. How many different lines of evidence allowed scientists to reconstruct the recent history of natural selection on sticklebacks in freshwater lakes?

Test Yourself

1. What does the shaded area in Figure 8.5B indicate?
   a. A large and significant change in mean beak size for the finch population
   b. Directional selection on beak size
   c. The year selection changed directions
   d. All of the above
   e. None of the above
2. Why did Hopi Hoekstra and her colleagues reject the hypothesis that light colored coats evolved just once in oldfield mice?
   a. Because the light coat color would only result with mutations to both melanocortin-1 receptor (*Mc1r*) and *Agouti*
   b. Because light-colored mice on the Atlantic Coast lacked the alleles of *Mc1r* that produced light coat color on the Gulf Coast
   c. Because even though the mice were phenotypically similar, they were actually distinct species
   d. Because oldfield mice cannot migrate that far
3. Why is gene flow important to evolution?
   a. Because it can prevent populations from becoming isolated
   b. Because it can change the frequencies of alleles within a population
   c. Because it can influence genetic drift
   d. Because it can minimize variation across habitats
   e. All of the above
4. George Harper and David Pfennig found that natural selection favors mimicry where the ranges of scarlet kingsnakes and coral snakes overlap. How far from that border did the fitness of mimicry alleles drop to 0?
   a. Between 0–100 km inside of the border
   b. Between 80–150 km outside of the border
   c. Almost 200 km outside of the border
   d. Right at the border (approximately 0 km)
   e. The mimicry alleles did not affect fitness

5. How did Arthur Weis and Warren Abrahamson determine that gall formation in host plants was at least partially heritable?
   a. They compared the mean gall size of different female's offspring while controlling for genetic differences among the plants.
   b. They bred distinct fly lineages so some only produced large galls and some only produced small galls.
   c. They isolated gall formation alleles using QTL.
   d. All of the above
   e. None of the above
6. Which of the following statements about gall formation is false?
   a. The optimal gall diameter for this species of fly is approximately 20 mm.
   b. Downy woodpeckers are more efficient predators of large galls than small galls.
   c. Parasitoid wasps cause strong directional selection on fly gall size.
   d. Only flies in galls that are approximately 20 mm survive to reproduce.
   e. Parasitoid wasps vary in their ability to detect fly galls.
7. Why have studies of lake sticklebacks been so informative to understanding the evolutionary response of populations?
   a. Because the lakes were so close to where scientists lived they could observe individual sticklebacks for extended periods of time to see how successful they were
   b. Because populations in the lakes are isolated from each other, and each population experienced natural selection independently of other populations
   c. Because the sticklebacks in lakes were easier to catch than sticklebacks that live in the oceans
   d. Because the populations were all related originally, scientists could observe genetic drift in the founder populations within each new lake
8. Which of the following statements about the low-*Eda* allele is true?
   a. The low-*Eda* allele arose in the marine ancestors of sticklebacks.
   b. The low-*Eda* allele can remain at low frequency in saltwater populations because it is recessive.
   c. The low-*Eda* allele is favored in one habitat type and not favored in another.
   d. Two copies of the recessive low-*Eda* allele can confer a selective advantage in freshwater habitats.
   e. All of the above
9. What is genetic linkage?
   a. Associations among alleles that result because loci are physically close to each other on a chromosome
   b. The linkages of codon triplets within DNA that form amino acids
   c. The linking of chromosome pairs that occurs before gametes form during meiosis
   d. Associations that arise when a gene regulates the expression of other genes that are downstream from its location
   e. None of the above

10. Given your experience with selective agents, why might natural selection favor smaller body size in introduced cane toads?
    a. Because cane toads were significantly larger shortly after their introduction than cane toads 10 years later
    b. Because cane toads exhibit variation in size
    c. Because cane toad predators were likely larger and more resistant in their native region
    d. Because cane toads can only eat small prey in the new habitat
    e. Natural selection does not favor smaller body size in cane toads.
11. According to Figure 8.28B, how has trophy hunting affected horn length of bighorn sheep?
    a. Harvested bighorn sheep have shorter horns and are less fit than unharvested bighorn sheep.
    b. Horn length in bighorn populations has been reduced by 20 cm.
    c. Bighorn sheep with small horns are more commonly harvested than bighorn sheep with large horns.
    d. The variation in horn length within bighorn populations has declined over time.
    e. Horn length varies among individuals.

Contemplate

Can the three conditions for evolution by natural selection be broadly applied to fashion in the United States? If so, would you expect to see similar patterns of variation in space and time?	Do humans act as agents of selection on diseases?

9 The History in Our Genes

Check Your Understanding

1. Which of the following is NOT a mutation that potentially alters DNA of future generations?
    a. Chromosome fusion
    b. Deletion
    c. Insertion
    d. Point mutation
    e. Morphogen
2. Are homologous traits ancestral or derived traits?
    a. They are ancestral because homologous traits are inherited from a common ancestor.
    b. It depends on the clade of interest within the phylogeny.
    c. They are derived because homologous traits define a monophyletic clade.
    d. None of the above
3. Are synapomorphies important when developing a phylogeny?
    a. Synapomorphies are not important when developing phylogenies because synapomorphies are traits shared by all the members of a species, population, or gene family, and they provide little useful information.
    b. Synapomorphies are not important when developing phylogenies because they are produced by convergent evolution or evolutionary reversal and as such, can confound the phylogeny.
    c. Synapomorphies are important when developing phylogenies because they are a lineage of tetrapods that gave rise to mammals.
    d. Synapomorphies are important when developing phylogenies because they are the shared derived traits that can distinguish groups of species, populations, or genes from other groups sharing different characters.

Learning Objectives for Chapter 9

Add important definitions and notes next to each learning objective for this chapter to help guide your understanding.

Learning Objective	Important Definitions	Notes
Discuss how coalescence can help us understand phylogenetic relationships.		
Explain how and why gene trees can be different from species trees.		
Describe the methods scientists use to construct phylogenetic trees.		
Discuss the evidence used to determine the origin of tetrapods, Darwin's finches, humans, and HIV.		
Explain the neutral theory of evolution, and discuss how the theory is used to deduce the timing of evolutionary events and the history of natural selection.		
Explain how phylogenetic approaches can assist in discovering disease-causing genes.		
Compare and contrast the evolution of genome size in bacteria and eukaryotes.		

Identify Key Terms

Match terms and definitions by filling in the blank to the right of the term with the appropriate letter.

1. **Bootstrapping** ____ 2. **Coalescence** ____ 3. **Distance-matrix methods** ____ 4. **Gene tree** ____ 5. **Genomics** ____ 6. **Incomplete lineage sorting** ____ 7. **Maximum likelihood and Bayesian methods** ____ 8. **Maximum parsimony** ____ 9. **Microsatellites** ____ 10. **Molecular clock** ____ 11. **Neighbor joining** ____ 12. **Non-synonymous (replacement) substitutions** ____ 13. **Orthologous genes** ____ 14. **Paralogs** ____ 15. **Purifying selection (also called negative selection)** ____ 16. **Synonymous substitutions** ____	a. The branched genealogical lineage of homologous alleles that traces their evolution back to an ancestral allele b. A distance method for reconstructing phylogenies. This method identifies the tree topology with the shortest possible branch lengths given the data. c. Procedures for constructing phylogenetic trees by clustering taxa based on the proximity (or distance) between their DNA or protein sequences. These methods place closely related sequences under the same internal branch, and they estimate branch lengths from the observed distances between sequences. d. The case when the history of a gene differs from the history of the species carrying the gene. e. The study of the structure and function of genomes, including mapping genes and DNA sequencing. The discipline unites molecular and cell biology, classical genetics, and computational science. f. The process by which, looking back through time, the genealogy of any pair of homologous alleles merges in a common ancestor. g. A method used to determine time based on base-pair substitutions. They use the rates of molecular change to deduce the divergence time between two lineages in a phylogeny, for example. They work best when they can be "calibrated" with other markers of time, such as fossils with known ages and placements. h. Approaches used to estimate parameter values for a statistical model. They are used in phylogeny reconstruction to find the tree topologies that are most likely, given a precise model for molecular evolution and a particular data set. i. A statistical method for estimating the strength of evidence that a particular branch in a phylogeny exists j. Homologous genes separated by a speciation event k. Noncoding stretches of DNA containing strings of short (1–6 base pairs), repeated segments. The number of repetitive segments can be highly polymorphic, and for this reason these stretches are valuable genetic characters for comparing populations and for assigning relatedness among individuals (DNA fingerprinting). l. Homologous genes that arise by gene duplication. Together they form a gene family. m. Substitutions that do not alter the amino acid sequence of a protein. Because these substitutions do not affect the protein an organism produces, they are less prone to selection and often free from selection completely. n. Substitutions that alter the amino acid sequence of a protein. They can affect the phenotype and are therefore more subject to selection. o. A statistical method for reconstructing phylogenies which identifies the tree topology that minimizes the total amount of change, or the number of steps, required to fit the data to the tree. p. A type of selection that removes deleterious alleles from a population. It is a common form of stabilizing selection (Chapter 7).

Link Concepts

Fill in the bubbles with the appropriate terms.

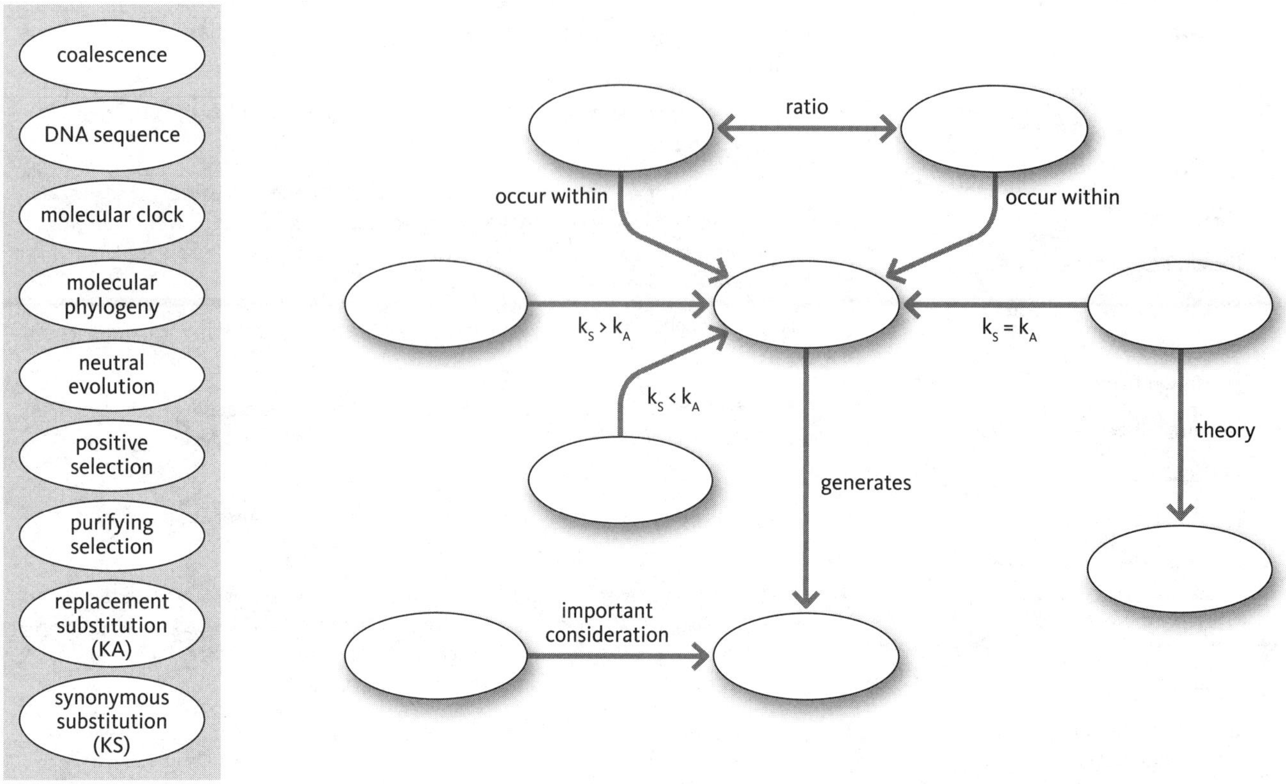

Develop your own concept map that explains the evolution of HIV, using the concepts HIV, SIV, mutation, chimpanzee, human, amino acid, and protein.

Key Concepts

Fill in the blanks for the key concepts from the chapter.

9.1 Coalescing Genes

It is possible to trace the genealogies of ____________ back through time, reconstructing when mutations generated ____________ and how these alleles subsequently spread.

9.2 Gene Trees and Species Trees

Gene trees occasionally have ____________ unlike those of the species that carry them. This is one reason that scientists often use information from several genes when they infer ____________ from molecular data.

9.3 Methods of Molecular Phylogenetics

Phylogenetic trees are actually ____________ about the relationships among species or groups of individuals.

____________ help scientists sift through volumes of molecular evidence to determine the ____________ hypothesis or hypotheses.

9.4 Four Case Studies in Molecular Phylogeny

Constructing phylogenies is often a process of ____________ evidence. Scientists can test the ____________ of phylogenetic ____________ developed with one line of evidence using other, ____________ lines of evidence to draw conclusions.

9.5 Natural Selection versus Neutral Evolution

The neutral theory of molecular evolution describes the ____________ of nucleotide sequence evolution under the forces of ____________ and random ____________ in the absence of selection.

The neutral theory ____________ that neutral mutations will yield nucleotide substitutions in a population at a rate ____________ to the rate of ____________, regardless of the size of the population.

As long as mutation ____________ remain fairly constant through time, neutral variation should ____________ at a steady rate, generating a molecular ____________ that can be used to date events in the distant past.

Both ____________ and ____________ selection leave distinctive signatures in nucleotide or amino acid sequences that can be detected using statistical tests.

9.6 Detecting Genes

By aligning ____________ of different species, scientists discovered that segments of ____________ could contain ____________ that had not previously been detected. The discovery has helped to identify new genes important for physiology and ____________ .

9.7 Genome Evolution

Bacteria typically have relatively ____________ genomes made up mostly of genes, whereas eukaryotes have genomes that ____________ greatly in size.

As more and more ____________ are sequenced, our understanding of genome evolution is ____________ rapidly. Scientists are discovering ways to answer more and more important ____________ about the role that ____________ and architecture play in the origins of our ____________ as well as in other eukaryotes.

Interpret the Data

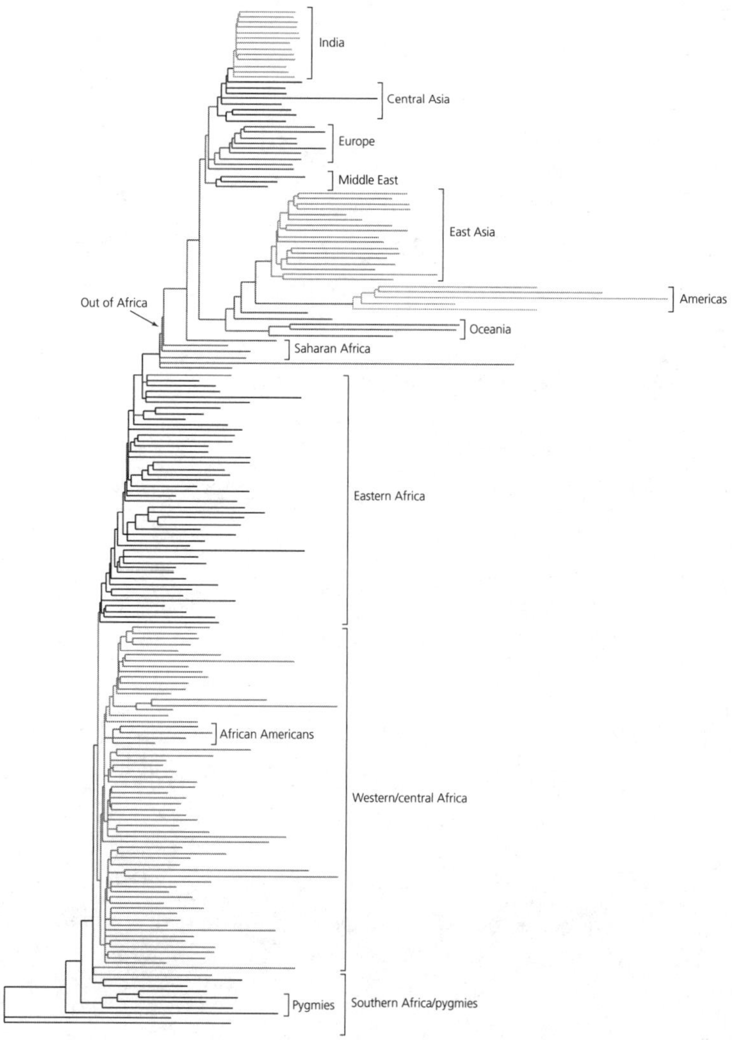

Sarah Tishkoff and colleagues developed this phylogeny of humans based on mitochondrial DNA (Figure 9.11).

- What group shares the most recent common ancestor with the group of humans that colonized the Americas?

- According to Tishkoff et al. (2009), the longer branch lengths for groups of humans in the Americas, Oceania, and Pygmy, and some of the hunter-gatherers, indicate high levels of genetic drift. Why might those groups have experienced higher levels of genetic drift than groups that remained in Africa?

Games and Exercises

More Fun with Phylogenies

In Chapter 4, using some simple characters, you were able to develop a cladogram showing the relationships of cartoon beetles. Although it may have seemed a bit silly to do with cartoons, scientists use similar methods to examine the relationships among genes, species, and even languages! In fact, Walter Fitch and Charles Langley were able to plot out the evolutionary changes in a gene that codes for a protein called cytochrome c. Cytochrome c is necessary for energy transfer in cells and is found in most, if not all, Eukaryotes. Fitch and Langley worked out the mutations that had accumulated in 17 lineages, including humans and horses, and found that the mutations became fixed with clock-like regularity. With this exercise developed by Beth Kramer (and added to ENSI website in 2002 http://www.indiana.edu/~ensiweb/lessons/mol.bio.html), you can compare the relatedness of a variety of Eukaryotes—including humans—based on cytochrome c amino acid sequences.

Start by comparing the amino acid sequences of cytochrome c between pairs of species. Use horses, donkeys, whales, chickens, penguins, snakes, moths, yeast, and wheat (see the table on the next page). You aren't comparing the DNA sequences themselves—codons are associated with the specific amino acids (see Chapter 5), and you are comparing amino acids.

Make sure you examine the entire molecule. Each protein sequence has 103–112 amino acids. The sequences start in the top half and continue onto a second line in the bottom half. Count the total number of differences between species pairs (the shaded amino acids are the same in all species) and record them in the table on the next page. Note also that some species have amino acids in locations that other species don't—those also count as differences.

Remember, you only have to fill out one half of the table because the halves mirror each other.

	Horse	Donkey	Whale	Chicken	Penguin	Snake	Moth	Yeast	Wheat
Horse	0								
Donkey		0							
Whale			0						
Chicken				0					
Penguin					0				
Snake						0			
Moth							0		
Yeast								0	
Wheat									0

Use the data in the table below to make a cladogram. The most closely related species have the fewest differences in amino acid sequences. Examine the table and determine which species are most closely related. Use the tree at the top of the next page and place species according to their relationships. Start with the two most closely related. Add those species to the shortest lines in the tree to indicate they are the most closely related. Is there another species pair that is closely related? Add them to the next shortest lines for the same reason. Continue to add species based on their relative relatedness until all have been placed.

	10 20 30 40 50
Amino Acid Number	1 2 3 4 5 6 7 8 9 0 1 2 3 4 5 6 7 8 9 0 1 2 3 4 5 6 7 8 9 0 1 2 3 4 5 6 7 8 9 0 1 2 3 4 5 6 7 8 9 0 1 2 3 4 5 6 7 8 9
Human, Chimpanzee	- - - - - - - - G D V E K G K K I F I M K C S Q C H T V E K G G K H K T G P N L H G L F G R K T G O A P G V S Y T A A
Rhesus monkey	- - - - - - - - G D V E K G K K I F I M K C S Q C H T V E K G G K H K T G P N L H G L F G R K T G O A P G Y S Y T A A
Horse	- - - - - - - - G D V E K G K K I F V Q K C A Q C H T V E K G G K H K T G P N L H G L F G R K T G O A P G F T Y T D A
Donkey	- - - - - - - - G D V E K G K K I F V Q K C A Q C H T V E K G G K H K T G P N L H G L F G R K T G O A P G F S Y T D A
Common zebra	- - - - - - - - G D V E K G K K I F V Q K C A Q C H T V E K G G K H K T G P N L H G L F G R K T G O A P G F S Y T D A
Pig, Cow, Sheep	- - - - - - - - G D V E K G K K I F V Q K C A Q C H T V E K G G K H K T G P N L H G L F G R K T G O A P G F S Y T D A
Dog	- - - - - - - - G D V E K G K K I F V Q K C A Q C H T V E K G G K H K T G P N L H G L F G R K T G O A P G F S Y T D A
Gray whale	- - - - - - - - G D V E K G K K I F V Q K C A Q C H T V E K G G K H K T G P N L H G L F G R K T G O A V G F S Y T D A
Rabbit	- - - - - - - - G D V E K G K K I F V Q K C A Q C H T V E K G G K H K T G P N L H G L F G R K T G O A V G F S Y T D A
Kangaroo	- - - - - - - - G D V E K G K K I F V Q K C A Q C H T V E K G G K H K T G P N L N G L F G R K T G O A P G F T Y T D A
Chicken, Turkey	- - - - - - - - G D I E K G K K I F V Q K C S Q C H T V E K G G K H K I G P N L H G L F G R K T G O A E G F S Y T D A
Penguin	- - - - - - - - G D I E K G K K I F V Q K C S Q C H I V E K G G K H K I G P N L H G I F G R K T G O A E G F S Y T D A
Peking duck	- - - - - - - - G D V E K G K K I F V Q K C S Q C H T V E K G G K H K T G P N L H G L F G R K T G O A E G F S Y T D A
Snapping turtle	- - - - - - - - G D V E K G K K I F V Q K C S Q C H T V E K G G K H K T G P N L N G L I G R K T G O A E G F S Y T E A
Rattlesnake	- - - - - - - - G D V E K G K K I F S M K C S T C H T V E E G G K H K T G P N L H G L F G R K T G O A V G Y S Y T A A
Bullfrog	- - - - - - - - G D V E K G K K I F V Q K C V Q C H T C E K G G K H K V G P N L Y G L I G R K T G O A A G F S Y T D A
Tuna	- - - - - - - - G D V A K G K K I F V Q K C V Q C H T V E N G G K H K V G P N L W G L F G R K T G O A E G Y S Y T D A
Screwworm fly	- - - - G V P A G D V E K G K K I F V Q R C V Q C H T V E A G G K H K V G P N L H G L F G R K T G O A A G V A Y T N A
Silkworm moth	- - - - G V P A G N A E N G K K I F A Q R C A Q C H T V E A G G K H K V G P N L H G F Y G R K T G O A P G F S Y S N A
Tomato hoen "worm"	- - - - G V P A G N A D N G K K I F A Q R C A Q C H T V E A G G K H K V G P N L H G F F G R K T G O A P G F S Y S N A
Wheat	A S F S E A P P G N P D A G A K I F K T K C P Q C H T V D A G A G H K Q G P N L H G L F G R Q S G T T A G Y S Y S A A
Rice	A S F S E A P P G N P K A G E K I F K T K C P Q C H T V D K G A G H K Q G P N L N G L F G R Q S G T T P G Y S Y S T A
Baker's yeast (Fungus)	- - - T E F K A G S A K K G A T L F K T R C A L C H T V E K G G P H K V G P N L H G I F G R H S G O A Q G Y S Y T D A
Candida yeast (Fungus)	- - P A P F E 0 G S A K K G A T L F K T R C A E C H T I E A G G P N K V G P N L H G I F S R H S G O A Q G Y S Y T D A
Neurospora (Fungus)	- - - - G F S A G D S K K G A N L F K T R C S E C H T E G G N L T Q K I G P A L H G L F G R K T G O V D G Y A Y T D A

(continued from above)

	60 70 80 90 100 110
Amino Acid Number	0 1 2 3 4 5 6 7 8 9 0 1 2 3 4 5 6 7 8 9 0 1 2 3 4 5 6 7 8 9 0 1 2 3 4 5 6 7 8 9 0 1 2 3 4 5 6 7 8 9 0 1 2
Human, Chimpanzee	N K N K G I I W G E D T L M E Y L E N P K K Y I P G T K M I F V G I K K K E E R A D L I A Y L K K A T N E
Rhesus monkey	N K N K G I T W G E D T L M E Y L E N P K K Y I P G T K M I F V G I K K K E E R A D L I A Y L K K A T N E
Horse	N K N K G I T W K E E T L M E Y L E N P K K Y I P G T K M I F A G I K K K T E R E D L I A Y L K K A T N E
Donkey	N K N K G I T W K E E T L M E Y L E N P K K Y I P G T K M I F A G I K K K T E R E D L I A Y L K K A T N E
Common zebra	N K N K G I T W K E E T L M E Y L E N P K K Y I P G T K M I F A G I K K K T E R E D L I A Y L K K A T N E
Pig, Cow, Sheep	N K N K G I T W G E E T L M E Y L E N P K K Y I P G T K M I F A G I K K K G E R E D L I A Y L K K A T N E
Dog	N K N K G I T W G E E T L M E Y L E N P K K Y I P G T K M I F A G I K K T G E R A D L I A Y L K K A T K E
Gray whale	N K N K G I T W G E E T L M E Y L E N P K K Y I P G T K M I F A G I K K K G E R A D L I A Y L K K A T N E
Rabbit	N K N K G I T W G E D T L M E Y L E N P K K Y I P G T K M I F A G I K K K D E R A D L I A Y L K K A T N E
Kangaroo	N K N K G I T W G E D T L M E Y L E N P K K Y I P G T K M I F A G I K K K G E R A D L I A Y L K K A T N E
Chicken, Turkey	N K N K G I T W G E D T L M E Y L E N P K K Y I P G T K M I F A G I K K K S E R V D L I A Y L K D A T S K
Penguin	N K N K G I T W G E D T L M E Y L E N P K K Y I P G T K M I F A G I K K K S E R A D L I A Y L K D A T S K
Peking duck	N K N K G I T W G E D T L M E Y L E N P K K Y I P G T K M I F A G I K K K S E R A D L I A Y L K D A T A K
Snapping turtle	N K N K G I T W G E E T L M E Y L E N P K K Y I P G T K M I F A G I K K K A E R A D L I A Y L K D A T S K
Rattlesnake	N K N K G I T W G D D T L M E Y L E N P K K Y I P G T K M V F T G L K K K K E R T D L I A Y L K E A T A K
Bullfrog	N K N K G I T W G E D T L M E Y L E N P K K Y I P G T K M I F A G I K K K G E R Q D L I A Y L K S A C S K
Tuna	N K S K G I V W N N D T L M E Y L E N P K K Y I P G T K M I F A G I K K K G E R Q D L V A Y L K S A T S -
Screwworm fly	N K A K G I T W Q D D T L F E Y L E N P K K Y I P G T K M I F A G L K K P N E R G D L I A Y L K S A T K -
Silkworm moth	N K A K G I T W G D D T L F E Y L E N P K K Y I P G T K M V F A G L K K A N E R A D L I A Y L K E S T K -
Tomato hoen "worm"	N K A K G I T W Q D D T L F E Y L E N P K K Y I P G T K M V F A G L K K A N E R A D L I A Y L K Q A T K -
Wheat	N K N A A V E W E E N T L Y D Y L L N P K K Y I P G T K M V F P G L K K P Q D R A D L I A Y L K K A T S S
Rice	N K N M A V I W E E N T L Y D Y L L N P K K Y I P G T K M V F P G L K K P Q E R A D L I S Y L K E A T S -
Baker's yeast	N I K K N V O W D E N N M S E Y L T N P K K Y I P G T K M A F G G L K K E K D R N D L I T Y L K K A C E -
Candida yeast	N K R A G V E W A E P T M S D Y L E N P K K Y I P G T K M A F G G L K K A K D R N D L V T Y M L E A S K -
Neurospora (Fungus)	N K 0 K G I T W D E N T L F E Y L E N P K K Y I P G T K M A F G G L K K D K D R N D I I T F M K E A T A -

Amino Acid Symbols
A = Alanine
C = Cysteine
S = Aspartic acid
E = Glutemic acid
F = Phenylalanine
G = Glycine
H = Histidine
I = Isoleucine
K = Lysine
L = Leucine
M = Methionine
N = Asparagine
P = Proline
Q = Glutamine
R = Arginine
S = Serine
T = Threonine
V = Valine
W = Tryptophan
Y = Tyrosine

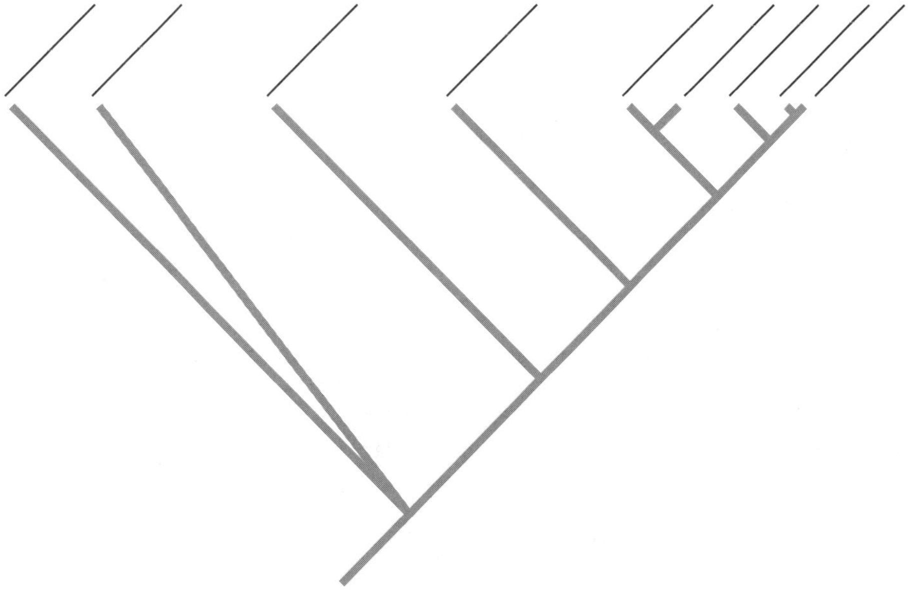

- Why do more closely related organisms have more similar cytochrome c?

Now try adding the cytochrome c sequence difference data for humans.

	Horse	Donkey	Whale	Chicken	Penguin	Snake	Moth	Yeast	Wheat
Human									

- How confident are scientists in these relationships? Why or why not?

- Are there polytomous branches? Which?

- What other tools do they use to determine the relationships among these species?

Explore!

Bird Brains

WGBH and *NOVA Science Now* present the evidence for the evolution of the *FOXP2* gene, its effect on birdsong evolution, and what that means for understanding human speech. The site includes a Q&A with Ofer Tchernichovski, a scientist studying vocal learning in birds at the City College of New York, a matching game for birds and their songs, and a great audio slide show that shows how experts are starting to learn to speak "walrus."

http://www.pbs.org/wgbh/nova/nature/bird-brains.html

Jake Westhoff—Molecular Clocks

Jake Westhoff sings about molecular clocks, using them as a metaphor for human relationships in this YouTube video.

http://www.youtube.com/watch?v=PEUNTk75ESo

Pecking Away at Beak Evolution

Science Now discusses two important research reports from Arhat Abzhanov and colleagues and Ping Wu and colleagues that together illuminate the molecular basis of the species diversity of Darwin's finches on the Galápagos Islands.

http://news.sciencemag.org/sciencenow/2004/09/02-03.html?ref=hp

You can view the original research papers here:

"*Bmp4* and Morphological Variation of Beaks in Darwin's Finches."

http://www.sciencemag.org/cgi/content/full/305/5689/1462

"Molecular Shaping of the Beak"

http://www.sciencemag.org/cgi/content/full/305/5689/1465

A Beginner's Guide to Making a Phylogenetic Tree

In this video, Sandra Porter walks viewers through the steps to generate a phylogenetic tree from data available on the web. Using the multiple copies of 16S ribosomal RNA (rRNA) genes found in bacterial genomes, Porter ultimately develops a tree using the neighbor-joining method.

http://scienceblogs.com/digitalbio/2008/03/27/a-beginners-guide-to-making-a/

Creating Phylogenetic Trees from DNA Sequences

The Howard Hughes Medical Institute developed this "Click & Learn" to explain how DNA sequences can be used to generate and interpret phylogenetic trees. The "Click & Learn" includes insightful video clips from scientists, and you can download a worksheet to complete as you go through to check your understanding.

http://www.hhmi.org/biointeractive/creating-phylogenetic-trees-dna-sequences

Overcoming Misconceptions

The Coalescent IS an Individual

It's tempting to believe that if you can trace an allele all the way back to a coalescent, that coalescent must be the single individual in the population with the mutation, especially when people write about "mitochondrial Eve." But it's important to remember that tracing the allele back to the coalescent actually represents the *lineage* of the allele that later became fixed in the population. For example, each individual carries multiple, genetically distinct, versions of their mitochondrial genome within cells, and only a subset gets included in daughter cells each time they divide. Thus, when we reconstruct a lineage of mitochondrial DNA, we are working backwards to the origin of a particular mitochondrial allele that arose and eventually became fixed within a population. The so-called "mitochondrial Eve" would likely have carried multiple different mitochondrial alleles, only one of which was the new allele that later became fixed within the population.

Mutations Can Be Deleterious, Beneficial, and Even Neutral

Mutations are errors that arise when DNA is replicated, but just because they are errors doesn't always make them harmful. Most may be neutral, or synonymous substitutions. In fact, recent research on humans indicates that from 20 to 40 mutations occur in each gamete (http://www.nature.com/ng/journal/v43/n7/abs/ng.862.html)! New mutations lead to variation among individuals (and they generate genetic variation within populations). And while mutations may occur randomly, their persistence in a population is a result of the mechanisms of evolution, such as natural selection and genetic drift.

Tips and Relatedness

Think of a phylogenetic tree as a mobile, with each group of branches spinning freely at each node. The tips aren't locked in to where they appear on the page. Relatedness comes from the shared common ancestors, (i.e., the nodes).

Exactly What Do Straight Lines in Phylogenies Mean?

Phylogenies illustrate relationships among groups of organisms (e.g., genes, populations, or species), and the straight lines may or may not confer meaning, depending on the type of tree. In trees designed simply to illustrate relationships among groups, branch length is often just a function of the order of the tree. But it's a mistake to think the long straight branches indicate that a species hasn't changed. As Chapter 5 explains, mutations are occurring all the time. Purifying selection tends to remove deleterious mutations from populations, but other mutations can exist simply because they are neutral (e.g., if they are synonymous substitutions). Neutral and positive mutations can accumulate in a population's genome, even though no lineages ever split. Evolutionary biologists often draw trees with branches that do indicate the amount of change in a lineage over time, however. In these trees, the branch length is proportional to the amount of change that has been uncovered.

Time Is Relative—When It's Looked at Correctly

No matter how a phylogeny is drawn, reading time simply depends on the location of the common ancestor. For example,

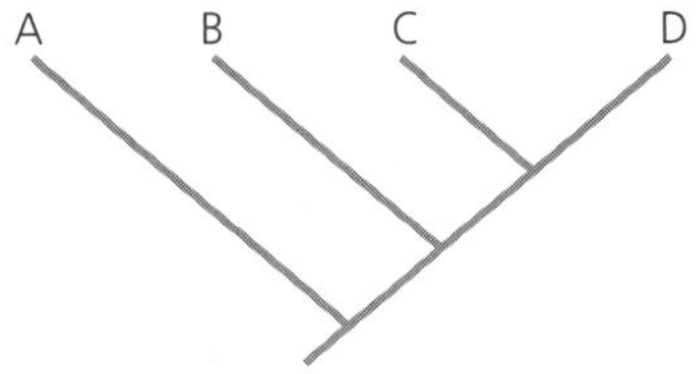

time in this style of tree is represented from bottom to top—the extant (living) species or groups are listed at the top, and each successive node down represents an earlier and earlier common ancestor.

In this style,

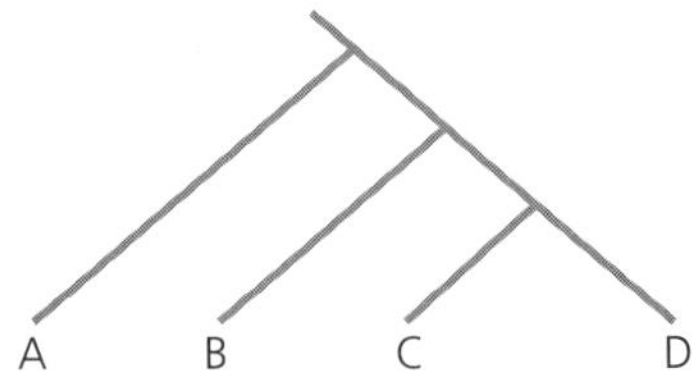

time is just the opposite of the figure above—the extant (living) species or groups are listed at the bottom, so each successive node up represents an earlier and earlier common ancestor.

In this style,

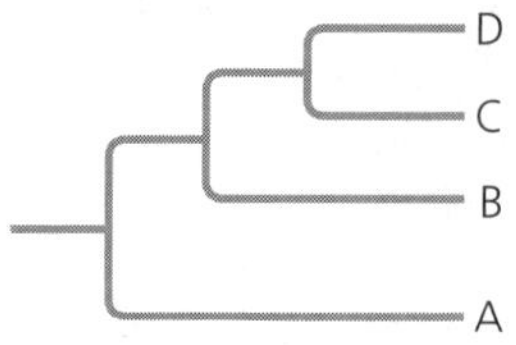

the extant (living) species or groups are listed at the right, so each successive node to the left represents an earlier and earlier common ancestor.

And in this style,

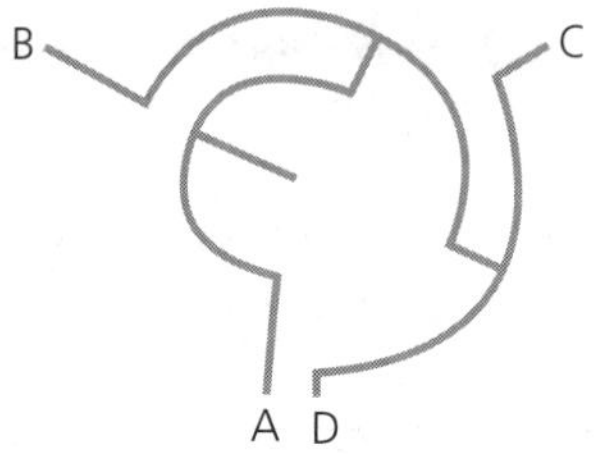

time is depicted as concentric circles. Extant species are on the outside, and earlier and earlier common ancestors can be found closer and closer to the center.

Which of the following is a true statement?

a. The coalescent represents the lineage of an allele that arose and ultimately persisted within a population.
b. Mutations are always deleterious.
c. A straight line in a phylogeny means that the species has not changed over time.
d. The oldest species can be found at the top of any phylogeny.

Go the Distance: Examine the Primary Literature

Wolfgang Enard, Molly Przeworski, Simon E. Fisher, Cecilia S. L. Lai, Victor Wiebe, Takashi Kitano, Anthony P. Monaco, and Svante Pääbo sequenced complementary DNAs that encode the FOXP2 protein in a variety of species, including humans. They investigated the molecular history of the *FOXP2* gene in humans and our common ancestors.

- In the clade including orangutans, how many synonymous substitutions occurred in the lineage leading to humans? How many replacement substitutions?

- Why do they argue that natural selection has shaped the *FOXP2* gene during recent human evolution?

- What type of method did they use to determine the timing of branching events?

Enard, W., M. Przeworski, S. E. Fisher, C. S. L. Lai, V. Wiebe, et al. 2002. Molecular Evolution of *FOXP2*, a Gene Involved in Speech and Language. *Nature* 418 (6900): 869–72. doi:10.1038/nature01025. http://www.nature.com/nature/journal/v418/n6900/abs/nature01025.html.

Delve Deeper

1. Do scientists use different lines of evidence to support and test phylogenetic hypotheses developed with molecular data?

2. How can molecular data be used to understand natural selection in the past?

Test Yourself

1. What is meant by coalescence?
   a. The process of tracing homologous alleles back to a common ancestor
   b. The history of an allele within a population
   c. The point in the history of a genetic locus when it becomes polymorphic
   d. All of the above
   e. None of the above
2. What is a gene tree?
   a. A tool used by molecular biologists to trace the lineage of a species
   b. A branched genealogical lineage of homologous alleles
   c. A graphic illustration of coalescence
   d. Both b and c
   e. All of the above

3. In Figure 9.5B, why would species 2 be considered more closely related to species 3 than to species 1 if scientists only sampled this single gene?
   a. Because more alleles can be traced to the common ancestor of species 2 and species 3 than to species 1
   b. Because species 2 branches from species 3 before species 1 does
   c. Because species 2 and species 3 share a more recent common ancestor than either shares with species 1 in terms of that single gene's alleles
   d. Because if scientists only sampled that single gene, they would miss the gene that actually identifies the species
   e. Species 2 is more closely related to species 3 because they share a more recent common ancestor than with species 1
4. Which method(s) for reconstructing phylogenies incorporates information about variation in mutation rates among genetic loci?
   a. Maximum likelihood methods
   b. Bayesian methods
   c. Bootstrapping methods
   d. Both a and b
   e. Both b and c
5. Why do scientists use bootstrapping for the analyses of phylogenetic trees?
   a. Because bootstrapping identifies the true relationships of organisms within a tree
   b. Because bootstrapping improves their data so scientists can make stronger claims
   c. Because scientists need to use statistical analyses when they are not sure of the truth
   d. Because other methods of developing phylogenetic trees are not very accurate
   e. Because scientists it can provide a test of the likelihood of the branching events they've proposed
6. Which of the following statements does Sarah Tishkoff's phylogenetic analysis of human mitochondrial DNA (Figure 9.10) illustrate?
   a. Humans that colonized India are no more distantly related to humans that colonized the Americas than humans that colonized the Middle East.
   b. Humans that colonized India, Central Asia, Europe, the Middle East, East Asia, the Americas, and Oceania represent a monophyletic clade.
   c. Humans that colonized the Americas share their most recent common ancestor with humans in East Asia.
   d. All of the above
   e. None of the above
7. Which of the following statements about the phylogenetic relationships among finches is TRUE?
   a. The Cocos finch shares a more recent common ancestor with Darwin's finches than with other finch species.
   b. The vegetarian finch is more closely related to the tree finches than to the ground finches.
   c. Tiaris bicolor is the common ancestor of all of Darwin's finches.
   d. *C. olivacea* is more closely related to *C. fusca* than to Darwin's finches.
   e. All of the above are true statements.

8. What proportion of the human genome has no known function?
   a. 98.8 percent of our genome does not encode proteins and has no known function.
   b. Most of the 98.8 percent of our genome that does not encode proteins has no known function, but scientists continue to discover new segments that are important.
   c. Scientists don't actually know what proportion of our genome has no known function because they are finding new methods of uncovering important elements.
   d. 1.2 percent of our genome does not encode proteins and has no known function.
   e. Every element of the human genome has a function.
9. Why do scientists consider APOAV a conserved gene?
   a. Because it occurs so far away from other APO genes
   b. Because mice evolved before humans
   c. Because mice produce fewer triglycerides than humans
   d. Because of the high level of homology between mice and humans
   e. Because it encodes a protein important to lipid regulation

Contemplate

What differences and similarities in DNA would you expect to find if our human ancestors interbred?	Could you examine the evolution of different languages using phylogenies? If so, what would the coalescent represent?

10 Adaptation

From Genes to Traits

Check Your Understanding

1. Which of the following statements about genetically controlled traits is true?
    a. Interactions among alleles at different genetic loci (epistasis) can affect the expression of a trait, such as height, in different ways from individual to individual.
    b. A single genotype may produce different phenotypes depending on the environment.
    c. A single mutation to a regulatory gene can affect many phenotypic traits.
    d. All of the above
    e. b and c only
    f. None of the above
2. Which effects of a mutation contribute most to the phenotypic resemblance among relatives: additive effects, dominant effects, or epistatic effects?
    a. Additive effects
    b. Dominant effects
    c. Epistatic effects
    d. All are equally important to heritability.
    e. None are important.
3. What is a cis-acting regulatory element?
    a. Stretches of DNA that are located far away from a focal gene (e.g., on another chromosome) that influence the expression of that gene
    b. Stretches of DNA located near a focal gene that influence the expression of that gene
    c. A non-coding region of the genome that can be found either immediately upstream (adjacent to the promoter region), downstream, or inside an intron
    d. All of the above
    e. b and c only
    f. None of the above

Learning Objectives for Chapter 10

Add important definitions and notes next to each learning objective for this chapter to help guide your understanding.

Learning Objective	Important Definitions	Notes
Explain how mutations to a regulatory network may affect development of an organism.		
Describe two examples where proteins were co-opted for other functions.		
Compare and contrast how scientists discovered the role of gene duplication in the evolution of citrate eating in *E. coli* and the evolution of venom in snakes.		
Explain why *Hox* genes are considered part of the "genetic toolkit."		
Explain how location and timing of the expression of developmental genes influences limb and beak development.		
Describe three important steps in the evolution of complex traits, such as the vertebrate eye.		
Distinguish between the outcomes of a mutation to a gene with pleiotropic effects and those of a mutation to a gene without pleiotropic effects within a regulatory network.		
Analyze the imperfections of a familiar complex adaptation.		
Explain the relationship between homology, convergent evolution, and parallel evolution.		

Identify Key Terms

Match terms and definitions by filling in the blank to the right of the term with the appropriate letter.

1. **Antagonistic pleiotropy** f
2. **Complex adaptations** h
3. **Deep homology** j
4. **Gene recruitment** i
5. **Novel traits** g
6. **Orthologs** b
7. **Parallel evolution** c
8. **Paralogs** a
9. **Promiscuous proteins** d
10. **Regulatory networks** e

a. Homologous genes that arise by gene duplication. Together they form a gene family.

b. Homologous genes separated by a speciation event (as opposed to homologous genes, produced by gene duplication, that are both possessed by the same species).

c. Independent evolution of similar traits, starting from a similar ancestral condition

d. Proteins capable of carrying out more than one function, such as catalyzing reactions of different substrates

e. Systems of interacting genes, transcription factors, promoters, RNA, and other molecules. They function like biological circuits, responding to signals with outputs that control the activation of genes during development, the cell cycle, and the activation of metabolic pathways.

f. Occurs when a mutation with beneficial effects for one trait also causes detrimental effects on other traits

g. Traits that arise *de novo* (i.e., not inherited from an ancestor) within a lineage and have no obvious counterparts (homologs) in related lineages

h. Suites of coexpressed traits that together experience selection for a common function. Phenotypes are considered complex when they are influenced by many environmental and genetic factors, and when multiple components must be expressed together for the trait to function.

i. The co-option of a particular gene or network for a totally different function as a result of a mutation. The reorganization of a preexisting regulatory network can be a major evolutionary event.

j. A condition that occurs when the growth and development of traits in different lineages result from underlying genetic mechanisms (e.g., regulatory networks) that are inherited from a common ancestor.

Link Concepts

Fill in the bubbles with the appropriate terms.

- bat wings
- cricket hind legs
- finch beaks
- ~~human eyes~~
- leg loss in snakes
- mutation
- paralogs
- promiscuous proteins
- regulatory elements
- snake venom
- wing loss in ants

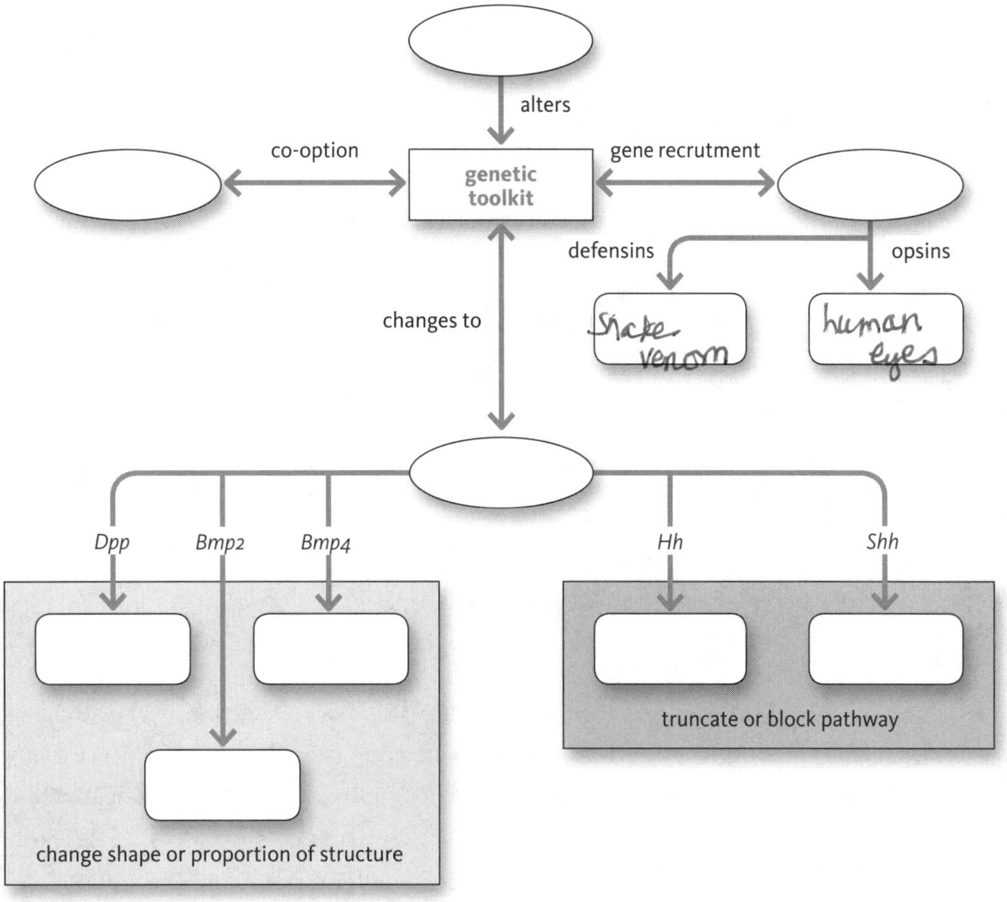

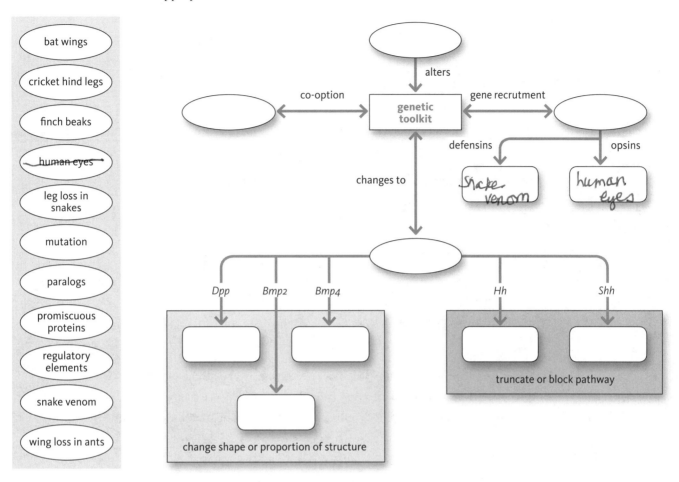

Develop your own concept map that explains how evolution imposes constraints on adaptations like eyes. Think about mutations, pleiotropy, genetic toolkits, homology, gene networks, and lineages.

Key Concepts

Fill in the blanks for the key concepts from the chapter.

10.1 Cascades of Genes

Genes rarely function in ________—the expression of a gene is influenced by a network of ________ and ________ with RNA and other gene products. These ________ may be important building blocks in the evolution of complex adaptations.

________ can arise when existing genes or processes are expressed in new ________ contexts (e.g., activating a genetic pathway at a new time or in a new tissue during development).

10.2 Generating Innovations

Proteins are promiscuous when they are able to perform functions other than their primary task. This ability can serve as a starting point for the evolution of novel traits, if proteins that evolved primarily for one function become coopted to another function.

Gene duplication facilitates divergence in gene function because multiple copies can take on new and different tasks. Since they are "extra" copies, duplicated genes are released from selection acting on the performance of the original gene. They are able to accumulate mutations faster and are often co- opt to serve new functions.

10.3 The Origin of New Adaptations in Microbes

Microbes are ideal organisms to examine the evolution of ________ because they are so diverse, and they reproduce rapidly. Studying these organisms gives scientists valuable insight into how the co-option of ________ and ________ can lead to novel traits.

10.4 The History of Venom: Evolving Gene Networks in Complex Organisms

Many novel traits (such as snake venoms) seem so complex that it can be hard to imagine how the ________ could have evolved. But research has shown they can evolve through a series of ________ events and ________ of proteins originally involved with other body functions. The complex adaptations we see today are the result of the combination of these processes with ________ and ________.

10.5 The Genetic Toolkit for Development

Hox genes and other ____________ genes participate in regulatory ____________ that demarcate the geography of developing animals, determining the ____________ locations and sizes of body parts. These networks comprise an ancient "genetic toolkit" inherited by ____________ with bilateral body symmetry.

10.6 The Deep History of Limbs

Subtle changes in the patterns of ____________ of ____________ genes can dramatically alter ____________. For example, ____________ the expression of a patterning ____________ can result in complete loss of a structure (as in the loss of limbs in snakes and of wings in ants).

10.7 Sculpting Adaptations

The genetic toolkit can be ____________ relatively quickly in evolutionary time, producing extraordinary diversity among even closely related species.

10.8 Evolving Eyes

The vertebrate eye is the result of a ____________ and ____________ evolutionary history: ____________ link vertebrates to organisms that lived 650 million years ago, and so too do the mechanisms for co-opting regulatory networks.

Very ____________ adaptations, such as the vertebrate eye, can evolve through a series of small ____________. ____________ link vertebrates to organisms that lived 650 million years ago, and so too do the mechanisms for co-opting ____________.

10.9 Constraining Evolution

Antagonistic pleiotropy can constraining the directions of evolution because it causes some phenotypes to be unsuccessful even when mutations may be beneficial.

10.10 Building on History: Imperfections in Complex Adaptations

Natural selection often retools the form and function of characters ____________ within a population, leading to adaptation that are far from perfect, but that still provide fitness advantages.

10.11 Convergent Evolution

The deep homology of ______________ may help explain many cases of ______________ evolution.

Examining traits that ______________ can shed light on how each lineage arrived at the shared phenotype—the genetic and developmental underpinnings of similar adaptations. It can reveal ______________ to a common challenge as well as the ______________ imposed by history within each lineage.

Interpret the Data

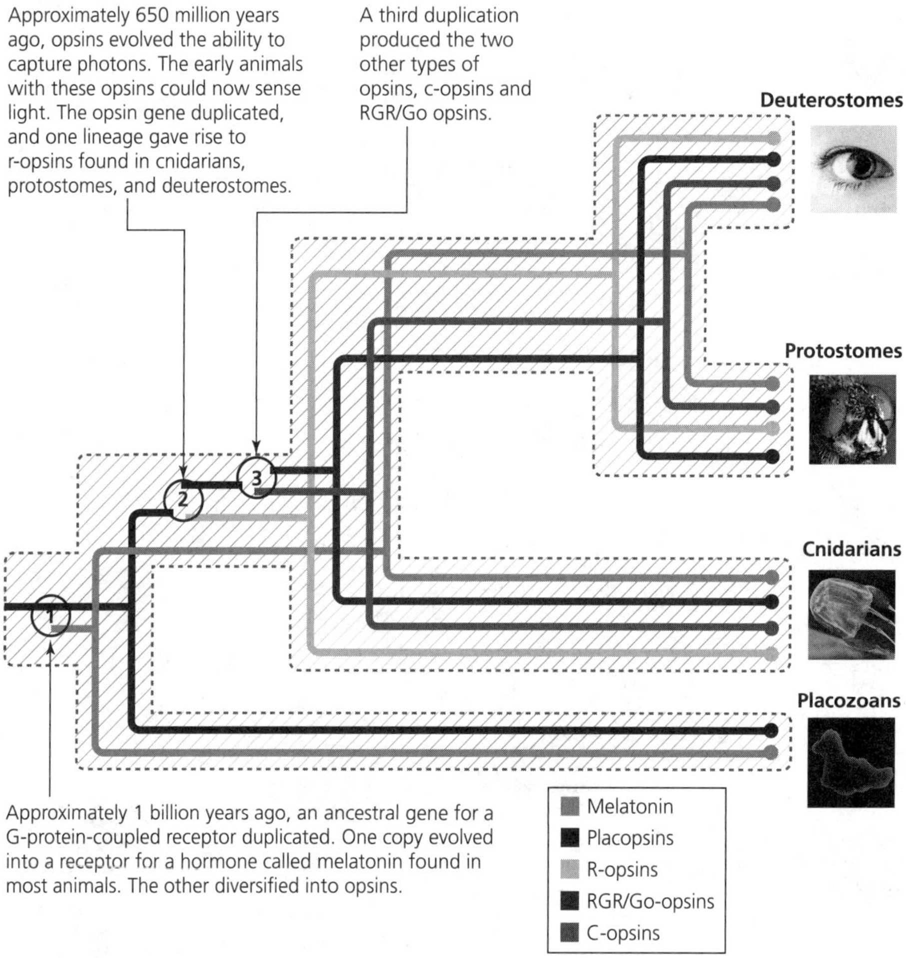

Approximately 650 million years ago, opsins evolved the ability to capture photons. The early animals with these opsins could now sense light. The opsin gene duplicated, and one lineage gave rise to r-opsins found in cnidarians, protostomes, and deuterostomes.
A third duplication produced the two other types of opsins, c-opsins and RGR/Go opsins.
Deuterostomes
Protostomes
Cnidarians
Placozoans
1
2
3
Approximately 1 billion years ago, an ancestral gene for a G-protein-coupled receptor duplicated. One copy evolved into a receptor for a hormone called melatonin found in most animals. The other diversified into opsins.
Melatonin
Placopsins
R-opsins
RGR/Go-opsins
C-opsins

Gene duplication events were crucial to the evolution and diversification of opsins so important for human vision. Scientists have been able to trace the origins of opsins by examining similar proteins across taxa. Opsins evolved from a family of proteins known as G-protein-coupled receptors (GPCRs). These GCPRs are found in all animals (and even fungi).

- Are all opsins used in vision?

- How many types of opsins can be found in the deuterostomes (a group of animals that includes humans)?

- Are cnidarians sensitive to light? Support your argument.

- What mutation was necessary for the evolution of opsins? Why was that important?

Games and Exercises

Paint by Numbers

As scientists began to explore genome sequences in various model organisms, they discovered amazing patterns! For example, in genes that influence the shapes of our bodies, like *Hox* genes, they discovered segments about 180 bases long that were virtually identical across organisms. They named these segments homeoboxes, and these homeoboxes translate into protein sequences that are 60 amino acids in length. Regulatory genes containing these homeoboxes (called *Hox* genes) act as switches, controlling suites of other genes in cascades that ultimately influence how we develop. These homeobox genes are strong evidence for the shared ancestry among bilateral organisms.

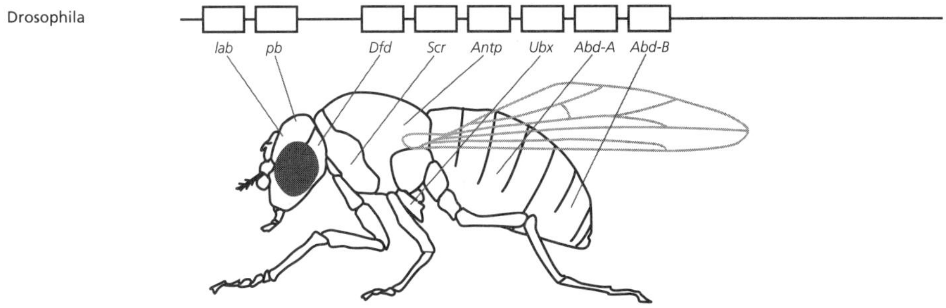

Seeing some of the homologies scientists have discovered is as easy as paint by numbers. In the following diagram, the *Hox* genes are lined up, literally, from head to tail on the chromosomes. Start with the *Drosophila* genes: *labial* (*lab*), *proboscipedia* (*pb*), *deformed* (*Dfd*), *sex combs reduced* (*Scr*), *antennapedia* (*Antp*), *ultrabithorax* (*Ubx*), and *abdominal-A* (*Abd-A*). Simply choose different colors for each of the boxes (genes) indicated on the *Drosophila* chromosome and color each one. Color the homologous boxes for the *Hox* genes on the four human chromosomes with the corresponding colors from the *Drosophila* chromosome.

Match these important developmental genes with the areas where they are expressed in the *Drosophila* larva, the *Drosophila* adult, the mouse larva, and the human larva. Color each "domain of expression" with the color corresponding to the genes.

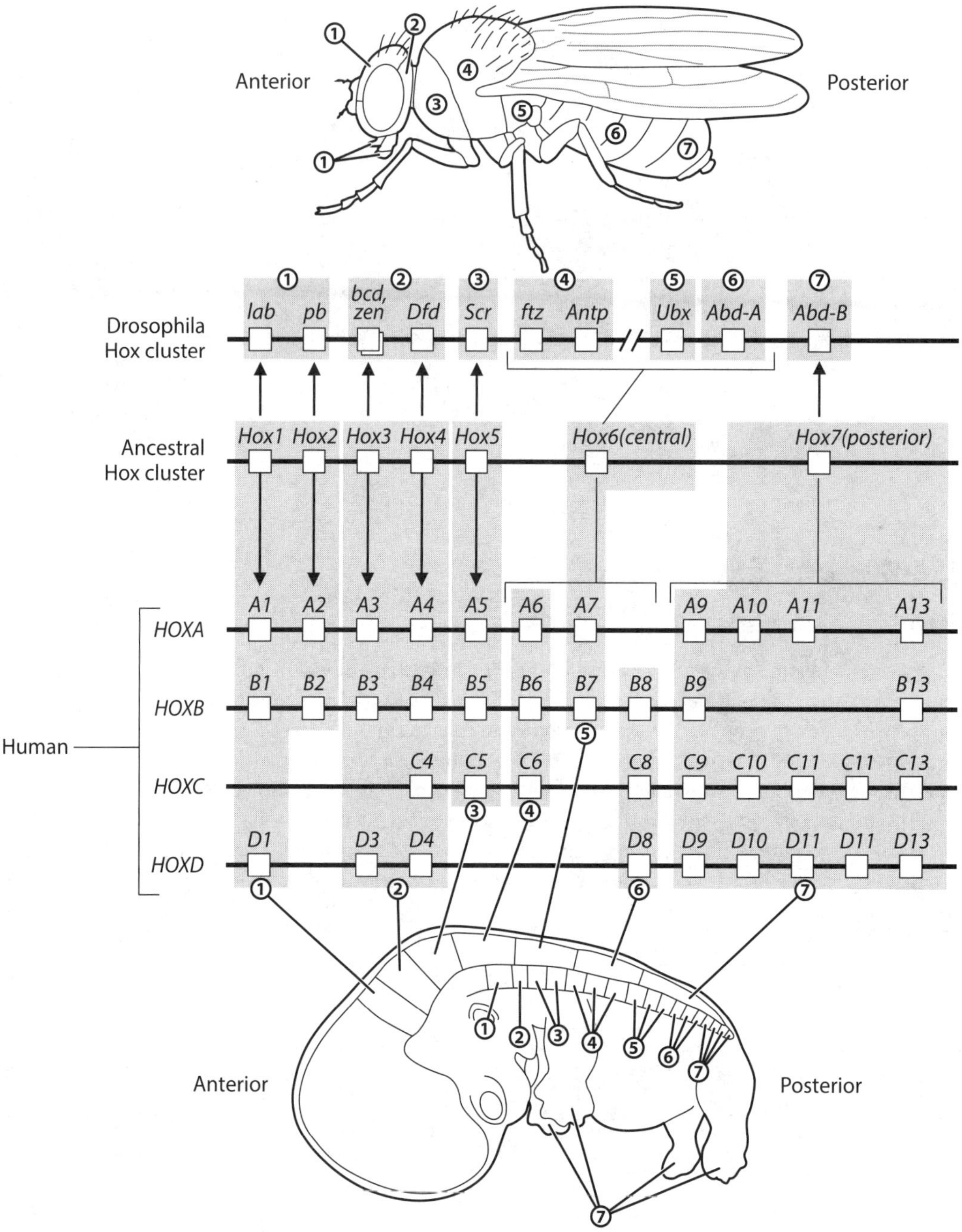

- The diagram indicates that *Drosophila Hox* genes lie on a single chromosome in a single cluster, and mouse and human *Hox* genes lie in clusters on four different chromosomes. What kind of mutation(s) likely produced this diversity?

- *Hox* genes are also expressed in the limbs of Stage 19 human embryos (specifically, 5'*HoxA* and *HoxD* genes). Which ortholog found in *Drosophila* gave rise to *Hox* genes governing human limb development?

Complex Adaptations and Selection

Werner Heim developed this activity using regular playing cards to help illustrate how complex adaptations can arise given cumulative, nonrandom natural selection. You can do this activity alone or with friends, but make sure to complete both versions of the activity.

You will need a deck of cards. Separate out each of the suits, choose one suit, and shuffle it thoroughly. Shuffling represents the probability of a functional mutation, and each round represents time. So you'll need to tally the number of times you shuffle the cards on a piece of paper. For round 1, shuffle the cards, then look at the top card. If the top card is the ace, you can start an "organism" stack. If not, shuffle the cards for round 2, round 3, and so on, until the top card is the ace. Once you've started the "organism" stack, the objective is to build up the mutations through the entire suit of cards by shuffling and recording rounds. If the top card is the next card needed for the construction of the organism, place it face up on the organism stack, shuffle, and repeat. If it's not, then shuffle the cards again. Continue shuffling and placing appropriate cards on the organism stack until you get to the king.

- How many rounds did it take to complete the complex adaptation? What do you think would happen if you repeated the experiment? How many times would you have to repeat the experiment to get the right answer?

- How does this simulation compare with the cumulative mutations necessary for the evolution of a complex adaptation?

- Compare the evolution of citrate utilization with being forced to start with the ace. What might the outcome be in terms of citrate feeding if the ace wasn't the first mutation?

Now repeat the procedure, but this time, there is no "organism" stack—you don't get to work with cumulative mutations. Shuffle the cards thoroughly. Keep track of the number of rounds, just like in the previous experiment. For the first round, examine all of the cards. Are they in order from ace to king? If not, shuffle the cards and examine the cards again. Continue shuffling and examining until the order of cards comes out in order from ace to king.

- How many rounds did it take to complete the complex adaptation? How many rounds did you try? How long do you think it would take to complete the complex adaptation simply by shuffling cards?

Adapted from Heim, Werner G. 2002. Natural Selection among Playing Cards. *American Biology Teacher* 64 (4): 276–78. doi:http://dx.doi.org/10.1662/0002-7685(2002)064[0276:NSALC]2.0.CO;2

Also available from the ENSI website

http://www.indiana.edu/~ensiweb/lessons/ns.cum.l.html

Explore!

Let's Shake Wings

Alexander Vargas and John Fallon examined development in chicken embryos and found evidence that may clarify the relationship between dinosaurs and birds.

http://news.sciencemag.org/sciencenow/2005/01/11-01.html?ref=hp

Quantity Is Key for *Hox* Genes

Hox proteins apparently aren't restricted to single tasks—they may actually be able to switch places fairly easily. The amount of proteins made, however, may be the key to developmental patterns.

http://news.sciencemag.org/sciencenow/2000/02/14-02.html?ref=hp

Evolutionary Development: Chicken Teeth—Crash Course Biology #17

In this Crash Course video, Hank Green introduces evo-devo, *Hox* genes, and the process of development. Find out how about an experiment where scientists caused eyes to develop on the legs of fruit flies, and how mutations can result in chickens with teeth!

http://www.youtube.com/watch?v=9sjwlxQ_6LI

Evolution: Great Transformations

PBS *Evolution* shows how scientists discovered the genetic toolkit. Scientists were able to switch the patterning genes for eyes between fruit flies and mice and show that not only did the two species use the same mechanism for gene expression, they also used the same gene.

http://www.pbs.org/wgbh/evolution/library/03/4/l_034_04.html

Evolution: Darwin's Dangerous Idea

Zoologist Dan-Erik Nilsson developed a model for the evolution of the eye. This video segment from PBS *Evolution* shows the important evolutionary steps that transformed eyes and the organisms that possess functional eyes at these various stages.

http://www.pbs.org/wgbh/evolution/library/01/1/l_011_01.html

Overcoming Misconceptions

Different Cell Types of an Individual Carry the Same DNA

If a cell is a diploid, nucleated cell, it contains all the organism's chromosomes, half from the female and half from the male. With only a few exceptions, it doesn't matter whether those cells are kidney cells, muscle cells, brain cells, or skin cells—they all inherit the same suite of genes. Instead, the genes within these cell types are differentially regulated, whether that regulation occurs as functions within the cell or outside of it. In fact, the differences among cell types arise primarily because different sets of genes are expressed in different cell types.

Complexity in Evolution Is Explainable

People who are unfamiliar with evolutionary biology may find it impossible to believe that complex adaptations could have evolved. Darwin himself recognized the reluctance that readers of *On the Origin of Species* might have to such ideas. After all, the complexity of biological structures had long been considered one of the most compelling arguments that life had been directly created by a designer.

The Reverend William Paley (1743-1805) had famously argued that looking at complexity in nature was no different from coming across a watch lying on the ground and inferring it had been built by a watch-maker.

But complex adaptations, like the human eye, can be explained by evolutionary theory. Many different lines of evidence (physiology, chemistry, genetics, phylogenetics, and such) show that mutations, gene duplication, and gene recruitment are all processes that can combine to produce incredible structures. Although the occurrence of mutations may be random—a matter of chance—their persistence in a population is definitely *not* random if the mutation affects the reproductive success of the individuals that have it. And *any* functional advance can affect the fitness of organisms—it doesn't have to be the *pinnacle of all complex adaptations* to be effective. As humans, we see human eyes as the best of all eye adaptations, but there are many different solutions to the problem of detecting light. Numerous organisms have different versions of light-sensing organs, from very simple to very complex, and some are more sophisticated and more sensitive than ours! These adaptations are all the result of many small changes over time, each small change adding a new level of functionality.

Genes Aren't Recruited Based on Need

Gene recruitment may sound goal-oriented, but genes aren't planning on recruiting other genes. Gene recruitment arises largely through random mutations and selection favors the associated fitness advantages. So *Sphingobium* did not actively recruit genes from other pathways. Mutations altered the regulatory networks and began producing proteins that helped break down PCP, enhancing the fitness of individuals with those mutations.

Evolution and the Nature of Science

Science is a process involving creativity and argument. Scientists propose explanations, based on their evidence, while other scientists test those explanations and argue and debate the evidence. The publication of *On the Origin of Species* by Charles Darwin, not surprisingly, triggered fierce public debates. But scientists soon came to agree with Darwin's argument that life had evolved, and later they came to understand how molecular biology made natural selection possible. They continue to debate each other today. When scientists meet at scientific conferences to present their research on evolution, the disputes can get intense. Evolutionary biologists argue about the best way to reconstruct evolutionary trees. They argue about what caused mass extinctions. They argue about the relative importance of natural selection and neutral evolution. But evolutionary biologists do not rehash long-settled subjects, such as the fact that life has existed for billions of years. Evolutionary biologists would consider it as pointless to re-argue these points as it would be for astronomers to revive debates about whether the Earth revolves around the Sun, or vice versa.

Despite this scientific consensus, a number of organizations and individuals claim that evolution is fundamentally untrue. They hold a wide range of views. Some claim that the Earth is only 6,000 years old, and that God created all life pretty much in its current form at the beginning of that time. Others don't object to the notion that life is in fact billions of years old. Instead, they claim that major features of biology are the result of something that they call "intelligent design"—a planned creation by some kind of intelligent agent they claim not to be able to identify. These claims are known collectively as "creationism" because they all explain the diversity of life as the result of creation by a supernatural being, rather than the result of natural processes such as natural selection.

Biologists reject creationism as a *scientific explanation* for life's diversity for several reasons. Many creationist arguments are not, in fact, arguments in favor of creationism. Instead, they are attempts to raise doubts about evolution. But the version of evolution they attack bears little resemblance to the *actual* science. For example, creationists often claim that evolution is a purely random process. Since organisms have many genes that encode many interacting proteins, the argument goes, the odds are astronomically tiny that pure randomness could produce a complex trait like the eye.

But natural selection is not a random process. Whether a particular organism gets a particular mutation is a matter of chance, but natural selection only spreads mutations through an entire population if they have some beneficial effect. Many lines of evidence show that natural selection can favor an entire series of mutations, which together give rise to new traits, including eyes.

Creationists also frequently claim that since fossils have not been found to document every intermediate stage in evolution, evolutionary biologists cannot claim that new species evolved from earlier ones. In reality, the process by which fossils actually form makes it inevitable that the fossil record is incomplete. What matters is the fact that the fossil record is consistent with the theory of evolution. Before the 1980s, for example, paleontologists had found no fossils of whales that could shed light on the early stages of their evolution. Based on their study of living whales, biologists predicted that whales descended from terrestrial mammals, most likely artiodactyls (a group that includes such hoofed animals as pigs, camels, and hippopotamuses—for more, see Chapter 1). Over the past three decades however, scientists have found not just one of these transitional fossils, but fossils from about 30 different species of early whales, documenting a 10-million-year transition. And these early whales show clear links to artiodactyls.

Creationism, by contrast, doesn't predict the pattern that we see in nature. It does not offer an explanation for why phylogenies derived from anatomy and from molecules are congruent, for example (see Chapter 4 for the case of tetrapod relationships). To say that such patterns are simply the work of God is to make a religious claim, not a scientific one. Scientists cannot evaluate a claim about the actions of a supernatural agent, since scientific theories explain phenomena that follow natural, repeatable patterns.

To invoke an intelligent designer instead of God does not make creationism any more scientific. The designer is either supernatural or natural. If the designer is supernatural, the claim of intelligent design is not scientific. If the designer is natural, it should be possible to make predictions about how it has produced life's diversity. But since we can't know the nature of the designer, we're told, this claim reaches a scientific dead end. For deeper treatments of these issues, see Miller (2008), Pennock (1999), and Scott (2008).

Which of the following is a true statement?

a. Only cells that make a specific gene product, like a protein, carry that gene.

b. Through natural selection and time, evolution can produce seemingly complex adaptations through gene duplication and recruitment.

c. Scientists do not debate the evidence for evolution; they accept the theory of evolutionary biology on faith.

d. Evolutionary biologists believe that pure randomness can produce complex adaptations, like the eye.

Citations

Miller, K. R. 2008. *Only a Theory: Evolution and the Battle for America's Soul.* New York: Viking Penguin.

Pennock, R. T. 1999. *Tower of Babel: The Evidence against the New Creationism.* Cambridge, MA: MIT Press.

Scott, E. C. 2008. *Evolution vs. Creationism: An Introduction.* Westport, CT: Greenwood Press.

Go the Distance: Examine the Primary Literature

Irma Varela-Lasheras and her colleagues continue work on the evolutionary constraints imposed by changes in the number of cervical vertebrae. They examine sloths and manatees, two mammal species that are exceptions to the seven-vertebrae rule—sloths with eight cervical vertebrae and manatees with six.

- What are homeotic transformations?

- What five predictions did the authors make about homeotic transformations?

- According to the authors, how can the pleiotropic constraints on body plan changes, including changes to the number of cervical vertebrae in mammals, be overcome?

- Why might this discovery be important?

Varela-Lasheras, I., A. Bakker, S. van der Mije, J. Metz, J. van Alphen, et al. 2011. Breaking Evolutionary and Pleiotropic Constraints in Mammals: On Sloths, Manatees, and Homeotic Mutations. *EvoDevo* 2 (1): 11. doi:10.1186/2041-9139-2-11. http://www.evodevojournal.com/content/2/1/11.

Delve Deeper

1.

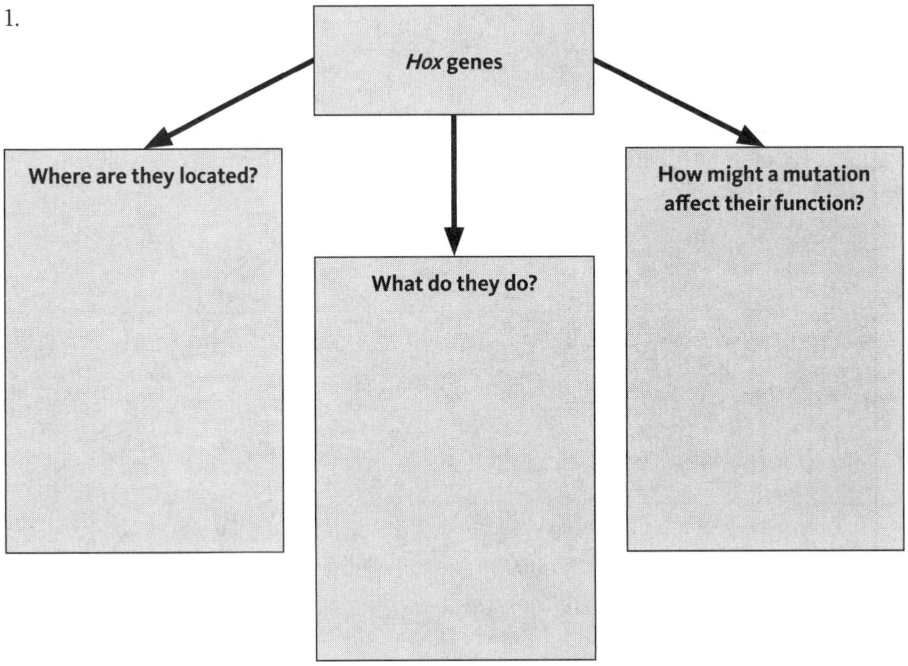

2. If the DNA in every cell within an organism includes all the *Hox* genes for the development of that organism, why doesn't every cell develop into a cell in the eye, for example?

Test Yourself

1. Why are gene networks important in the evolution of new traits?
   a. Because gene networks lead to the quantitative traits that scientists are interested in studying
   b. Because within a gene network, a single mutation can have cascading effects, leading to rapid evolution of novel traits
   c. Because gene networks act like self-contained "modules" that can be deployed in new developmental contexts, yielding novel traits
   d. Because mutations to upstream genes can truncate the entire pathway, dramatically affecting traits
   e. All of the above
2. Which of the following statements about Richard Lenski's *E. coli* populations is false according to the following figure (Figure 10.6)?

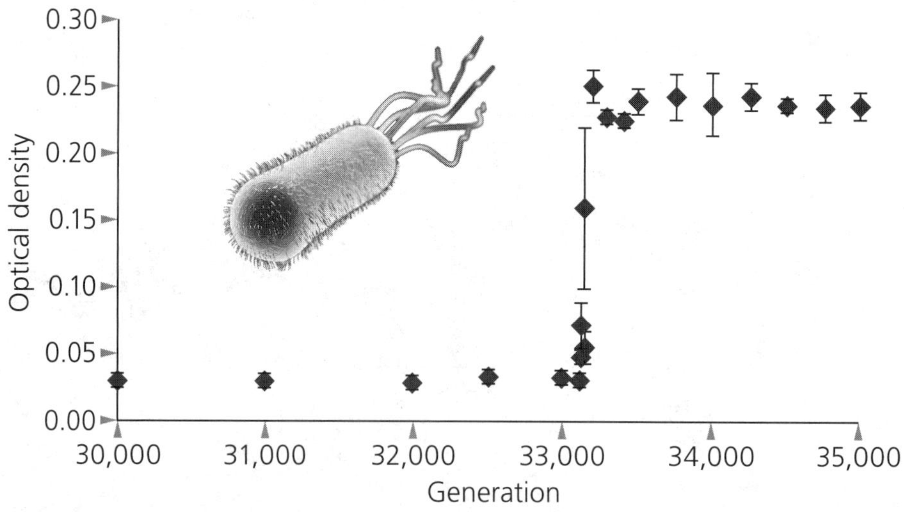

   a. Initially, only a small proportion of one lineage of *E. coli* could feed on citrate.
   b. Citrate feeding did not evolve in the lineage for 30,000 generations.
   c. The citrate feeding trait quickly came to dominate the *E. coli* lineage.
   d. One generation of *E. coli* showed significant variation in the citrate feeding trait.
   e. The frequency of citrate feeding was highly variable in all generations within the lineage of *E. coli.*

3. Mutations to which genes in the limb patterning network led to the loss of legs in snakes?
   a. None. Snakes don't have legs, so they don't have genes for legs.
   b. *Hedgehog* (*Hh*) and *sonic hedgehog* (*Shh*)
   c. *Bmp2* and *Dpp*
   d. *Hox* genes and *sonic hedgehog* (*Shh*)
   e. *Hedgehog* (*Hh*) and *Hox* genes

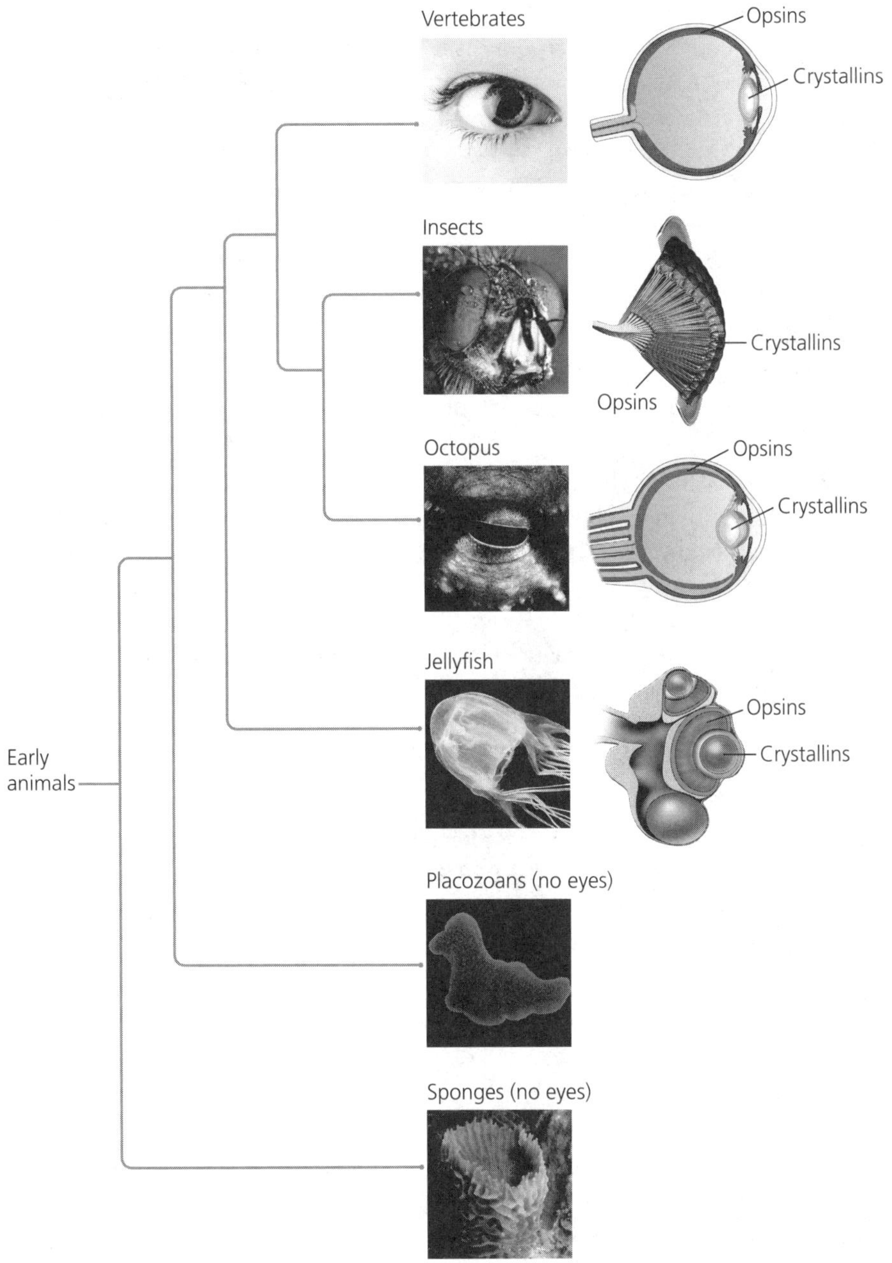

4. According to Figure 10.2 (above), which of the following statements is FALSE?
   a. Jellyfish eyes are more similar to octopus eyes than they are to insect eyes.
   b. The structure of an octopus eye and the structure of an insect eye are more similar to each other than to the structure of a human eye.
   c. The proteins in human eyes can be traced back to a common ancestor with octopi and insects.
   d. Complex eyes have evolved in several distinct lineages of animals.

5. Do scientists have evidence to support their hypotheses about the early evolution of eyes?
   a. No. Scientists have only a hypothesis based on the structure of proteins.
   b. No. The evidence for the evolution of eyes is inconsistent among lineages, such as bilaterians and cnidarians.
   c. Yes. Scientists know exactly how the eye evolved based on a series of steps involving mutations of different proteins.
   d. Yes. Scientists have evidence from phylogenies based on shared derived morphological structures as well as gene sequences and entire genomes.
6. How did scientists test the hypothesis that insect size is constrained by access to oxygen?
   a. They looked at insects of different sizes and compared how well they survived and reproduced.
   b. They controlled oxygen concentrations and determined how large insects could grow.
   c. They compared the sizes of insects in the fossil record when oxygen concentrations were higher to when oxygen concentrations were lower.
   d. They examined the fossil record of large insects and then looked for a factor that provided some explanation.
   e. They compared the sizes of the tubes insects use to get oxygen between large and small fossil insects.
7. Is the human eye a perfect adaptation?
   a. Yes. Because in combination with our brain's ability to overcome the blind spot caused by the optic nerve, the human eye allows humans to see what we need to see.
   b. Yes. Because the human eye evolved through modifications of ancestral eyes, natural selection favored mutations that improved the eye until it became a perfect adaptation.
   c. No. Because the human eye evolved through modifications of ancestral eyes, the optic nerve leaves the retina through the back of the eye creating a blind spot.
   d. No. The human eye can't be perfect because some people need glasses.
   e. None of the above
8. Why are the wolf and the Tasmanian wolf (Figure 10.27) considered examples of convergent evolution?
   a. Because even though they are from two different lineages of mammals, their genotypes converged on the same form and ecological function
   b. Because even though they were found on different continents, they have identical phenotypes
   c. Because they share a common ancestor
   d. Because their phenotypes converged on similar forms and ecological functions even though they are from two different lineages of mammals
   e. Because their adaptations arise from similar developmental underpinnings

Contemplate

Do you think placental and marsupial mammals faced different constraints in the course of their evolution?	If the human lineage shared a more recent common ancestor with octopus or squid, do you think we'd have the same constraints on our ability to see?

11 Sex

Causes and Consequences

Check Your Understanding

1. What is most important element necessary for natural selection to act?
   a. Individuals that can change to meet their needs
   b. Variation within a population
   c. Thousands of years
   d. All of the above
   e. None of the above
2. How does $R = h^2 \times S$ apply to artificial selection?
   a. In artificial selection, breeders manipulate the entire reproductive process of all individuals, so the response to selection is simply the strength of that selection, $R = S$.
   b. In artificial selection, breeders create very strong selection on traits, which can result in a rapid evolutionary response if the heritability of the selected traits is high.
   c. In artificial selection, breeders manipulate narrow sense heritability, h^2, by selecting for specific traits.
   d. $R = h^2 \times S$ is the breeder's equation and therefore only applies to artificial selection.
3. Is phenotypic plasticity heritable?
   a. No. Phenotypic plasticity is the capacity of an individual to change in response to the environment, and characteristics acquired during an individual's lifetime are not heritable.
   b. Yes. Phenotypic plasticity is the capacity of an individual to change in response to the environment, and this ability to change in response to need is the very basis of evolution.
   c. No. Phenotypes cannot be directly linked to any heritable component of genes, so phenotypic plasticity cannot be heritable.
   d. Yes. Individuals may vary in their plastic responses to the environment, and those responses can be heritable.

Learning Objectives for Chapter 11

Add important definitions and notes next to each learning objective for this chapter to help guide your understanding.

Learning Objective	Important Definitions	Notes
Compare and contrast the genetic consequences of sexual and asexual reproduction.		
Explain how differential investment by males and females in sexual reproduction can lead to sexual selection.		
Compare and contrast hypotheses developed to explain mate choice.		
Discuss the costs and benefits to males and females of different mating systems.		
Explain how sexual selection can act on traits that function after mating.		
Use mating systems to predict when sexual conflict will lead to antagonistic coevolution.		

Identify Key Terms

Match terms and definitions by filling in the blank to the right of the term with the appropriate letter.

1. **Anisogamy** ___	a. The burden imposed by the accumulation of deleterious mutations
2. **Certainty of paternity** ___	b. A difference in form between males and females of a species, including color, body size, and the presence or absence of structures used in courtship displays (elaborate tail plumes, ornaments, pigmented skin patches) or in contests (antlers, tusks, spurs, horns)
3. **Cryptic female choice** ___	c. A form of sexual selection that arises after mating, when females store and separate sperm from different males and thus bias which sperm they use to fertilize their eggs
4. **Direct benefits** ___	d. A mating system in which one male pairs with one female. It can be sexual or social. The sexual form is very rare and occurs when each male mates only with a single female, and vice versa. The social form occurs when a male and female form a stable pair bond and cooperate to rear the young, even if either or both partners sneak extra-pair copulations. The social form occurs in a few fish, insect, and mammal species, and in almost 90 percent of bird species.
5. **Fecundity** ___	e. The probability that a male is the genetic sire of the offspring his mate produces
6. **Genetic load** ___	f. The disadvantages of being a sexual rather than an asexual organism. Asexual lineages multiply faster than sexual lineages because all progeny are capable of producing offspring. In sexual lineages, half of the offspring are males who cannot themselves produce offspring. This limitation effectively halves the rate of replication of sexual species.
7. **Hermaphrodites** ___	g. A phenomenon seen in coevolving populations—to maintain relative fitness, each population must constantly adapt to the other. Leigh Van Valen borrowed the tale from Lewis Carroll's *Through the Looking-Glass* and compared the difficulties to those in biological arms races, such as between parasites and their hosts.
8. **Indirect benefits** ___	h. Occurs when members of the less limiting sex (generally males) compete with each other over reproductive access to the limiting sex (generally females). Often called male–male competition
9. **Intersexual selection** ___	i. The variance in fitness within a population. When there is no variance in fitness, there can be no selection; when there is large variance in fitness, there is a great opportunity for selection. In this sense, the opportunity for selection constrains the intensity of selection that is possible.
10. **Intrasexual selection** ___	j. The ratio of male to female individuals who are available for reproducing at any given time
11. **Leks** ___	k. A mating system where females mate (or attempt to mate) with multiple males
12. **Monogamy** ___	l. A form of sexual selection that arises after mating, when males compete for fertilization of a female's eggs
13. **Muller's ratchet** ___	m. The process by which the genomes of an asexual population accumulate deleterious mutations in an irreversible manner
14. **Operational sex ratio (OSR)** ___	n. Differential reproductive success resulting from the competition for fertilization, which can occur through competition among individuals of the same sex (intrasexual selection) or through attraction to the opposite sex (intersexual selection)
15. **Opportunity for selection** ___	o. Benefits that affect a particular female directly, such as food, nest sites, or protection
16. **Polyandry** ___	p. Benefits that affect the genetic quality of a particular female's offspring, such as male offspring that are more desirable to females
17. **Polygyny** ___	q. The evolution of phenotypic characteristics that confer a fitness benefit to one sex but a fitness cost to the other
18. **Red Queen effect** ___	
19. **Reproduction** ___	
20. **Sexual conflict** ___	
21. **Sexual dimorphism** ___	
22. **Sexual selection** ___	
23. **Sperm competition** ___	
24. **Twofold cost of sex** ___	

	r. The reproductive capacity of an individual, such as the number and quality of eggs or sperm. As a measure of relative fitness, this refers to the number of offspring produced by an organism.
	s. Sexual reproduction involving the fusion of two dissimilar gametes; individuals producing the larger gamete (eggs) are defined as female, and individuals producing the smaller gamete (sperm) are defined as male.
	t. A mating system where males mate (or attempt to mate) with multiple females
	u. Occurs when members of the limiting sex (generally females) actively discriminate among suitors of the less limited sex (generally males). Often called female choice
	v. Assemblages of rival males who cluster together to perform courtship displays in close proximity
	w. Individuals that produce both female and male gametes
	x. The formation of new individual organisms (offspring)

Link Concepts

Fill in the bubbles with the appropriate terms.

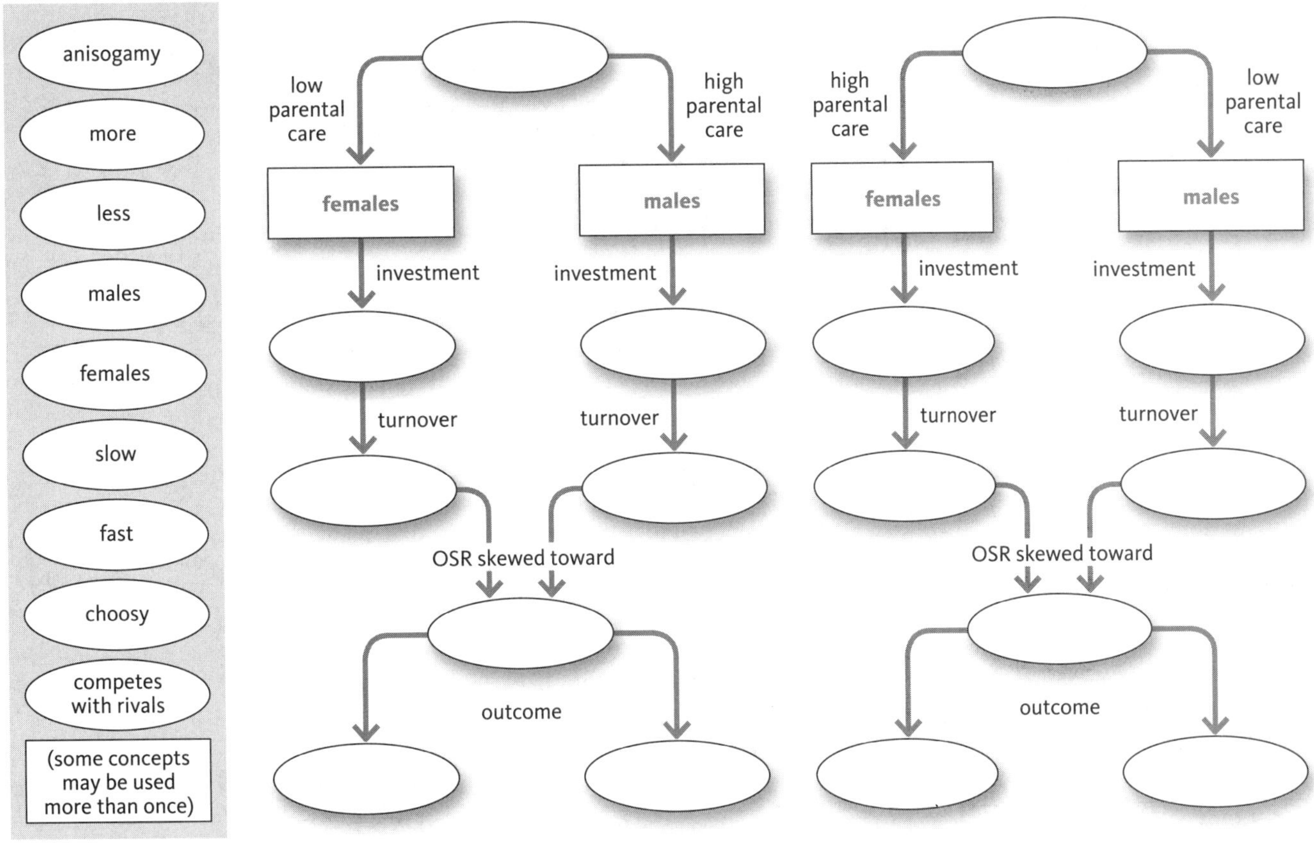

Key Concepts

Fill in the blanks for the key concepts from the chapter.

11.1 Evolution of Sex

Sex is the combining and mixing of ____________ during the formation of offspring. It involves two main processes: (1) ____________ halves the number of chromosomes during gamete formation, and (2) ____________ restores the original chromosome number as two gametes fuse to form a zygote.

Self-fertilization occurs in ____________ organisms when the two gametes fused in fertilization come from the ____________ individual. Although this is still sexual reproduction (because it incorporates ____________ and ____________), it yields offspring that are lacking in genetic variation.

Sexual reproduction may have evolved because sex creates new genetic ____________ within ____________ by mixing parental alleles into novel offspring ____________ while at the same time enabling the purging of ____________ alleles.

The Red Queen effect is used to explain the evolution of ____________, because sexually reproducing organisms are likely to fare better in the continual evolutionary arms races that can arise between species (e.g., between ____________ or ____________ and their ____________).

Ecological situations that require ____________ and ____________ evolution are likely to favor the evolution and maintenance of ____________.

11.2 Sexual Selection

Anisogamy is an important driver of behavior because it constitutes a difference between the ____________ in their relative investment in ____________. Because of anisogamy, males and females maximize their reproductive success in ____________ ways.

A female's reproductive success is often limited by the number of ____________ that she can produce and provision. Females with access to the most ____________ generally achieve the highest egg (and offspring) numbers and have the highest ____________ relative to other females.

Male reproductive success is often limited by the number of ____________ he can obtain. Males who mate with the most ____________ generally sire the greatest number of ____________ and have the highest ____________ relative to other males.

Biased operational sex ratios can generate strong ____________ selection because the ____________ sex (typically males) must compete over access to the ____________ sex (typically females).

Males often ____________ with each other over ____________ to females, and this behavior can generate strong sexual selection for large body size, weapons, and aggression.

Depending on the species, females may select males based on characteristics that benefit her ____________ (such as nutrients, nest sites, protection, parental care) or ____________ (e.g., high genetic quality transmitted to her offspring).

Females often choose males based on ____________ or details of their ____________ behavior.

The ____________ in reproductive success of males and females can influence the ____________ for sexual selection.

11.3 The Rules of Attraction

Females who choose males with attractive traits pass to their offspring ____________ influencing the expression of *both* the ____________ *and* the ____________. The resulting genetic correlation between preference and trait can lead to an escalating ____________ feedback cycle of ____________ of stronger preferences and larger display traits.

Although the traits that strike the females' fancy may originally be ____________, they do allow females to choose males based on ____________ that reliably ____________ the highest-quality, healthiest individuals from the rest of the lot.

____________ signals or display traits will be the most ____________ because they are difficult or impossible for those of poor ____________ quality to bluff.

11.4 The Evolution of Mating Systems

Mating systems evolve because of the ____________ and ____________ they confer on males and females. Who mates with whom can depend on the ____________ that arise through both inter- and intrasexual ____________. In some species, a combination of systems may be at play.

11.5 The Hidden Dimension of Sexual Selection

When females mate with multiple males, ____________ from the different males must ____________ for opportunities to fertilize the females' eggs.

Sometimes ______________ can control whose sperm they use to ______________ their eggs. In these species males may continue to court females even after they have finished ______________ .

11.6 Sexual Conflict and Antagonistic Coevolution

Sexual conflict can lead to ______________ and arms races between males and females. Mutations ______________ the reproductive interests of males select for ______________ in the females, and vice versa.

Interpret the Data

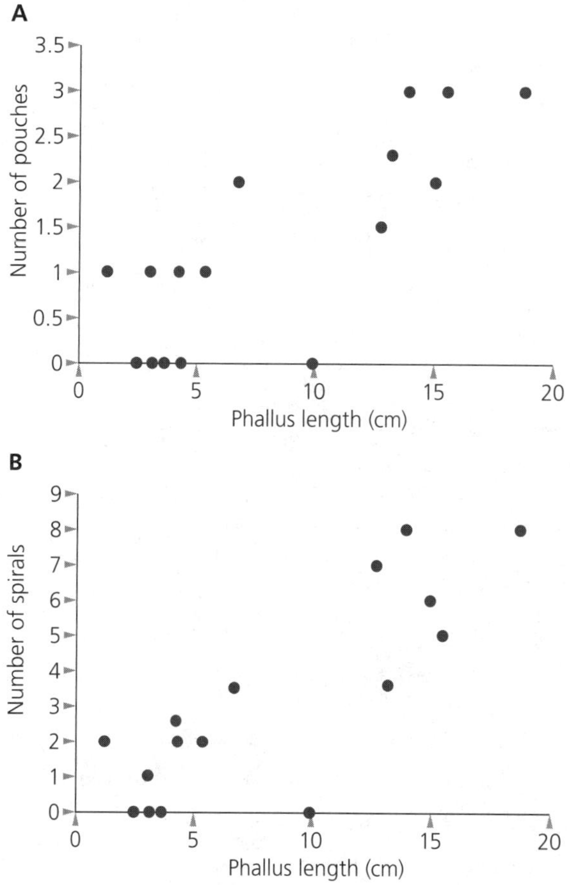

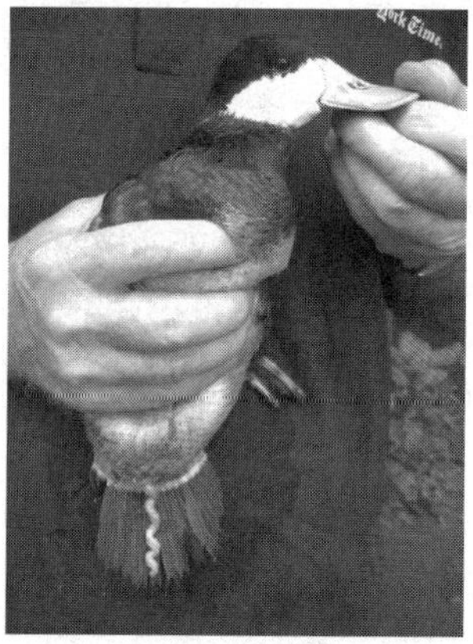

Patricia Brennan found that among ducks and other wading birds, species where males had long phalluses also had females with many pouches and spirals in their reproductive tracts. (Graph adapted from Brennan et al. 2007.) The mean length of the male's phallus in centimeters for each species is plotted along the x-axes, and mean number of pouches (top) and mean number of spirals (bottom) in the reproductive tracts of females for each species are plotted along the y-axes.

- What was the longest phallus found for a species?

- What was the maximum number of pouches found in females for a species?

- What was the maximum number of spirals found in females for a species?

- From these graphs, could you predict the number of pouches or spirals you would expect to find in a species where the male phallus length was 9 cm?

- Do the relationships between phallus length and number of pouches/spirals shown in the above graphs indicate that male phallus length causes female reproductive tract characteristics?

Games and Exercises

Mating Game

Live from the PBS *Evolution Library*, it's the Mating Game! Test your ability to pick the right mate based on the information the "bachelors" provide. The game is tricky, so read carefully!

http://www.pbs.org/wgbh/evolution/sex/mating/index.html

Mating Game II: Craigslist

Humans are evolving, too, and patterns in our mating preferences, although complex, can be identified. Go to Craigslist.com and examine the personal ads. Although the ads don't represent a true sample of mating preferences in our society, you can still summarize some of the characteristics. For example, how many ads are seeking long-term versus short-term relationships?

	males	females
18–20		
21–25		
26–30		

Develop a table and tally the number of ads according to gender. Based on your understanding of personal relationships, would you expect age to have an influence? Add age categories to your table. Think about the duration of relationships requested by the people that posted the ads (you probably need to make a new table).

- Is duration related to gender?

- Is duration related to age?

- Does gender affect the preference for a mate of a specific age?

- Could age be relevant to female reproductive potential?

- Compare ads for heterosexual and homosexual relationships.

- Does the sex of target audience affect the ads? How?

Develop a hypothesis about sexual selection in humans.

Now, go online and read the abstract for a research paper on human mating preferences:

http://onlinelibrary.wiley.com/doi/10.1046/j.1439-0310.2002.00757.x/abstract

- Does this research affect your hypothesis about age, gender, and duration of relationships in personal ads? Why or why not?

This activity is modified from an exercise developed by Janis Antonovics and Doug Taylor of the Biology Department at University of Virginia.

http://www.faculty.virginia.edu/evolutionlabs/home.html

Explore!

TheCornellLab of Ornithology
Exploring and Conserving Nature

In this video, The Cornell Lab shares the stunning diversity of plumage and courtship that evolved in one group of birds, the birds-of-paradise, and explains how these traits would have evolved through natural and sexual selection.

http://ed.ted.com/on/JLk7NshX

Birds-of-Paradise Project

The Birds-of-Paradise Project is a multi-media website that explores the evolution, diversity, and adaptations of the birds-of-paradise. The videos are excellent. Check out the introduction.

http://www.birdsofparadiseproject.org/

The Elaborate Weapons of Beetles—Yes, Beetles

This slideshow from the *New York Times* presents some of the incredible diversity of beetle weapons courtesy of Dr. Douglas Emlen and his colleagues at the University of Montana.

http://www.nytimes.com/slideshow/2009/03/23/science/032409-Armor_index.html

Animal Attraction by Virginia Morell

Bower birds are another amazing example of the lengths males go to to gain access to females. Male bower birds actually build elaborate structures, called bowers, that females inspect. If they find the display suitable, the male may be granted her favor. This *National Geographic* article explores the lengths males will go to, as well as the females' interests, leading to some of the fantastic traits we see in the animal world.

http://science.nationalgeographic.com/science/health-and-human-body/human-body/animal-attraction.html#page=1

Color-Obsessed Bowerbirds

Another awesome video from *National Geographic* showing just how obsessed Satin Bowerbirds are with the color blue.

http://video.nationalgeographic.com/video/weirdest-bowerbird

Sexual Cannibalism

Visit the Andrade Lab web page to learn about recent discoveries with redback spiders. The website also provides videos or links to videos that show how sexual cannibalism operates.

http://www.utsc.utoronto.ca/~mandrade/index_files/Media.htm

Wrestling Elephant Seals

This video from *National Geographic* gets up close and personal as an elephant seal "beach master" defends his territory from a would-be interloper.

http://video.nationalgeographic.com/video/seal_elephant_wrestling

Bugling Bull Elk in Montana

Listen to the haunting calls of bull elk and watch them spar on the National Bison Range in western Montana in this clip from Montana PBS' *The Backroads of Montana.*

http://watch.montanapbs.org/video/2070414704/ (scroll to 09:56)

A Database of Sex

The Tree of Sex Consortium developed a database to facilitate access information about sexual systems and sex chromosomes across taxa, including plants, fish, amphibians, non-avian reptilians, birds, mammals, and invertebrates.

http://www.nature.com/articles/sdata201415

Overcoming Misconceptions

Evolution Does NOT Provides Species with the Adaptations They "Need"

Sexually reproducing males and females certainly need to breed if they are to pass on their genes to future generations, but natural and sexual selection do not provide males or females with adaptations simply because they need them. Sexually selected traits arise from processes such as random mutations and the change of gene frequencies from one generation to the next. Sex is the mechanism that permits that transmission.

Courtship Behavior Has a Genetic Component

Individual organisms undergo change during their own lifetime, and some of these changes are adaptive—learning elaborate courtship displays, for example. But it's generally not possible for these traits to be passed down to successive generations, unless there's a genetic component. If you dyed your hair purple, and you found a mate that liked purple hair, you wouldn't expect your children to have purple hair. But, some males may have a preference for purple that has a genetic basis. This preference could be passed to your male offspring, who would likely choose mates with purple hair. Similarly, courtship behavior (and any behavior) seems like something an individual can adapt it to its environment. Clearly, those more successful in wooing a female would have more offspring. Behavior has a genetic component, though—even the capacity for flexible strategies is related to our underlying genetic architecture (see Chapter 16). So, the capacity to learn an elaborate courtship display *and* carry it out successfully (especially in light of the costs it may incur) can evolve. The costs and benefits may vary among different environments, and this variation will affect the fitness of individuals within those environments differently, potentially leading to a diversity of courtship behaviors.

Evolution Has NOT Made Living Things Perfectly Adapted to Their Environment

As impressive as some sexually selected adaptations may be, they are far from perfect. Selection can act only on what already exists, and it operates under tight constraints of physics and development. Bees often "mate" with flowers that only look superficially like female bees, but they still end up pollinating the flower. Humans have very large brains, for example, that have allowed us to become nature's great thinkers, but those big brains also make childbirth much more dangerous for human mothers than for other female primates.

Evolution Does NOT Happen for the Good of the Species

Just as evolution has no foresight, evolution cannot recognize what is best for a large group of organisms. Sexual selection operates at the level of genes and the reproductive success of individuals.

Which of the following is a true statement?

a. When an animal needs to change to fit the environment, it will evolve to fit the environment.

b. Evolution of sexual ornaments results because individuals need to make themselves more attractive to the opposite sex.

c. Sexually selected traits are examples of the perfect fit between males and females of a species.

d. Ornaments will evolve through sexual selection when individuals vary in a sexually selected trait, that variation is heritable, and that variation affects reproductive success.

Go the Distance: Examine the Primary Literature

Cosima Hotzy and Göran Arnqvist examine an intriguing aspect of sexual selection: when one sex physically injures members of the other sex. Spines on the genitalia of male seed beetles wound female seed beetles during copulation. They test the hypothesis that these spines are adaptive, providing a reproductive edge, so to speak.

- Did Hotzy and Arnqvist find evidence to support the idea that male genital spines are adaptive? Why or why not?
- Did they offer alternative hypotheses? What was it?
- Do their conclusions affect scientists understanding of sexual selection theory? Why or why not?

Hotzy, C., and G. Arnqvist. 2009. Sperm Competition Favors Harmful Males in Seed Beetles. *Current Biology* 19 (5):404–7. doi:10.1016/j.cub.2009.01.045.

http://www.cell.com/current-biology/abstract/S0960-9822%2809%2900617-4.

Delve Deeper

1. In Figure 11.27, the genus *Homo* is included. What does the position of the point indicate?

2. What's the difference between sex, reproduction, and copulation?

Test Yourself

1. All of the following are components of Darwinian "fitness" except:
   a. Seasonal seed set in a perennial plant.
   b. The probability of an individual living from one breeding season to the next.
   c. The rate of cell division of a bacterial strain.
   d. The number and quality of grandchildren produced by an individual.
   e. The number of sperm a male produces.
   f. None of the above; they all are components of fitness.
2. The computer simulation for host-parasite coevolution graphed in Figure 11.5 demonstrates the Red Queen Effect because:
   a. Host and parasite population sizes fluctuate over time.
   b. The frequency of parasite genotypes peak every 25 generations.
   c. The frequency of parasite infections tracks host populations with regularity.
   d. The graph shows how sex helps hosts evolve.
   e. The frequency of the parasite genotype that can infect the host genotype tracks the frequency of the host genotype.
3. According to Figure 11.13, fighting entails high costs to male red deer because:
   a. Males do not survive at the same rate as females.
   b. Females choose older males that are more capable of fighting.
   c. Females reproduce at a steady rate.
   d. All of the above
   e. None of the above
4. Which of the following is a type of sexual dimorphism?
   a. Males bigger than females
   b. Males smaller than females
   c. Males more brightly colored than females
   d. Females bigger than males
   e. All of the above are sexually dimorphic.
5. Which of the following is true?
   a. Polyandry is a mating system where many males mate with a single female.
   b. Polygyny is a mating system where many males mate with many females.
   c. Monogamy is a mating system where a male and female mate only with each other.
   d. Social and sexual monogamy are the same thing.
6. Females benefit from polyandrous mating systems because:
   a. Gaining access to the highest quality males.
   b. Gaining access to males with genes different from their own.
   c. The costs of fending off courting males may be too high.
   d. All of the above
   e. None of the above
7. Which adaptations are likely related to sperm competition?
   a. Large sperm
   b. Large sperm plugs
   c. Cooperative sperm
   d. All of the above
   e. None of the above

8. Fertilization success in male seed beetles is related to:
    a. The number of mates males have.
    b. The amount of sperm the male produces.
    c. The length of spines on their intromittent organ.
    d. The number of eggs in the female.
    e. The amount of time spent mating.
9. When monogamous female *Drosophila* flies were forced to mate with males reared in a polygynous mating system, what happened according to Figure 11.32?
    a. After 20 days, female survival rate was lower than females reared in a polygynous mating system.
    b. Control females were more successful at gaining access to males.
    c. Young females were more likely to mate than old females.
    d. Survival rate of males declined over time.

Contemplate

Can artificial selection replicate sexual selection? Check out the HBO Real Sports clip on Unnatural Selection https://www.youtube.com/watch?v=BVS5XjcATwY	**How might sexual conflict shape the reproductive strategies of male and female primates?**

12 After Conception
The Evolution of Life History and Parental Care

Check Your Understanding

1. How does sexual selection differ from natural selection?
   a. Sexual selection only acts on males, whereas natural selection acts on populations.
   b. Variation among the choosy sex is not necessary for sexual selection to operate; variation among all individuals is necessary for natural selection to operate.
   c. Sexual selection does not really differ from natural selection; both require heritable variation in traits that confer some fitness differences to individuals possessing those traits.
   d. No. Sexual selection does not really differ from natural selection, but the optimum trait value under sexual selection can often be at odds with the optimum trait value under natural selection.
   e. Both c and d
2. What is an/are important factor(s) influencing gene expression?
   a. The gene control region
   b. Cis- and trans-acting factors
   c. Internal environmental variables, such as hormone levels and microRNAs
   d. The external environment in which the organism develops, such as sunlight or food availability
   e. All of the above
   f. None of the above
3. How may polyandry benefit females?
   a. Females may be able to get the highest-quality genes possible for their offspring by mating with a number of males.
   b. Females may benefit because they accumulate nutrients or other resources offered by each of the males.
   c. Mating with multiple males may allow females to boost the health of their offspring by giving them a defense against a wider range of pathogens.
   d. Females may mate with multiple males simply because the costs of resisting courtship overtures are higher than the costs of mating multiple times.
   e. All of the above

Learning Objectives for Chapter 12

Add important definitions and notes next to each learning objective for this chapter to help guide your understanding.

Learning Objective	Important Definitions	Notes
Use natural selection and trade-offs to explain why organisms might not produce as many offspring as they possibly can.		
Compare and contrast the investments males and females make in reproduction and the life-history trade-offs associated with different strategies.		
Explain how different parental care strategies can lead to conflicts between parents, and between parents and their offspring.		
Explain how parental conflict can influence gene expression.		
Describe how senescence may arise as a result of life-history trade-offs.		
Analyze the evidence scientists have used to explain the evolution of menopause in humans.		

Identify Key Terms

Match terms and definitions by filling in the blank to the right of the term with the appropriate letter.

1. **Genomic imprinting** ____ 2. **Intralocus sexual conflict** ____ 3. **Life history** ____ 4. **Methylation** ____ 5. **Parent-of-origin effect** ____ 6. **Parent-offspring conflict** ____ 7. **Parental conflict** ____ 8. **Senescence** ____	a. Occurs when parents benefit from withholding parental care or resources from some offspring (e.g., a current brood) and invest in other offspring (e.g., a later brood). Conflict arises because the deprived offspring would benefit more if they received the withheld care or resources. b. Occurs when genes inherited from one or the other parent are silenced due to methylation. Imprinting can result in offspring who express either the maternal or paternal copy of the gene, but not both. c. An effect on the phenotype of an offspring caused by an allele inherited from a particular parent. d. A conflict between the fitness effects of alleles of a given locus on males and females e. The deterioration in the biological functions of an organism as it ages. f. Refers to the pattern of investment an organism makes in growth and reproduction. Life-history traits include an organism's age at first reproduction, the duration and schedule of reproduction, the number and size of offspring produced, and life span. g. The process by which methyl groups are added to certain nucleotides. Methylation alters gene expression, thereby reducing or eliminating the production of proteins or RNA molecules. h. When parents have an evolutionary conflict of interest over the optimal strategy for parental care

Link Concepts

Fill in the bubbles with the appropriate terms.

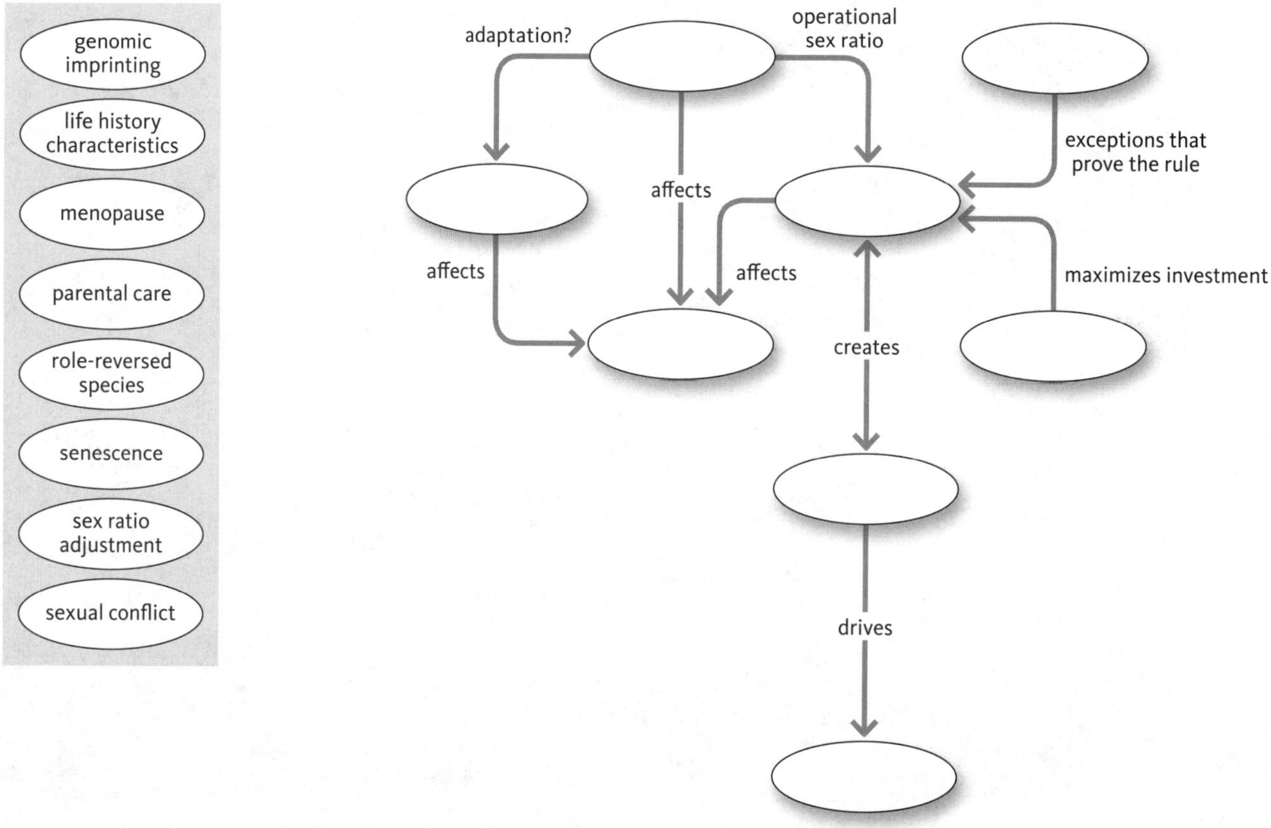

Key Concepts

Fill in the blanks for the key concepts from the chapter.

Life-history theory explores how the schedule and duration of ____________ in an organism's ____________ are shaped by natural selection. It helps explain ____________ in the age at which organisms begin reproducing, the size and number of offspring produced, the amount and type of parental care invested, and even the onset of senescence.

12.1 Selection across a Lifetime

____________ arise when allocation of resources to one life-history trait ____________ investment in another trait.

Investment in reproduction often comes at the expense of ____________ or ____________ .

Selection may favor alleles that are ____________ early in life, even if those same alleles are ____________ later on.

Investment in reproduction early in life often ____________ an individual's ability to breed later in life.

12.2 Parental Investment

____________ generally benefit more than ____________ from parental care of offspring. Exceptions to this rule, when they occur, provide exciting tests of ____________ theory.

Frequency-dependent selection can ____________ variation within populations, if the fitness of an allele or phenotype ____________ when it is common, and ____________ as it becomes rare. This may contribute to the persistence of two sexes (males, females) within populations.

The Trivers–Willard hypothesis predicts greater investment in male offspring by parents in ____________ condition, and greater investment in female offspring by parents in ____________ condition.

12.3 Family Conflicts

____________ is often rife with ____________ ; what is in the best interest of the female is often not the same as what is best for the male.

Offspring not only compete amongst themselves for parental care, ____________ between parents and offspring can lead to life-history ____________ .

12.4 Conflicts within the Genome

Parents sometimes have conflicting interests when it comes to how much a mother should ____________ in her offspring. Males benefit when they can ____________ a female's parental contribution to current offspring, but females benefit when they ____________ some of their resources for later offspring.

Parental conflict can lead to battles over the control of ____________ in offspring, through genomic imprinting.

12.5 Searching for Immortality Genes

Senescence is a by-product of ____________ for alleles that enhance growth and reproduction early in life. It arises because of a trade-off between investing in ____________ and investing in ____________ .

12.6 Menopause: Why Do Women Stop Having Children?

Menopause is likely a biological feature of humans that evolved at some point after our ____________ branched off from ____________ some 7 million years ago.

Interpret the Data

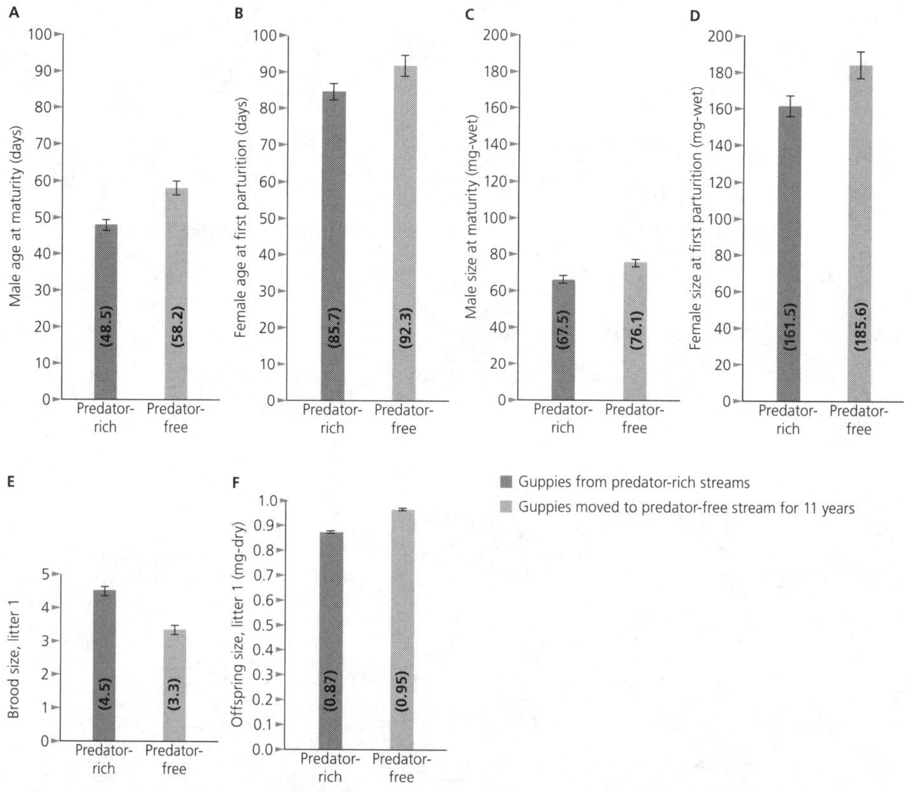

David Reznick and his colleagues developed an experiment to test predictions about the evolution of life-history traits in guppies. Because different populations of guppies experience different levels of predation (because different drainages have different predators of waterfalls that can exclude predators all together), they transplanted individuals from populations that lived with high predation into areas with very low predation. Then, 11 years later, they brought individuals from both populations into the lab. After two generations, they looked at the life history traits of the lab-reared guppies. The dark gray represent individuals from the original high-predation environments, and the light gray represent individuals that had been moved to low-predation environments.

- Which trait(s) decreased in the predator-free environments?
- In which populations were the females younger and smaller at first parturition (when they give birth)? Why might female age and size be life-history traits?
- Did males experience similar selection?
- Based on your understanding of natural selection and fitness differences, how many more generations do you think it would take for the population to be completely dominated by individuals with relatively long lives?

Games and Exercises

Ratios, Rates, and Proportions

A

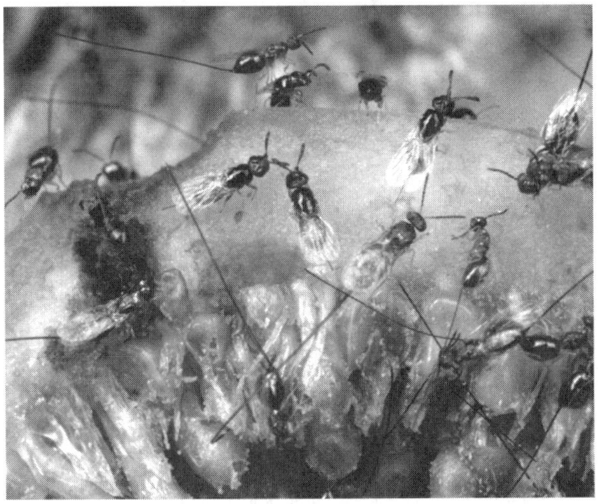

B

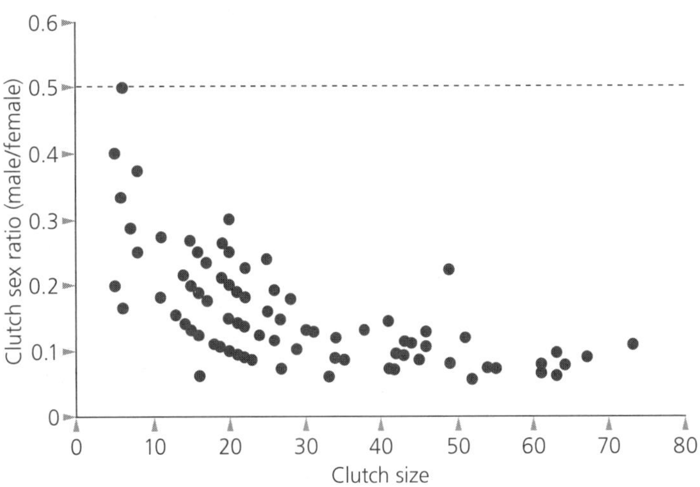

Female bees, wasps, and ants can manipulate the sex ratios of their offspring by determining which eggs are fertilized. Unfertilized eggs develop into males and fertilized eggs develop into females. Females benefit if they adjust their offspring sex ratios to produce just a few sons, and as many daughters as possible. Larger clutch sizes have more female bias than smaller clutch sizes.

- What does the sex ratio of 0.4 in the graph on the previous page mean?

A ratio describes the relationship between two numbers; it can be expressed as 4:5, 4 to 5, 4/5, or as a decimal, like 0.8. Ratios can be difficult to interpret, however. You have to know whether the ratio is part-to-whole or part-to-part—4 males for every 5 individuals (part-to-whole) or 4 males for every 5 females (part-to-part). When a ratio is expressed as 4/5 you can't divide unless you know whether it's part-to-whole or part-to-part because if there are 4 males of 5 total, the decimal form is 0.8. But if there are 4 males to 5 females, the decimal form would be rounded to 0.4. Usually, you are dealing with part-to-part ratios in science, but a ratio of 4 to 5 doesn't mean there are exactly 4 males and 5 females. The actual numbers could be 52 males and 65 females or 8 males to 10 females because all of those ratios are in proportion. 52:65 and 8:10 represent the same ratio. But, the order of the numbers matters—3:1 is not the same ratio as 1:3.

- What does the sex ratio of 0.1 in the graph on the previous page mean?

- How many males would you expect in a population of 75 wasps with a sex ratio of 0.2?

Here's a quick test of your mad math skills with these important science issues:

A scientist running an experiment where she has to dilute a phenol reagent with diH2O in a 2:3 ratio before use. She wants to make 450 milliliters of solution.

Milliliters of Reagent	Milliliters of diH2O	Milliliters of Solution
20	30	50
40	60	100

Use the table above to determine how many milliliters of diH2O the scientist needs. Sex ratios are usually expressed in the proportion that is male.

	Males	Population
Montana	47	181
Idaho	27	135
Nevada	18	142
Oregon	68	340

- What is the sex ratio for each population?

- Do any of the populations have equal sex ratios?

The CIA provides information on the sex ratios of countries throughout the world. They use the number of males for each female, and they break it down into five age groups: at birth, under 15 years, 15–64 years, 65 years and over, and for the total population. Here's data from the U.S., Germany, Tanzania and India.

	At Birth	0–14 Years	15–24 Years	25–54 Years	55–64 Years	65 Years and Over	Total Population
U.S.	1.05	1.05	1.05	1	0.97	0.77	0.97
Germany	1.06	1.06	1.04	1.03	0.97	0.76	0.97
Tanzania	1.03	1.02	1	1.01	0.99	0.76	0.99
India	1.12	1.13	1.13	1.06	1.08	0.91	1.08

From https://www.cia.gov/library/publications/the-world-factbook/docs/guidetowfbook.html (last accessed May 2015).

- What does the ratio at birth indicate for these four countries?

- How do sex ratios play out over time?

A rate is a little bit different than the ratio, it's a special ratio. It is a comparison of measurements that have different units, like cost (dollars per pound). A unit rate is a rate with a denominator of 1. A unit rate is actually a fraction that's understood. For example, 61 point mutations per birth means 61/1.

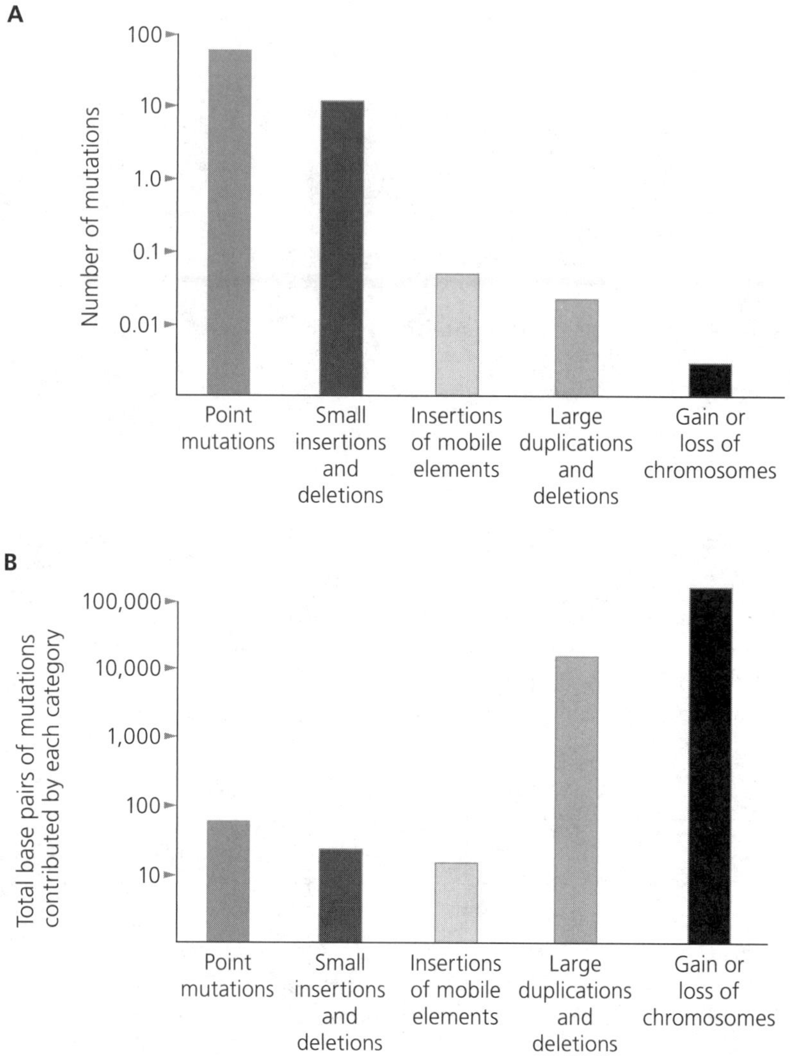

Figure 5.15a (above) shows the average number of mutations per live birth for each type of mutation.

- Roughly, what are the mutation rates for each category?

- So how often does a large duplication actually occur?

The BBC has a great interactive website designed to help people understand the relevance of ratios and proportions in their daily lives. Check out *Skillswise: English and Maths for Adults* here:

http://www.bbc.co.uk/skillswise/topic/ratio-and-proportion

Explore!

From Nature News: Male Pipefish Abort Embryos of Ugly Mothers

Read the news report about recent research on male pipefish and sexual selection from the journal *Nature*.

http://www.nature.com/news/2010/100317/full/news.2010.127.html

Eurasian Penduline Tits Nest Building

Eurasian penduline tits are known for more than just the elaborate deceptions that punctuate their mating system. They build incredibly elaborate nests as well. Watch this video by Jos Vroegrijk.

http://www.youtube.com/watch?v=_QqiWvmimg0

Pipefish Courtship

Watch this research video of Gulf Pipefish females courting males by Mark Currey, from Oregon State University.

http://www.youtube.com/watch?v=aZQ75Vr9b7M

Sand Goby Building a Nest Burrow

Watch this short video from ReefCollege of nest building in Sand Gobies.

http://www.youtube.com/watch?v=LS25go6E33I

Sex and the Wattled Jacana

This video from PBS *Evolution* shows the amazing adaptations of wattled jacanas. Dr. Stephen Emlen explains why the behaviors, often common in males of other species, evolved in this sex-role reversed species.

http://www.pbs.org/wgbh/evolution/library/01/6/l_016_04.html

Overcoming Misconceptions

Organisms Are Not Trying to Adapt

Trying is a human endeavor. We try to do better in school. We try to make our loved ones happy. We try to be good parents. It's pretty easy to assume that organisms are "trying," too, but we can't apply human motivations to other organisms (a symptom known as anthropomorphizing). Adaptations, like parental care, result from the process of natural selection, not because an organism is "trying" to raise more offspring. Evolution by natural selection results when individuals vary in some heritable way that confers a fitness advantage or disadvantage. If, for example, a mutation arises that causes a male to guard his nest more intensively than other males do, and his offspring survive at a greater rate as a result, not only would he have more offspring, but if the mutation is heritable, his offspring will be more likely to carry it as well. In the next generation, a greater proportion of the population will have that adaptation. And so on. And so on. The population will evolve because the behavior is adaptive, not because the individual is trying, and certainly not because that individual needed to adapt.

Evolutionary Theory May Be Incomplete, but It Does Explain the Diversity of Life

All of science is a work in progress, and evolutionary theory is no exception. For example, scientists don't understand cancer completely, and they're constantly improving their understanding of climate change. Smartphones are only a recent technological advance, and scientists are working to improve that technology, as well as studying how people use them and why. We can't expect to know all there is to know about a discipline before it has any value. Scientists are constantly making new discoveries, developing new hypotheses, creating new ideas. Evolutionary theory is supported by mounds and mounds of evidence from diverse disciplines. The evidence for some aspects has been debated for so long and is so strong that scientists have little doubt about the value of evolutionary theory to explain the history and diversity of life. And it explains so much more. Evolutionary theory is being applied to research in diseases, like cancer, to economics, to government, and even to artificial intelligence. But there are still many unanswered questions, and none of this would be possible if we settled with the easy way out, saying we don't know all there is to know. So, even though scientists are arguing and debating about different components of evolutionary theory, like hypotheses about menopause, that doesn't mean evolutionary theory our best explanation for the diversity of life.

Chance in Evolution

Chance is definitely a factor in evolution, but natural selection and other evolutionary mechanisms are definitely *not* random. Random mutations are the ultimate source of genetic variation, and these mutations can affect everything from gene products, like proteins, hormones, and other gene products, to regulatory mechanisms (cis- and trans-acting regulatory factors) and RNAs (microRNA and messenger RNA, for example). Natural selection, sexual selection, and genetic drift are the non-random processes that influence the relative frequency of the alleles that result from mutation. So, while part of the evolutionary process relies on randomness (see Section 5.2 for a discussion about mutation rates), shaping the diversity of life is decidedly not random.

Which of the following is a true statement?

a. Males that try to maximize their reproductive success over their lifetimes will be more successful than males that don't.

b. Females try to assess environmental conditions so they can adjust the numbers of male and female offspring they have.

c. Theories in science are generally incomplete, so they can't explain complex issues like the diversity of life.

d. Evolution is more than chance.

Go the Distance: Examine the Primary Literature

The question of menopause in humans is a topic of current debate in evolutionary biology. Is menopause adaptive or simply a life-history tradeoff? The mother hypothesis has been proposed as an explanation for this phenomenon, but few tests of the hypothesis had been published in the scientific literature until Dr. Mirka Lahdenperä and her colleagues presented their results in the journal *Evolution*.

- What data did Lahdenperä use to examine the predictions of the mother hypothesis and why?

- What was the relationship between age of breeding and likelihood of death in females?

- Did maternal mortality affect the growth and survival of offspring? Why or why not?

- What is the male longevity hypothesis?

Lahdenperä, M., A. F. Russell, M. Tremblay, and V. Lummaa. 2011. Selection on Menopause in Two Premodern Human Populations: No Evidence for the Mother Hypothesis. *Evolution* 65 (2):476–89. doi:10.1111/j.1558-5646.2010.01142.x. http://www.huli.group.shef.ac.uk/Lahdenpera2011Evolution pdf

Delve Deeper

1. How might parent-offspring conflict factor into sex-ratio adjustment in Seychelles warblers?

2. What predictions might you make about the timing of senescence in the different populations of guppies that Dr. Reznick studies?

Test Yourself

1. How do parents adjust the sex ratio of their broods?
    a. Parents cannot adjust the sex ratios of their broods.
    b. Sex ratio adjustment only occurs in social insects, where females control the sex of offspring by fertilizing (female) or not (male) their eggs.
    c. Sex ratio adjustment only occurs in role-reversed species, so males choose which gender survives.
    d. Parents selectively feed some offspring and not others, so only the most competitive gender survives.
    e. The mechanisms are largely unknown, but substantial evidence indicates that parents can and do adjust the sex ratios of their broods.

2. Are scientists just guessing about gene imprinting?
   a. Yes. Gene imprinting is just a theory, and therefore it is just a guess.
   b. Yes. Scientists don't really understand what genes are and what they do; they look for indirect evidence, but they are only guessing.
   c. Yes. Scientists have done a few experiments looking at how genes interact, but the evidence has not risen beyond the level of a guess.
   d. No. Gene imprinting has valid scientific context and initial experiments provide evidence to support Haig's hypothesis.
3. Which is the definition of senescence?
   a. The deterioration of the human mind as a person ages
   b. Alzheimer's disease
   c. The deterioration in the biological functions of an organism as it ages
   d. The tradeoff between reproducing and living longer
   e. None of the above
4. Why would some scientists argue that menopause simply results from life-history trade-offs?
   a. Because everything is a trade-off, and it makes sense that a female's inability to breed late in life must be a trade-off
   b. Because natural selection cannot act on individuals that no longer reproduce.
   c. Because genetic imprinting by males can cause females to become infertile
   d. Because the mechanism that leads to infertility in menopausal women (i.e., damage to a female's eggs over time) is the same in other species
   e. All of the above

Contemplate

Humans have longer lifespans then we used to. Do you think that has resulted from life-history tradeoffs? Have other life-history traits changed?	If sexual conflict is a coevolutionary arms race, what might be the limits on the evolution of genomic imprinting?

13 The Origin of Species

Check Your Understanding

1. What are clades?
    a. Groups made up of organisms and all of their descendants
    b. Hierarchies nested according to their synapomorphies
    c. Groups of organisms that comprise a taxonomic unit
    d. Groups of living organisms that share a phenotypic trait or character state
    e. Both a and b
    f. All of the above
2. Which of the following may explain the origin of a particular female mate preference?
    a. Male biased operational sex ratios
    b. Anisogamy
    c. Pre-existing sensory biases in females
    d. Sexual conflict
3. How accurate is the current estimate for the age of the Earth (4.568 billion years)?
    a. Not accurate at all because it is just an estimate
    b. Only slightly accurate because radiometric dating requires knowing decay rates, and no one has measured decay rates directly
    c. Somewhat accurate, but different dating techniques give different results
    d. Fairly accurate because scientists have been replicating the experiments and re-evaluating the evidence for decades

Learning Objectives for Chapter 13

Add important definitions and notes next to each learning objective for this chapter to help guide your understanding.

Learning Objective	Important Definitions	Notes
Compare and contrast the phylogenetic, biological, and general lineage species concepts.		
Discuss geographic and reproductive isolating barriers and their influence on gene flow.		
Differentiate between allopatric and sympatric speciation, and explain how isolating barriers contribute to these and other models of speciation.		
Discuss the types of evidence scientists use to evaluate different models of speciation and the challenges of studying the process of speciation.		
Explain why the rate of speciation may vary among organisms.		
Describe an example of cryptic species and discuss the significance of its discovery.		
Discuss the challenges of applying species concepts to bacteria and archaea.		

Identify Key Terms

Match terms and definitions by filling in the blank to the right of the term with the appropriate letter.

1. **Allopatry** ____
2. **Allopolyploidy** ____
3. **Bateson-Dobzhansky-Muller incompatibilities** ____
4. **Biological species concept** ____
5. **Cryptic species** ____
6. **Ecological speciation** ____
7. **Gametic incompatibility** ____
8. **General lineage species concept** ____
9. **Isolating barrier** ____
10. **Isolation by distance** ____
11. **Metapopulation** ____
12. **Parapatric speciation** ____
13. **Phylogenetic species concept** ____
14. **Postzygotic reproductive barriers** ____
15. **Prezygotic reproductive barriers** ____
16. **Reinforcement** ____
17. **Reproductive isolation** ____
18. **Speciation** ____
19. **Sympatry** ____

a. Occurs when populations are in the same geographic area

b. The evolution of new species within a spatially extended population that still has some gene flow

c. Lineages that historically have been treated as one species because they are morphologically similar but that are later revealed to be genetically distinct

d. A group of spatially separated populations of the same species that interact at some level (e.g., exchange alleles)

e. Occurs when sperm or pollen from one species fails to penetrate and fertilize the egg or ovule of another species

f. A pattern in which populations that live in close proximity are genetically more similar to each other than populations that live farther apart

g. Aspects of the genetics, behavior, physiology, or ecology of a species that prevent zygotes from successfully developing and reproducing themselves

h. The idea that species are the smallest possible groups whose members are descended from a common ancestor and who all possess defining or derived characteristics that distinguish them from other such groups

i. The idea that species are metapopulations of organisms that exchange alleles frequently enough that they comprise the same gene pool and therefore the same evolutionary lineage

j. Polyploidy (more than two paired chromosomes) resulting from interspecific hybridization. (If polyploidy arises within a species, it's called autopolyploidy.)

k. An aspect of the environment, genetics, behavior, physiology, or ecology of a species that reduces or impedes gene flow from individuals of other species. These barriers can be geographic or reproductive.

l. The increase of reproductive isolation between populations through selection against hybrid offspring.

m. Aspects of the genetics, behavior, physiology, or ecology of a species that prevent sperm from one species from fertilizing eggs of another species. These barriers reduce the likelihood that a zygote will form.

n. The evolution of reproductive barriers between populations by adaptation to different environments or ecological niches.

o. Occurs when populations are in separate, non-overlapping geographic areas (i.e., they are separated by geographic barriers to gene flow)

p. Occurs when reproductive barriers prevent or strongly limit reproduction between populations. The result is that few or no genes are exchanged between the populations.

q. Genetic incompatibilities in hybrid offspring arising from epistatic interactions between two or more loci

r. The idea that species are groups of actually (or potentially) interbreeding natural populations that are reproductively isolated from other such groups

s. The evolutionary process by which new species arise. This process causes one evolutionary lineage to split into two or more lineages (cladogenesis).

Link Concepts

Fill in the bubbles with the appropriate terms.

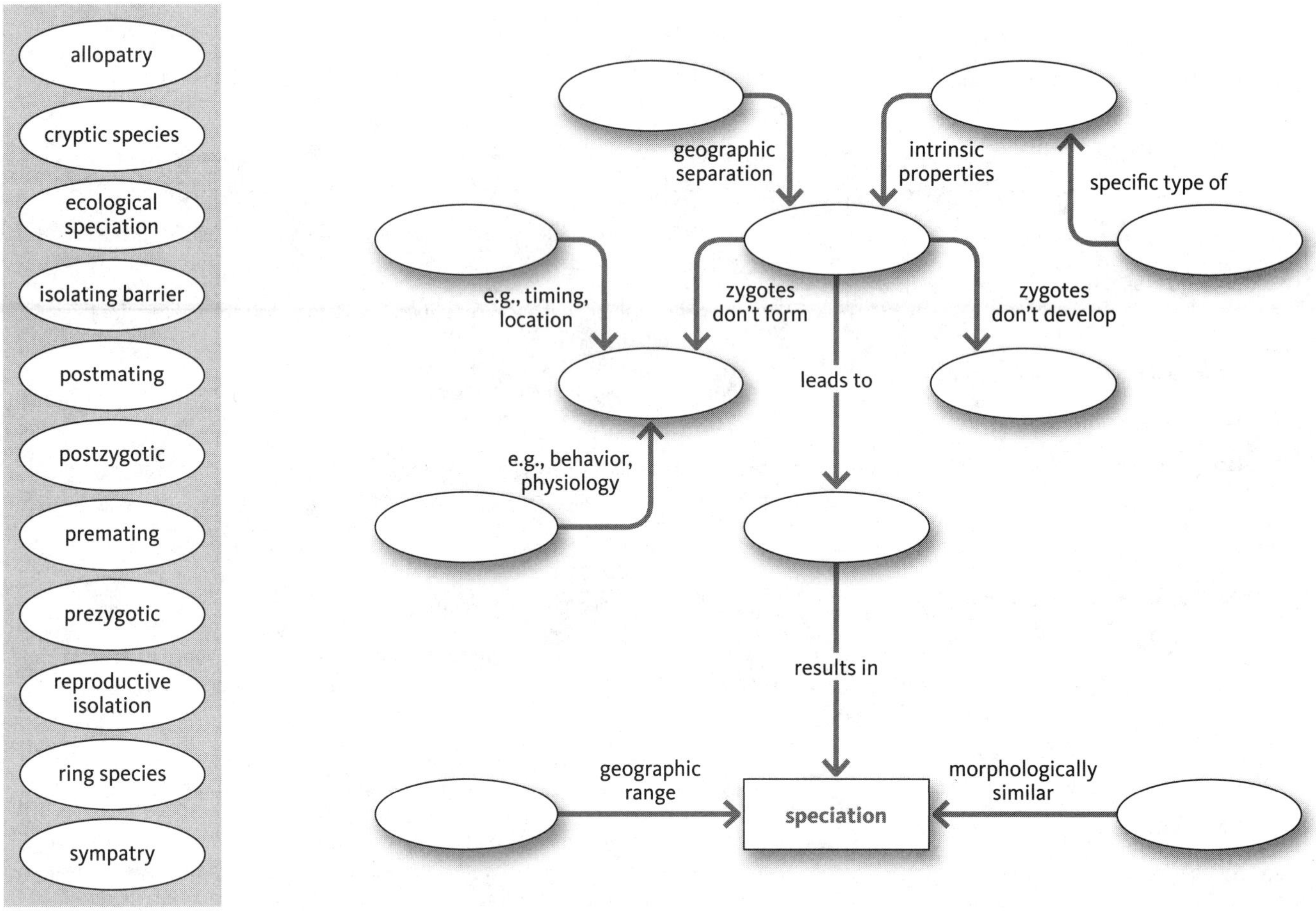

Key Concepts

Fill in the blanks for the key concepts from the chapter.

13.1 What Is a Species?

As scientists gain more insight into the mechanism of ______________ and the relationships between ______________ and ______________ across the diversity of life on our planet, they are developing better ways of thinking about how to define ______________ .

13.2 Barriers to Gene Flow: Keeping Species Apart

Geographic barriers to gene flow are features of the environment that ______________ separate populations from each other. They are important for all types of species, regardless of the ______________ that applies.

Reproductive barriers to gene flow are ____________ of organisms that reduce the likelihood of ____________ between individuals of different populations.

Separation of populations in ____________ can reduce the likelihood they will exchange ____________ . In these cases, divergent behavior (e.g., habitat preference, spawning time) acts as a ____________ to gene flow.

Floral traits can act as reproductive barriers to ____________ when their ____________ causes them to attract different pollinators.

____________ between eggs and sperm can be important barriers to gene flow between ____________ , causing ____________ sperm to be more likely to fertilize eggs than ____________ sperm.

Genetic incompatibilities can be important ____________ to gene flow between populations if they cause ____________ offspring to be sterile or to perform poorly ____________ to other individuals in the population.

13.3 Models of Speciation

New species can form when two or more ____________ become geographically isolated from each other (____________). Allopatric populations will begin to diverge ____________ because they independently experience ____________ , ____________ , and ____________ . Eventually, they may become sufficiently divergent that they no longer would or could ____________ , even if the physical barrier disappeared.

____________ can arise even when populations live in sympatry—they are not geographically separated from each other and can thus exchange migrants. New ____________ accumulate ____________ in these populations anyway because individuals rarely ____________ .

Isolation by distance occurs because individuals tend to mate with individuals from the same or nearby ____________ , resulting in ____________ of alleles across the geographic range of a species.

Speciation can arise as a by-product of ____________ of populations to ____________ habitats or resources.

13.4 Testing Speciation Models

Speciation is often a complex process, and ____________ of speciation are based on currently available ____________ . As scientists understand more and more about ____________ and their interactions, they are gaining valuable insight to the processes leading to these important ____________ in lineages.

Scientists have been able to test ____________ about the processes of speciation using ____________ from geology, DNA, and phylogenetic analyses, as well as with mating and breeding experiments.

13.5 The Speed of Speciation

The speed of speciation can ______________ normously among ______________ . In plants, interspecific hybridization and allopolyploidy can generate new species ______________ . In birds and mammals, however, ______________ can take millions of years.

13.6 Uncovering Hidden Species

Cryptic species may ______________ from ancestral populations without evolving easily ______________ morphologies.

Identifying cryptic species is important to measures of ______________ .

13.7 The Puzzle of Microbial "Species"

Microbial organisms present particular challenges to species concepts developed for ______________ because of the ______________ in rates and kinds of exchange of genetic material.

New models for species concepts in microbiology may provide valuable insight into the evolutionary history of ______________ organisms.

Interpret the Data

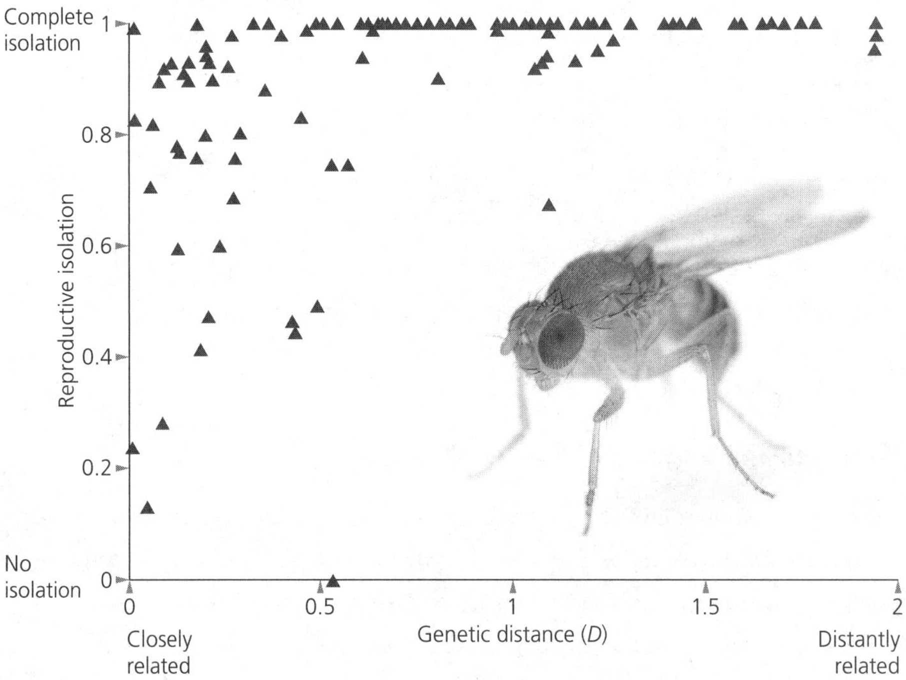

This graph shows how reproductive isolation evolved among species of *Drosophila*. The genetic distance (D) between two species increases with time. (Adapted from Coyne and Orr 2004.)

- Were *Drosophila* species pairs more likely to hybridize when they had a genetic distance < 0.5 or >0.5?

- Did *Drosophila* species pairs hybridize even if they weren't closely related?

- If it takes roughly a million years for the genetic distance (D) to reach a value of 1, when would you consider *Drosophila* species to be reproductively isolated?

Games and Exercises

TheCornellLab of Ornithology
Exploring and Conserving Nature

Singing a Song of Speciation

A number of species concepts have been proposed, and no one definition likely fits all species. For example, the biological species concept considers species as groups of actually (or potentially) interbreeding natural populations that are reproductively isolated from other such groups. The phylogenetic species concept considers species as the smallest possible groups whose members are descended from a common ancestor and who all possess defining or derived characteristics that distinguish them from other such groups. For many scientists who focus on determining the limits delineating one species from another, such as conservation geneticists concerned with species and subspecies designations relevant to management, the nuances of the different species concepts are critically important. For other scientists focusing on the mechanistic processes by which species evolve—the forces affecting populations, and the cohesion of these evolutionary units—details of the various species concepts are less relevant, and often the biological species concept suffices (since this definition focuses on the means by which new barriers to gene flow can arise).

The Cornell Lab of Ornithology developed this multimedia activity to illustrate some of the difficulties with our concepts of species (http://orb.birds.cornell.edu/orb/wp-content/uploads/2013/05/InstructorGuide-SpeciesConcepts.pdf). Eastern and western meadowlarks and blue-winged and golden-winged warblers are all recognized as distinct species. Examine the video and songs of these four species.

The first pair of species is the meadowlarks:

http://macaulaylibrary.org/video/415063

http://macaulaylibrary.org/video/435410

The second pair of species is the warblers:

http://macaulaylibrary.org/video/436872

http://macaulaylibrary.org/video/435383

- How different are the species in each pair?

- Do they look alike?

- Would you predict that the two meadowlark species might hybridize?

- What about the warblers?

Download the Raven Viewer (http://macaulaylibrary.org/raven-viewer) to analyze the bird songs using visual representations of the audio. Once the viewer is downloaded, use the visualize icon (it looks like a mini-waveform) to examine each species' song.

- Can you see any quantifiable differences in the birds' songs?

- Which species pair has greater differences, meadowlarks or warblers?

The lab has a number of songs of individuals of each species available.

- What kind of variation among individuals can you see?

Now examine the species accounts for eastern and western meadowlarks and blue-winged and golden-winged warblers at AllAboutBirds.org.

http://www.allaboutbirds.org/guide/eastern_meadowlark/id

http://www.allaboutbirds.org/guide/Western_Meadowlark/id

http://www.allaboutbirds.org/guide/Blue-winged_Warbler/id

http://www.allaboutbirds.org/guide/Golden-winged_Warbler/id

- How much geographic overlap is there between eastern and western meadowlarks?

- What about blue-winged and golden-winged warblers?

- What do the species accounts say about hybridization? Are you surprised given your predictions about how similar or different the species pairs look? What about the differences in their songs? What can you conclude about our understanding of species concepts based on this activity?

For more information about ring species and speciation, see Darren Irwin's *actionbioscience* page at http://www.actionbioscience.org/evolution/irwin.html.

ENSI also has a great activity to help you see how biogeography can inform speciation using different subspecies of the California salamander. You can examine the different traits of populations and map where they've been found on a topographic map of California. It's a great visual exercise!

http://www.indiana.edu/~ensiweb/lessons/step.sp.html

Explore!

The Speciation Song

Cheryl Van Buskirk and Cameron G. Brown have written and animated a song about speciation. Watch the *Speciation Song* here:

http://www.youtube.com/watch?v=WDPsZPKSEFg

Van Buskirk provides a little more explanation of the process here:

http://www.youtube.com/watch?v=PKb8Yi5xzhE

Connecting Concepts: Interactive Lessons in Biology

The University of Wisconsin-Madison produced this animated module on species and speciation. You can test your understanding of species and speciation, then apply that understanding to two real-world examples.

http://ats.doit.wisc.edu/biology/ev/sp/sp.htm

Speciation: Of Ligers and Men—Crash Course Biology #15

Hank Green explains speciation in this crash course in biology. He discusses the kinds of barriers that can arise that lead to reproductive isolation—prezygotic and postzygotic, allopatric and sympatric speciation, and hybridization.

http://www.youtube.com/watch?v=2oKlKmrbLoU

Hummingbird Species in the Transitional Zones

PBS *Evolution* "Darwin's Dangerous Idea" explores the hummingbirds and their habitats on the east slope of the Andes Mountains in Ecuador. Biologists studying these amazing birds are discovering that ecological differences can drive speciation. Small changes in bill length can lead to big changes in evolutionary lineages.

http://www.teachersdomain.org/resource/tdc02.sci.life.evo.hummingbird/

Salamander Ring Species

PBS *NOVA* "Evolution in Action" offers a great video about *Ensatina* salamanders in a classic example of what may prove to be a ring species:

http://video.pbs.org/video/1300397304/

Speciation in Real Time

Explore recent research on speciation in the central European blackcap. This website includes a video from National Evolutionary Synthesis Center (NESCent).

http://evolution.berkeley.edu/evolibrary/news/100201_speciation

Speciation: An Illustrated Introduction

In this animated video, the Cornell Lab uses the biological species concept to explain the diversity of birds, and species in general. The video includes stunning images of the birds-of-paradise.

http://ed.ted.com/on/R4PiqPn1

New Tools for Separating Geographical and Ecological Isolation

Hot off the press—scientists have developed new statistical techniques for examining genetic differentiation due to geographical isolation versus ecological isolation.

http://onlinelibrary.wiley.com/doi/10.1111/evo.12193/abstract

Anole Lizards: An Example of Speciation

This BioInteractive from the Howard Hughes Medical Institute explores how scientists understand speciation in these Caribbean lizards, including geographic isolation, ecological adaptation, and reproductive isolation. The site includes an animation, a virtual lab, and an activity where students can generate phylogenies from DNA data.

http://www.hhmi.org/biointeractive/anole-lizards-example-speciation

Overcoming Misconceptions

Evolution Is NOT Entirely Random

Chance is certainly an important component of evolution. Mutations—harmful, beneficial, and even neutral—are random, and within an individual they arise because of chance. But whether a particular mutation spreads throughout a population is *not* solely due to chance, especially if that mutation affects the reproductive success of the individuals that have it. Natural selection is definitely not random; it does act on the raw heritable variation that arises by chance. In fact, the nonrandomness of evolution by natural selection is amazingly apparent. Whales and fish occupy similar ecological niches in the same physical environment space. They face similar selection pressures, and the result has been convergence on similar shapes. The natural world is full of evidence for the nonrandom nature of evolution.

Evolution Can and Has Been Observed

Evolutionary biologists have been observing evolution for centuries, in artificial breeding, in the lab, and even in the wild. Artificial breeding may not seem like evolution, but the process has definitely changed the frequency of alleles over time. The result has been an amazing diversity, and over time these groups may or may not be able to interbreed. In the lab, scientists have used model organisms, like *Drosophila,* to show sympatric speciation, and even to show when reproductive isolation is likely to occur. In fact, in the lab scientists have been able to show many mechanisms that can cause gene frequencies to change over time, including predation and sexual selection. In the wild, evolutionary biologists have observed the evolution of mouse coat color, allopolyploidy in plants, and the evolution of resistance to pesticides and herbicides, for example.

Observing evolution isn't limited to just what we see before our eyes. Observation in evolution, and science in general, uses indirect evidence as well, and that indirect evidence is overwhelming. More importantly, evolutionary theory makes predictions about what we expect to see, and scientists can test those predictions with both direct and indirect evidence. We can predict where we would expect to find common ancestors in geologic formations; we can predict when life-history traits should differ among populations; and we can predict how new pathogens that affect human health may evolve.

Evolution Is NOT a March of Progress

People often ask, if humans evolved from apes, then why are there still apes? No evolutionary biologist would ever claim that humans evolved *from* modern apes—humans and apes share a common ancestor. The lineage of that common ancestor gave rise to both human and ape lineages, and that is why evolutionary biologists consider apes our closest living relatives.

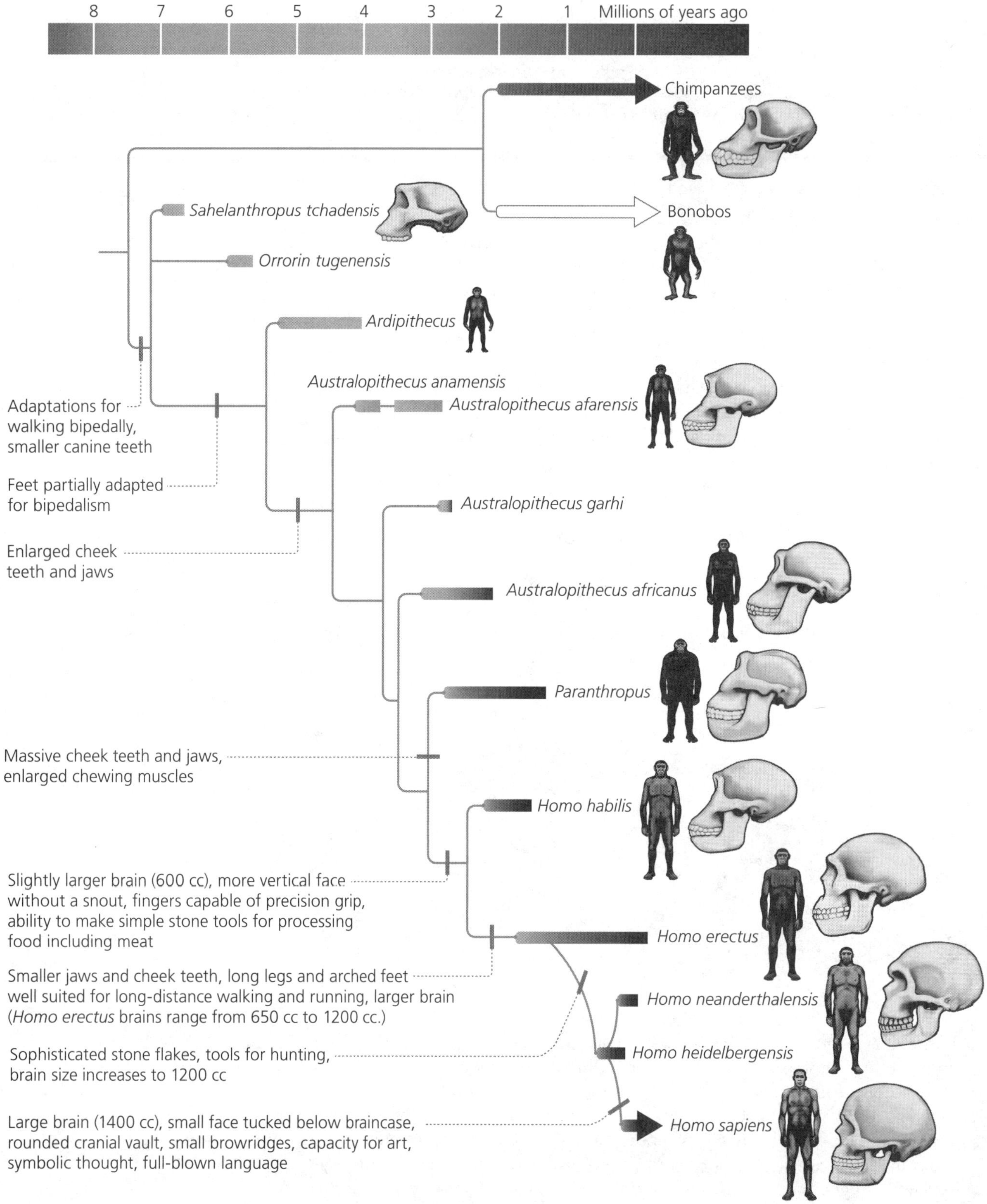

People often identify the "March of Progress" illustration as a "true" depiction of the relationship between modern apes (chimps) and humans, but evolution rarely occurs in a straight line. Phylogenetic trees are much better tools for visualizing how lineages split and diverge.

And nothing in evolutionary theory suggests that when one group splits from another, the first group has to go extinct to make way for the new group. Think about allopatric speciation—the rise of a geographic barrier, such as a river changing course, can lead to speciation in the group of individuals isolated by the new course. That doesn't mean the original group has to go extinct for the two groups to coexist in time (or even space if the geographic barrier is removed in the future).

So the common ancestor of apes and humans may have gone extinct, but apes and humans (as well as all related species) can coexist because lineages split and change. Some lineages go extinct, some do not, and evolutionary biologists use the clues left behind from this great diversity as evidence to reconstruct our complex past.

Transitional Fossils Are Everywhere

Scientists expect gaps in the fossil record because the fossil record preserves only a tiny fraction of the history of life, and discovering fossils of new species is a slow, arduous process. Since the publication of the *On the Origin of Species,* however, thousands and thousands of new fossils have been discovered, including many fossils of human ancestors.

So what counts as a *transitional* fossil? To evolutionary biologists, it's a fossil that has some characteristics of one lineage, and some characteristics of another, and perhaps some characteristics of its own lineage. It is an intermediate—a position that depends on the categories being compared. So *Archaeopteryx* is a transitional fossil between dinosaurs and birds if, like scientists, you understand that dinosaurs and birds are very loose categories. No paleontologist expects to find a fossil that is directly transitional between *Tyrannosaurus rex* and *Cardinalis cardinalis* (the northern cardinal). But, scientists have started looking for fossils with characteristics the two groups share—like feathers—and they have found a ton of new fossils (literally) that have features that make them intermediate—or transitional—to birds and dinosaurs. Likewise, the new fossil hominids that are being discovered regularly function as transitional lineages between apes and humans.

Which of the following is a true statement?

a. Random mutations are important to evolution because they provide the raw material for natural selection.

b. Only direct observations of change over time can be used as evidence of evolution.

c. When one lineage splits from another, the first usually goes extinct.

d. Scientists have not found the transitional fossils necessary to support the theory of evolution.

Go the Distance: Examine the Primary Literature

The neotropical skipper butterfly, *Astraptes fulgerator,* was first described in 1775, and it was considered a common species that was variable and wide ranging. In 2004, Paul Hebert and his colleagues examined *Astraptes fulgerator* to understand biodiversity in this biologically important part of the world, and they found that *Astraptes fulgerator* was likely more than a single, generalist species—it was 10 distinct species.

- What evidence did Hebert and his colleagues use to evaluate the differences among skipper species?

- Was only a single type of evidence used to develop their phylogenetic tree?

- Why do Hebert and his colleagues suggest that adults are so similar in coloration?

Hebert, P. D. N., E. H. Penton, J. M. Burns, D. H. Janzen, and W. Hallwachs. 2004. Ten Species in One: DNA Barcoding Reveals Cryptic Species in the Neotropical Skipper Butterfly *Astraptes fulgerator. Proceedings of the National Academy of Sciences* 101 (41): 14812–17. doi:10.1073/pnas.0406166101. http://www.pnas.org/content/101/41/14812.

Delve Deeper

1. Explain the concept of ring species with this semantic map by describing what happens, why it happens, and when it happens.

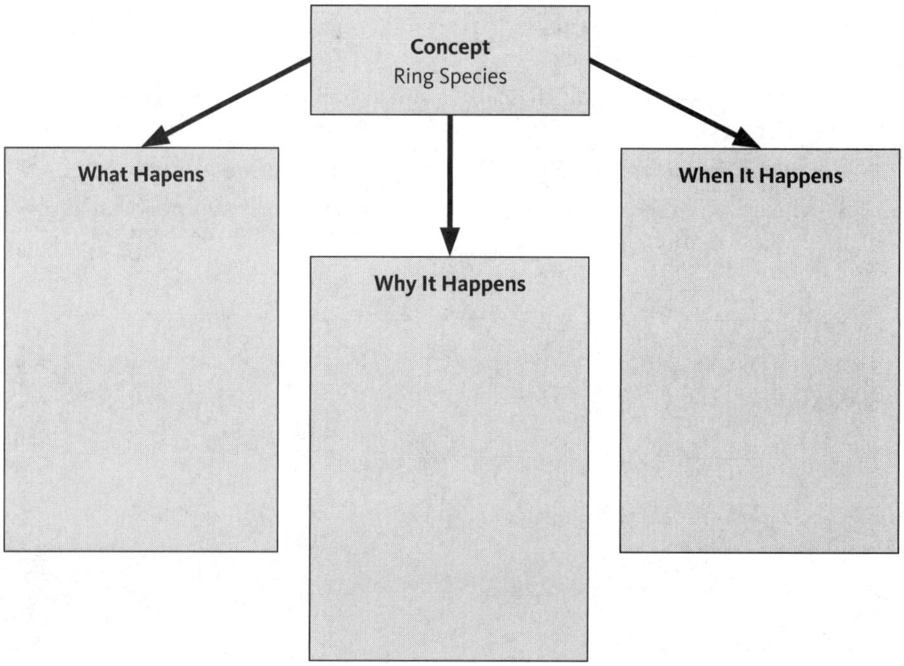

Test Yourself

1. According to Figure 13.4B, how are *Monostrea* corals reproductively isolated?
   a. They live in different habitats.
   b. One species spawns in the morning, the other in the evening.
   c. The distribution of spawning times is completely non-overlapping.
   d. Species release their gametes for different lengths of time.
   e. The gametes are dispersed by ocean currents at different rates.
2. Are reproductive isolating barriers more important in allopatric speciation or sympatric speciation?
   a. Allopatric speciation
   b. Sympatric speciation
   c. Reproductive isolating barriers are important in both allopatric and sympatric speciation.
   d. Only geographic isolating barriers are important in speciation.
3. According to Figure 13.13, which lineages of *Laupala* crickets are most closely related?
   a. *L. kona* and *L. molokaiensis*
   b. *L. oahuensis* and *L. tantalis*
   c. *L. cesarina* and *L. eukolea*
   d. *L. hualalai* and *L. cesarina*
   e. *L. cesarina* and *L. oahuensis*
4. Which of the following statements about Figure 13.16 is TRUE?
   a. Only *Drosophila* species pairs that were closely related hybridized.
   b. *Drosophila* species pairs that had a genetic difference of 0.51 were more likely to hybridize than other species pairs.
   c. *Drosophila* species pairs are more likely to hybridize when the genetic distance is < 0.5 than >0.5.
   d. All species pairs of *Drosophila* will likely hybridize at some point, even distantly related species.
   e. The genetic distance of *Drosophila* species pairs does not influence hybridization.
5. What are cryptic species?
   a. Metapopulations of organisms that exchange alleles frequently enough that they comprise the same evolutionary lineage but are almost indistinguishable morphologically.
   b. Groups of organisms that are genetically distinct and do not interbreed, but are almost indistinguishable morphologically.
   c. The smallest possible groups whose members are descended from a common ancestor and who all possess defining or derived characteristics that distinguish them from other such groups, even though they are almost indistinguishable morphologically to humans.
   d. Groups of organisms that have converged on similar adaptations so much so that they are almost indistinguishable.

6. Why is horizontal gene transfer an important factor in defining microbial species?
   a. Because horizontal gene transfer, along with rapid diversification, undermines any universal species concept
   b. Because homologous recombination could be considered microbial "sex"
   c. Because horizontal gene transfer transforms microbial genomes into mosaics difficult to classify
   d. All of the above
   e. None of the above
7. What are scientists discovering as they learn more and more about the genomic structure of microbes (Figure 13.20)?
   a. *E. coli* lineages are highly unpredictable.
   b. *E. coli* is actually many cryptic species.
   c. Among microbe "species," *E. coli* is more diverse than others.
   d. Within a single "species," microbial strains may not be that closely related genetically.
   e. Microbe "species" are highly adaptable because of the immense amounts of heritable variation.

Contemplate

What are the costs and benefits of speciation?	How might language serve as an isolating barrier in human populations?

14 Macroevolution

The Long Run

Check Your Understanding

1. What is speciation?
   a. The process by which new species arise
   b. Cladogenesis
   c. When one lineage splits into two or more lineages
   d. All of the above
   e. a and c only
2. Why don't scientists agree on a single definition of species?
   a. Because research methods can dictate which definition is most useful
   b. Because they have not discovered the true definition of a species
   c. Because different scientists have different philosophies about defining species
   d. All of the above
   e. a and b only
   f. a and c only
3. Why isn't the fossil record a complete record of life on Earth?
   a. Because organisms eat other organisms
   b. Because conditions have to be just right in order to preserve fossils
   c. Because wind and rain can erode fossils from the substrate
   d. Because rock-bearing fossils can be difficult to access
   e. All of the above
   f. a and b only
   g. c and d only

Learning Objectives for Chapter 14

Add important definitions and notes next to each learning objective for this chapter to help guide your understanding.

Learning Objective	Important Definitions	Notes
Compare and contrast the processes involved in macroevolution and microevolution and the patterns that result from these processes.		
Evaluate the effects on total species diversity when origination and extinction rates vary.		
Evaluate the kinds of evidence needed to distinguish between dispersal events and vicariance events in the fossil record.		
Explain how paleontologists analyze the fossil record to reconstruct macroevolutionary patterns.		
Explain what an adaptive radiation is and what kinds of opportunities can give rise to adaptive radiations.		
Compare background extinctions with mass extinctions, and provide an operational definition of mass extinction.		
Describe two abiotic factors potentially responsible for mass extinctions.		
Evaluate the evidence for human influence on biotic and abiotic factors affecting biodiversity, and discuss whether those influences may lead to another mass extinction.		

Identify Key Terms

Match terms and definitions by filling in the blank to the right of the term with the appropriate letter.

1. **Adaptive radiations** ~~e~~ f
2. **Anagenesis** e
3. **Background extinction** j
4. **Biogeography** d
5. **Dispersal** i
6. **Macroevolution** A
7. **Mass extinction** B
8. **Microevolution** l
9. **Punctuated equilibria** K
10. **Standing diversity** C
11. **Turnover** h
12. **Vicariance** g

a. Evolution occurring above the species level, including the origination, diversification, and extinction of species over long periods of evolutionary time

b. A statistically significant departure from background extinction rates that results in a substantial loss of taxonomic diversity

c. The number of species (or other taxonomic unit) present in a particular area at a given time

d. The study of the distribution of species across space (geography) and time

e. Wholesale transformation of a lineage from one form to another. In macroevolutionary studies, this process is considered to be an alternative to lineage splitting or speciation.

f. Evolutionary lineages that have undergone exceptionally rapid diversification into a variety of lifestyles or ecological niches

g. The formation of geographic barriers to dispersal and gene flow, resulting in the separation of once continuously distributed populations

h. The disappearance (extinction) of some species and their replacement by others (origination) in studies of macroevolution. This rate is the number of species eliminated and replaced per unit of time.

i. The movement of populations from one geographic region to another with very limited return exchange, or none at all

j. The normal rate of extinction for a taxon or biota

k. A model of evolution that proposes that most species undergo relatively little change for most of their geologic history. These periods of stasis are punctuated by brief periods of rapid morphological change, often associated with speciation events.

l. Evolution occurring within populations, including adaptive and neutral changes in allele frequencies from one generation to the next

Link Concepts

Fill in the bubbles with the appropriate terms.

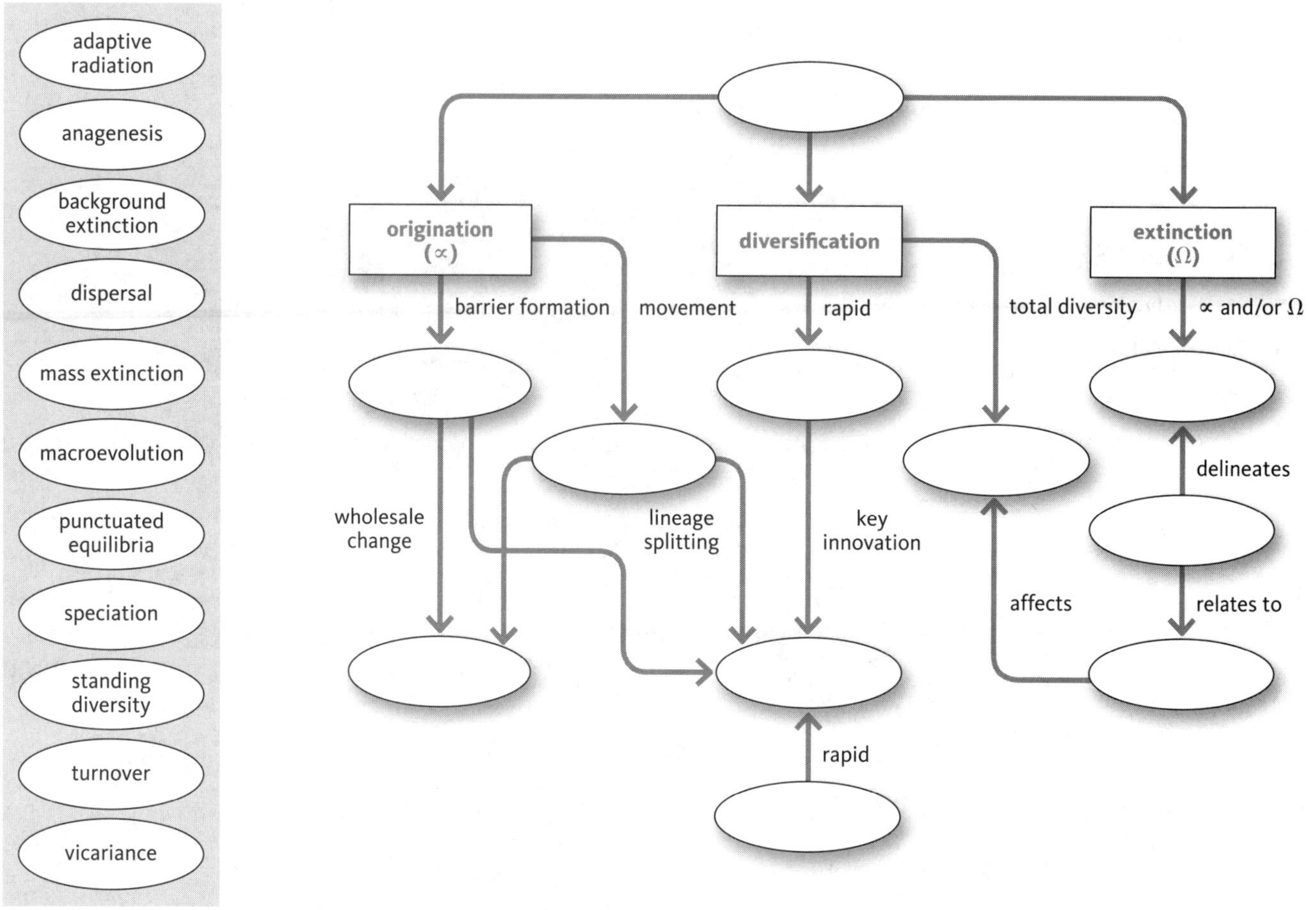

Key Concepts

Fill in the blanks for the key concepts from the chapter.

14.1 Biogeography: Mapping Macroevolution

Biogeography is a highly interdisciplinary field that explores the roles of __________ and __________ in explaining the __________ of species in nature.

Dispersal and vicariance explain __________ patterns of taxa. __________ occurs when a taxon crosses a preexisting barrier, like an ocean. __________ occurs when a barrier interrupts the preexisting range of the taxon, preventing __________ between the now separated populations.

14.2 The Drivers of Macroevolution: Speciation and Extinction

Scientists use the rates of ____________ and ____________ of species documented in the fossil record to examine the history of life on Earth.

____________ occur when the fossil record indicates a lineage split into ____________ distinct clades. The clade may be a single ____________, or a higher ____________.

An ____________ occurs when the last member of a clade ____________. Trilobites, for example, were a clade containing many species; the entire trilobite ____________ became extinct 250 million years ago.

14.4 The Drivers of Macroevolution: Changing Environments

Because the fossil record is ____________, examining macroevolutionary ____________ over time is challenging. Statistical analyses can ____________ for known biases and help scientists make and ____________ predictions about the processes that shaped the observed patterns.

14.5 Adaptive Radiations: When α Eclipses Ω

Most adaptive radiations have a common theme: the absence of established ____________ for the ____________ within an environment. Undercontested resources permitted ancestral populations to flourish and ____________ to increasingly specialized and localized subsets of those available resources and/or habitats, leading to diversification and speciation.

Sometimes ____________ properties of a lineage create ecological opportunity. ____________ can transform how organisms interact with their environments in ways that take them into new and undercontested habitats or permit them to exploit novel ways of life. These opportunities can trigger explosive subsequent ____________ and adaptive radiation.

14.6 The Cambrian Explosion: Macroevolution at the Dawn of the Animal Kingdom

Adaptive radiations are basically high rates of ____________ that occur in an area in a relatively short period of geologic time. The Cambrian Explosion likely resulted because a developmental innovation at the ____________ level allowed lineages to radiate and occupy a tremendous diversity of new ecological opportunities.

14.7 Extinctions: From Background Noise to Mass Die-Offs

Extinction is a common event in the history of life, well documented in the fossil record. Scientists can determine the ____________ of this process and use that background extinction rate to examine how ____________ from that rate affects the diversity of life on Earth.

14.8 The "Big Five" Mass Extinctions

The Big Five mass extinctions had different causes and affected different kinds of organisms. There is no reason that explains all mass extinctions.

14.9 Macroevolution and Our "Sixth Mass Extinction"

A single extinction may not have significant effects on an ecosystem, but we should be concerned about the impact that may result from a sixth mass extinction.

Cascading effects

Interpret the Data

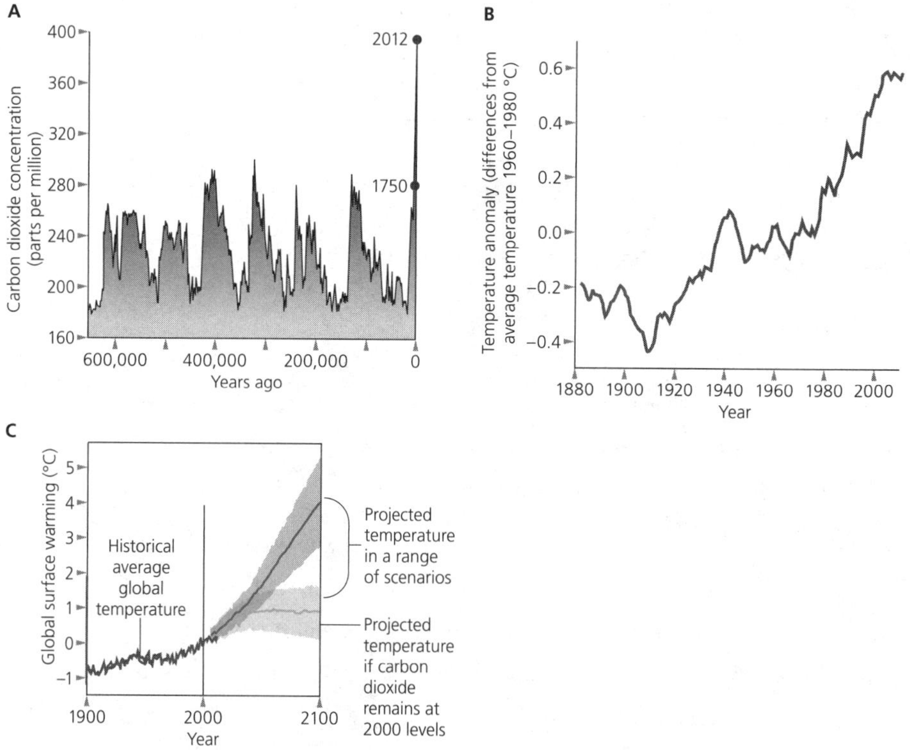

The graphs from Figure 14.27 show some of the data scientists are using to understand climate change. Graph A plots the concentration of carbon dioxide over time, and Graph B plots temperature variation. Graph C shows some of the projections made by climate models for global surface warming by the year 2100.

- According to Graph A, how many times has the Earth experienced warming in the last 600,000 years?

- Prior to 2012, what was the highest carbon concentration the Earth experienced?

- According to Graph B, what is the baseline against which other temperature deviations are being measured? Why might a scientist choose this method to make comparisons?

- Does the variation in projected temperatures for the year 2100 mean that scientists don't know what they are talking about?

Games and Exercises

How Big Is a Billion?

We live in very short time frames (for example, the original iPhone was released in 2007!). As a result, the "deep time" over which evolution has taken place can be exceptionally difficult to comprehend. Here's a quick visual trick that can help you grasp exactly how long a billion years is. All you need is 10 quarters (100 quarters would be awesome!), a ruler, and a calculator. Use your ruler to determine how tall a quarter is when it is lying flat on a table (you can use English or metric, whichever is more comfortable to you). Stack the 10 quarters, one on top of another in a nice, neat stack. Use your ruler to determine how tall the stack is—it should be 10 times as tall as a single quarter. (If you have 100 quarters, see if you can stack them and measure how tall that stack is.) Now use your calculator to fill in this table:

# of Quarters	Height	
1		
10		
100		
1000		How tall are you? (60 inches = 5 feet or 152 cm)
10,000		
100,000		About 1/10 of a mile or 1 ¾ football fields (0.16 km)
1,000,000		How far do you live from your school? (63,360 inches = 1 mile or 1.61 km)
10,000,000		
100,000,000		
1,000,000,000		How far do you live from Anchorage, Alaska?

Adaptive Radiation in the Hawaiian Honeycreepers

Hawaiian honeycreepers are an amazing group of birds. The ancestors of this group first colonized the volcanic islands of Hawaii about 5 million years ago. They diversified into dramatically different forms—identifiable largely by their diverse bills and feeding habits (Figure 11.11). To get a feel for how natural selection could act so rapidly to lead to such different forms, you can do a little experiment with bill shapes and foraging efficiency.

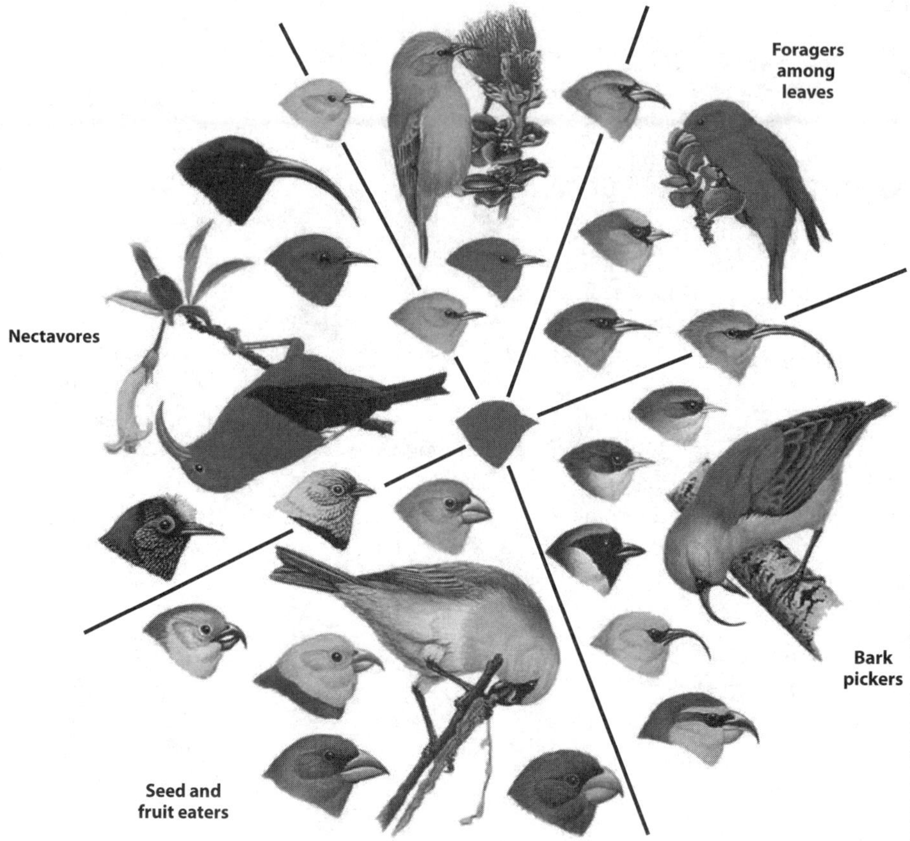

ADAPTED FROM LOSOS AND RICKLEFS (2009).

You will need a pair of regular pliers, a pair of needle-nosed pliers, a pair of scissors, some peanuts or walnuts in the shell, sunflower seeds (in the shell), uncooked rice, a bottle of water (make sure the neck is somewhat narrow), a bowl, and a small glass. The nuts, seeds, rice, and water represent food types. Place some nuts, seeds, and rice on a flat table and open the bottle of water and place the glass nearby. Use each tool with each type of food and see how easy it is to "eat" by transferring the food from the table to the bowl (or small glass for the water). Time your trials. Give yourself one minute with each, and fill out the data table below. Make it harder by actually cracking the nuts or seeds.

Beak Type	**# Peanuts or Walnuts**	**# Sunflower Seeds**	**# Rice**	**Height of Water in Glass**
Regular pliers				
Needle-nosed pliers				
Tweezers				
Medicine dropper				

- Did any single beak type work best for all foods? If not, which type worked best for which foods?

- How could different beak types evolve from a common ancestor through natural selection?

- If the Hawaiian Islands differed radically in the types of habitats available (say islands to the west are desert-like and islands to the east are rainforest-like), would you predict that different groups of birds would exist on different islands? Why?

Now look at some specific honeycreepers. See if you can match their bill types with the types of food they can access.

Kona grosbeak

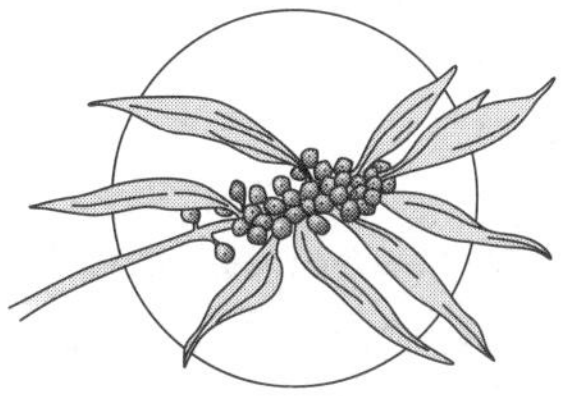

Naio seeds (in a very hard shell)

Kaua'i 'akialoa

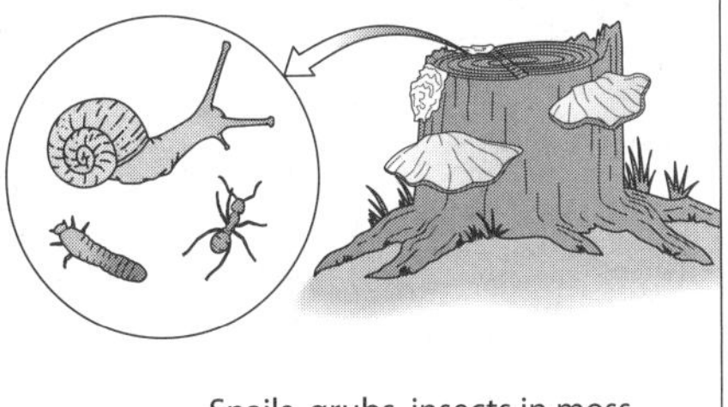

Snails, grubs, insects in moss

po'ouli

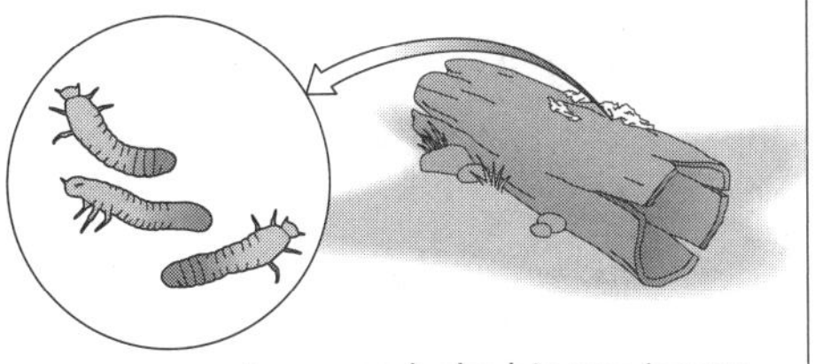

Larvae under bark in tree, insects under lichens

'akiapōlā'au

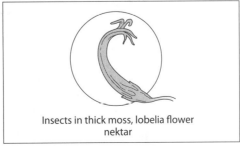

Insects in thick moss, lobelia flower nektar

These four bill forms represent some of the diversity in the honeycreeper radiation.

- Kona grosbeaks used their thick bills to crack open hard seeds, such as the naio. Unfortunately, the Kona grosbeak went extinct, even though the naio is quite abundant.
- The Kaua'i 'akialoa foraged for insects that lived in deep mats of moss. Its unusually long bill also allowed it to access nectar from long tubular flowers, like lobelia. Prior to their extinction, different species of 'akialoa (and two fossil species) were found on different islands in the Hawaiian chain, including Kaua'i.
- The po'ouli used its small bill to feed on snails, grubs, and insects. The po'ouli is likely extinct as well—the last known birds were observed on Maui in 2004.
- The akiapola'au still inhabits the island of Hawaii, but its numbers are plummeting. Like a woodpecker, it uses its sharp beak to forage for insects.

The Hawaiian Islands offered a diverse, but fragile, environment for this extraordinary radiation of birds, but they are a cautionary tale. Humans introduced predators and mosquito-borne diseases to the island, and also destroyed Hawaii's pristine habitat. As a result, less than half of the species first discovered there still remain.

Adapted from *Eruption!*, an electronic field trip produced by Ball State University (http://electronicfieldtrip.org/volcanoes/teachers/classroom_honeycreeper.html).

Explore!

100 Greatest Discoveries: Continental Drift

Alfred Wegener was a meteorologist and geophysicist. This short clip from *How Stuff Works* illustrates some of the biological evidence Wegener used when he developed his ideas about continental drift.

http://geography.howstuffworks.com/29267-100-greatest-discoveries-continental-drift-video.htm

Earth's Paleogeography—Continental Movements through Time

This cool animation created by Dr. Ron Blakey at Northern Arizona University shows the history of the continents on Earth and points out when the major radiations and extinctions of organisms occurred—you can see what the land masses of the continents looked like when these major changes in biodiversity were occurring!

http://www.youtube.com/watch?v=GNmUd43pabg

Mass Extinctions

Hank Green explains the Big Five mass extinctions and offers clues to what could be the sixth.

http://www.youtube.com/watch?v=FlUes_NPa6M

Mass Extinction

In this edition of *NOVA Science* Now, Neil deGrasse Tyson explores the causes of the Permian Mass Extinction 250 million years ago and its effects on life on Earth.

http://www.pbs.org/wgbh/nova/earth/mass-extinction.html

Facts of Evolution: Speciation and Extinction

In this comprehensive video, Best of Science and the Science Channel use the fossil record to show that extinction has always been a part of the Earth's history, but so has speciation. More importantly, they use current evidence to show how speciation arises and the direct and indirect evidence for the evolution of the diversity of life.

http://www.youtube.com/watch?v=T5kumHLiK4A

Scientists Gauge Ancient Die-Off of Pacific Birds

Humans have often been the cause of massive extinctions. For example, the Pacific islands were home to a huge diversity of birds before the arrival of humans, but the total loss couldn't be determined because of the incomplete nature of the fossil record. Now a new study estimates that nearly 1000 species went extinct—about 1/10th of the world's species.

http://news.sciencemag.org/plants-animals/2013/03/scientists-gauge-ancient-die-pacific-birds?ref=em

Ancestor of All Placental Mammals Revealed

Aided by the wealth of mammal fossils that have been discovered and a phylogeny based on anatomy, physiology, and genetics, scientists have inferred what the common ancestor of placental mammals probably looked like.

http://news.sciencemag.org/plants-animals/2013/02/ancestor-all-placental-mammals-revealed?ref=em

Gut Microbes Can Split a Species

For a whole new take on speciation, scientists examined the microbes harbored in the guts of organisms to see if differences in biodiversity at that scale affect speciation at the larger scale.

http://news.sciencemag.org/evolution/2013/07/gut-microbes-can-split-species

Beaks Keep Avian Lovers Apart

And now for an indirect effect of natural selection on bill size and shape—morphology influences sexual selection and speciation!

http://news.sciencemag.org/2001/01/beaks-keep-avian-lovers-apart

ScienceShot: Big Smash, Dead Dinos

Did dinosaurs go extinct because of an asteroid or was the asteroid just the coup de grâce (the death blow) to a group that was fading fast? New research using a high-resolution radiometric dating technique provides some pretty convincing evidence for one of the hypotheses.

http://news.sciencemag.org/2013/02/scienceshot-big-smash-dead-dinos?ref=em

Let There Be Mammals

When did the radiation of placental mammals occur? Did it occur only after the disappearance of the dinosaurs, or did it begin while dinosaurs were still around? This research paper from *Science* presents some striking new evidence that living placental mammals originated and radiated after the Cretaceous.

http://app.aaas-science.org/e/er?s=1906&lid=26839&elq=c49daf2845264f4b919e607a3ffc81e7

Or read the *Science Perspective* by Anne D. Yoder:

http://app.aaas-science.org/e/er?s=1906&lid=26798&elq=c49daf2845264f4b919e607a3ffc81e7

Overcoming Misconceptions

Microevolution and Macroevolution Are Two Sides of the Same Coin

Microevolution and macroevolution can be distinguished by the patterns they describe, but underneath it all, the mechanisms leading to those patterns are the same. Microevolution describes the genetic variation within populations, and how that variation came about through mutation, selection, and drift. Microevolution, or changes in allele frequencies over time, has been well documented, and given enough time, microevolutionary processes can produce enough differences between populations that they can be considered separate species.

Macroevolution, on the other hand, describes origination, diversification, and extinction of groups of organisms over time. The periods of time necessary for macroevolutionary patterns to appear can be unfathomably long, especially when our own experiences are usually just measured in decades. Although such long periods of time make direct observation of these changes unlikely, macroevolutionary patterns are very apparent in the fossil record. More importantly, volumes of indirect evidence, especially in the burgeoning field of evolutionary development, indicate that given enough time, microevolutionary changes can accumulate and translate into macroevolutionary change.

Gaps in the Fossil Record and Evolution

The fossil record is not complete, and we can't ever expect it to be. As Chapter 3 illustrates, fossilization is a complex process, and fossilization events that preserve the diversity of life, even as just snapshots in time, are extremely rare. When Darwin introduced natural selection as a process for producing the diversity of organisms on Earth, scientists were just beginning to be able to predict fossil locations. As a component of evolutionary theory, natural selection and common ancestry gave scientists an additional framework for making predictions about where and when to locate transitional fossils.

In fact, since Darwin's time, scientists have discovered many transitional fossils (including transitional fossils within the human lineage, between land mammals and modern whales, and between dinosaurs and modern birds). But because of the nature of the fossil record, they don't expect to find *every* transition. And more importantly, missing transitional fossils doesn't negate the overwhelming evidence for evolution.

Which of the following is a true statement?

a. Macroevolutionary changes are not related to microevolutionary changes.

b. Microevolutionary processes may produce changes within lineages, but they do not lead to new species.

c. The fossil record is a complete record of life on Earth.

d. Scientists have discovered many fossils that show the transition between dinosaurs and birds and between our primate ancestors and modern humans.

Go the Distance: Examine the Primary Literature

Peter Mayhew and his colleagues examine biodiversity patterns in marine invertebrates during the Phanerozoic Eon (an eon is a time period longer than an era; the Phanerozoic Eon includes the Paleozoic, Mesozoic, and Cenozoic eras). An important component of their research included controlling for the bias historically inherent in paleontological samples.

- According to the authors, what causes bias in studies of paleodiversity?

- Why did Mayhew and his colleagues argue that this sampling bias must be considered?

- What did Mayhew and his colleagues find?

Mayhew, P. J., M. A. Bell, T. G. Benton, and A. J. McGowan. 2012. Biodiversity Tracks Temperature over Time. *Proceedings of the National Academy of Sciences* 109: 15141–45. http://www.pnas.org/content/early/2012/08/27/1200844109.abstract.

Delve Deeper

1. Are punctuated equilibria and Darwin's theory of natural selection at odds in evolutionary theory?

2. How is the term "Cambrian Explosion" misleading?

Test Yourself

1. What is a turnover rate?
   a. The total number of originations and extinctions in a given interval of time
   b. The number of species present in a particular area at a given time
   c. α minus Ω divided by α
   d. Originations minus extinctions
   e. Extinctions minus originations
2. Which of the following statements about biogeography is FALSE?
   a. Biogeography is study of the distribution of species across space and time.
   b. Alfred Russel Wallace was responsible for many early ideas about biogeography.
   c. Darwin did not believe in vicariance.
   d. The discovery of plate tectonics provided a new line of evidence for patterns in biogeography.

3. Studying the Earth's biodiversity over time is not a straightforward scientific process. Which of the following is NOT a problem for macroevolutionary studies?
   a. Fossil bearing rocks differ in area and volume exposed masking complex geographical patterns of biodiversity.
   b. Patterns observed in the fossil record cannot be tested using statistical methods.
   c. Groups in the fossil record used to study species diversity over time may not be monophyletic.
   d. Scientists often must use taxonomic units other than species, and these higher order taxa are not necessarily comparable.
   e. All of the above are problems for macroevolutionary studies.
4. Punctuated equilibria is a hypothesis about the evolution of species. According to this hypothesis, why should macroevolutionary biologists expect abrupt breaks in the fossil record?
   a. Because the processes that form fossils are inconsistent through time, so there will always be gaps in the record.
   b. Because significant changes in phenotypes may occur within small isolated populations on the fringes of a species' range, leading to rapid speciation and little fossil evidence.
   c. Because species are defined as substantial morphological shifts within the fossil record.
   d. Macroevolutionary biologists should not expect abrupt breaks in the fossil record, because entire populations transform from one form to another in a process called anagenesis.
   e. Macroevolutionary biologists should not expect abrupt breaks in the fossil record. They should expect to find enough fossils to demonstrate a smooth transition from one species to another over time.
5. What key innovation led to an adaptive radiation of insects?
   a. Evolution of flowering plants
   b. Jointed exoskeletons
   c. Wings
   d. Complex eyes
   e. Predators that changed the fitness landscape
6. Which of the following statements about the Cambrian Explosion is FALSE?
   a. Changing oxygen levels in the Earth's oceans may have created ecological opportunities that organisms were able to exploit.
   b. The emergence of the genetic toolkit may have allowed the evolution of new developmental pathways leading to a diversity of body forms.
   c. The Cambrian Explosion was not really an explosion because the complex body plans of animal taxa evolved from precursors—they did not simply appear.
   d. The sudden appearance of sponges marks the beginning of the Cambrian Explosion.
   e. Molecular clocks indicate that the common ancestor of all animals lived about 800 million years ago.

7. How are mass extinctions defined?
   a. As a statistical departure from background extinction rates
   b. As any large number of species that disappears at a point in the stratigraphic column
   c. As the extinction of many species in a short period of time
   d. As the extinction of many species in a short period of time minus Lazarus, Elvis, and Zombie taxa
8. Why do scientists argue that climate change may be leading to a mass extinction (Figure 14.27)?
   a. Because according to models incorporating a variety of scenarios, projected temperatures could rise 2–3 times temperatures observed in 2000
   b. Because four of the big five mass extinctions were related to climate changes
   c. Because the average temperature has increased by 0.8 in the last 80 years
   d. Because although carbon dioxide concentrations clearly cycle, in the last 100 years they have become 100 parts per million greater than the highest concentrations recorded for the last 600,000 years
   e. All of the above

Contemplate

Based on your understanding of vicariance and dispersal, what might you predict about the relationships between South American plants and African plants (you might want to use Figure 11.7 as a guide)?	**Could the central insights scientists gained from punctuated equilibria be applied to evolution of the English language? How?**

15 Intimate Partnerships

How Species Adapt to Each Other

Check Your Understanding

1. Which of the following statements are depicted by this phylogeny?

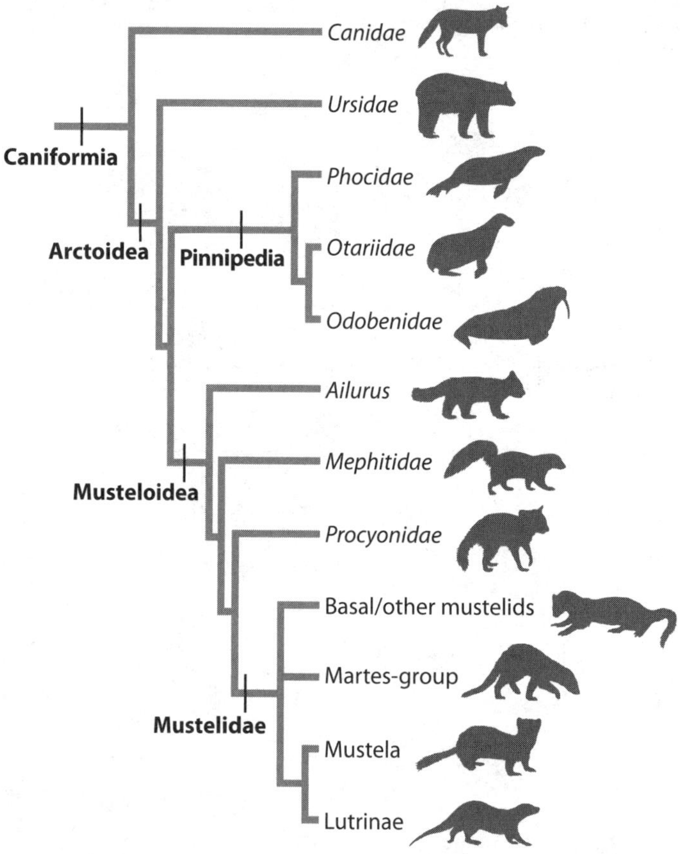

Flynn, J. J., J. A. Finarelli, S. Zehr, J. Hsu and M. A. Nedbal. 2005. Molecular Phylogeny of the *Carnivora (Mammalia)*: Assessing the Impact of Increased Sampling on Resolving Enigmatic Relationships. *Systematic Biology* 54(2): 317–337.

a. Otters (*Lutrinae*) evolved from martins (Martes group).

b. The ancestors of otters (*Lutrinae*) became gradually more "otter-like" over time.

c. Living otters (*Lutrinae*) represent the end of a lineage of animals whose common ancestor was wolf-like.

d. Otters (*Lutrinae*) share a common ancestor with *Odobenidae* (walruses).

e. All of the above are depicted by this phylogeny.

2. What is a species?
   a. Groups of actually (or potentially) interbreeding natural populations that are reproductively isolated from other such groups.
   b. The smallest possible groups whose members are descended from a common ancestor and who all possess defining or derived characteristics that distinguish them from other such groups.
   c. Metapopulations of organisms that exchange alleles frequently enough that they comprise the same gene pool, and therefore, the same evolutionary lineage.
   d. All are valid definitions of a species.
   e. Scientists don't know what a species actually is.
3. What is a mobile genetic element?
   a. A bacterium or viral parasite
   b. A gene with no known function
   c. Types of DNA that can move around in the genome
   d. Noncoding sections of DNA that function in alternative splicing

Learning Objectives for Chapter 15

Add important definitions and notes next to each learning objective for this chapter to help guide your understanding.

Learning Objective	Important Definitions	Notes
Define coevolution.		
Describe reciprocal selection and relate that process to the geographic mosaic theory of coevolution.		
Compare and contrast the outcomes of coevolution when antagonistic relationships are between two species versus among several to many species.		
Differentiate between negative and positive frequency-dependent selection and how they function in coevolutionary relationships.		
Compare and contrast the coevolution dynamics of Müllerian and Batesian mimicry.		
Explain how coevolution can promote diversification.		
Describe the evolution of mitochondria.		
Explain how retroviruses alter host genomes and affect host fitness.		

Identify Key Terms

Match terms and definitions by filling in the blank to the right of the term with the appropriate letter.

1. **Batesian mimicry** i
2. **Coevolution** ____
3. **Coevolutionary escalation or coevolutionary arms race** B
4. **Diversifying coevolution** ____
5. **Endosymbionts** A
6. **Geographic mosaic theory of coevolution** ____
7. **Müllerian mimicry** C
8. **Reciprocal selection** ____
9. **Retrovirus** E

a. Mutualistic organisms that live within the body or cells of another organism

b. Selection that occurs in two species, due to their interactions with one another. This process is the critical prerequisite of coevolution.

c. Occurs when harmless species resemble harmful or distasteful species, deriving protection from predators in the process

d. Reciprocal evolutionary change between interacting species, driven by natural selection

e. An RNA virus that uses an enzyme called reverse transcriptase to become part of the host cells' DNA. The virus that causes AIDS, the human immunodeficiency virus (HIV), is one type.

f. An increase in genetic diversity caused by the heterogeneity of coevolutionary processes across the range of ecological partners

g. A theory that proposes that the geographic structure of populations is central to the dynamics of coevolution. The direction and intensity of coevolution varies from population to population, and coevolved genes from these populations mix together as a result of gene flow.

h. Occurs when species interact antagonistically in a way that results in each species exerting reciprocal directional selection on the other. As one species evolves to overcome the weapons of the other, it, in turn, selects for new weaponry in its opponent.

i. Occurs when several harmful or distasteful species resemble each other in appearance, facilitating the learned avoidance of predators

Link Concepts

Fill in the bubbles with the appropriate terms.

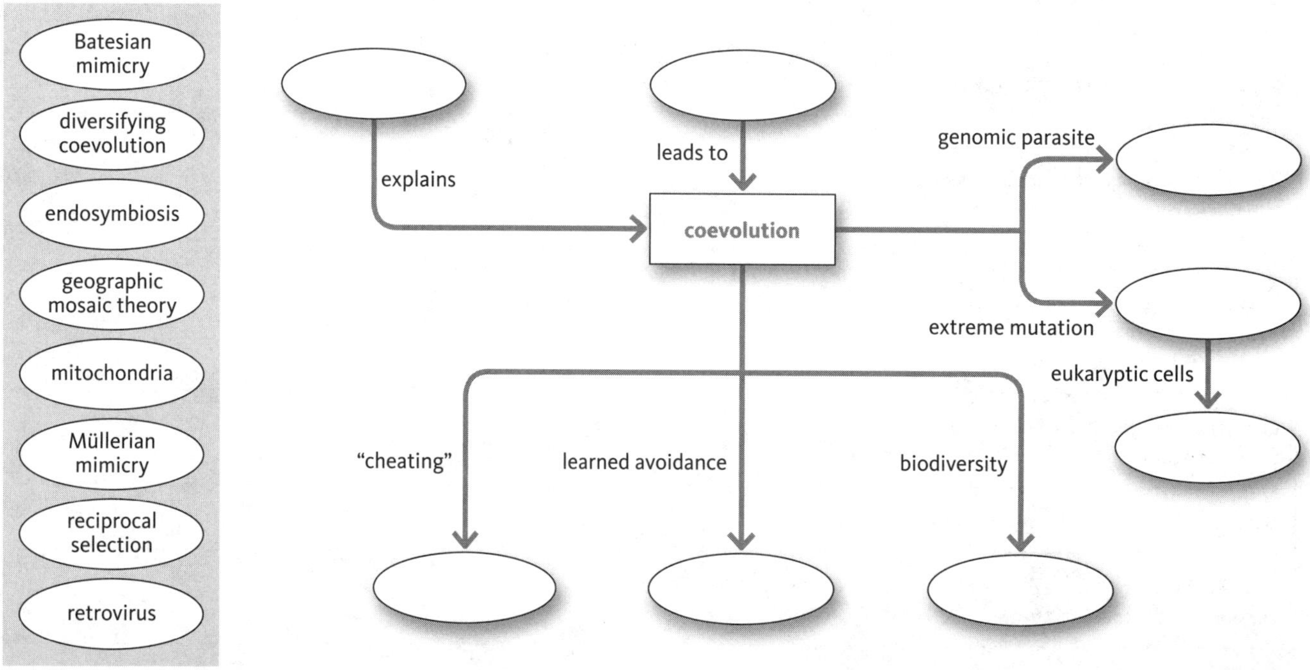

Key Concepts

Fill in the blanks for the key concepts from the chapter.

15.1 The Web of Life

Each species in a coevolutionary relationship exerts ____________ on the others, thereby affecting each other's ____________ .

15.2 Variation and Populations: The Building Blocks of Coevolution

The intensity and specificity of ____________ among ____________ species may ____________ across landscapes, and this spatial mosaic of interactions can ____________ over time.

Host and parasite (or pathogen) populations may generate ____________ frequency-dependent selection on each other, a ____________ interaction that can ____________ genetic variation within both populations.

Predator populations sometimes switch between several different prey ____________ , resulting in sequential or alternating bouts of pairwise ____________ between multiple species.

Escalated and rapid ____________ of weapons and defenses (an "arms race") results when species interact ____________ so that each generates ____________ selection on the other.

Interactions among species that begin as strongly antagonistic sometimes become less antagonistic over time, as each species ____________ in ways that ____________ the intensity of the interaction for the other.

When mutualistic species exert ____________ frequency-dependent selection on each other, the rapid coevolution that results can be very similar to ____________ , only without the antagonism.

Mutualistic interactions are vulnerable to the invasion of ____________ . ____________ can spread rapidly within a population, resulting in the ____________ of the mutualism or the evolution of defenses, such as sanctions.

15.3 Coevolution as an Engine of Biodiversity

Coevolutionary interactions can increase ____________ as genetic diversity ____________ among populations of each species within a geographic ____________ .

Highly specialized coevolutionary interactions make species highly ____________ on each other. If one species becomes ____________ , the other may be more likely to ____________ as well.

15.4 Endosymbiosis: How Two Species Become One

Mitochondria and plastids are ____________ bacteria that ____________ with their hosts until they became organelles, rather than free-living organisms.

15.5 Invasion of the Genomic Parasites

Endogenous ____________ and ____________ have coevolved with their host genomes.

Interpret the Data

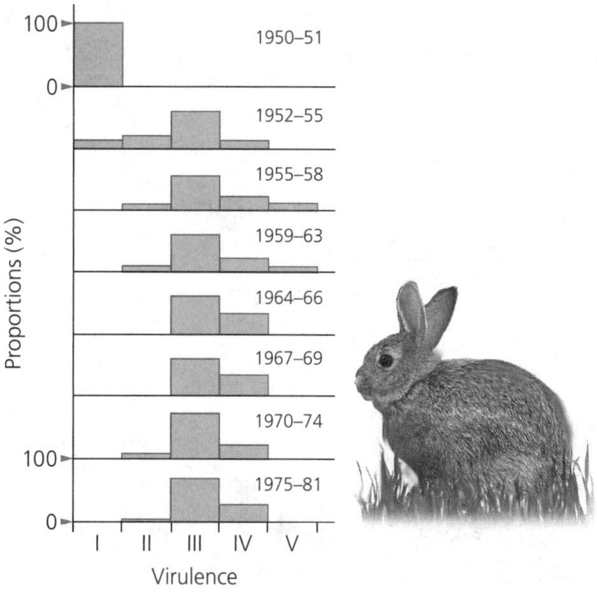

Figure 12.8 shows the changes in virulence in myxoma virus over time. Myxoma virus was introduced into Australia to control the exploding rabbit population. Virulence is measured as "virulence grades": I (deadliest) to V (mildest). Each bar represents a proportion (0–100 percent) of the total virus population during the specific time period (the scales have been left off to make the graphs easier to read).

- What was the most common virulence grade in 1951? In 1952?

- Explain how the virulence grades could change so quickly between 1951 and 1952?

- How stable is the intermediate grade of virulence in the myxoma virus?

Games and Exercises

Plants and Their Pollinators

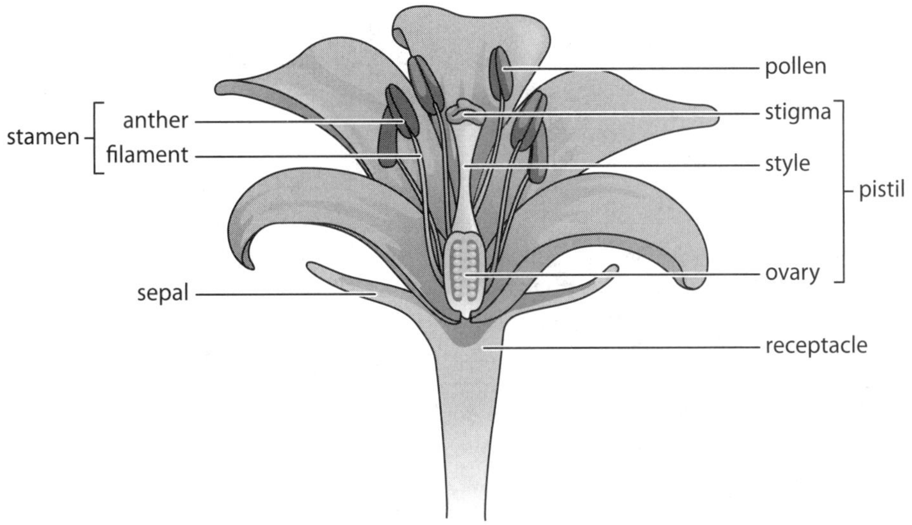

Pollination is incredibly important for sexual reproduction in plants (see Chapter 9 to learn how important sexual reproduction is in evolution). Pollination is simply the transfer of pollen grains from the stamen of one flower to the stigma of the same or another flower. Coevolution of plants and their pollinators has resulted in an amazing diversity of flowers and pollinators, each with adaptations that favor traits in the other. You can examine these adaptations at work simply by going outside and making careful observations of pollination in progress.

Start by finding a patch of similar looking flowers—a square meter would be a standard size you could observe easily and replicate elsewhere. Spend 10 minutes observing pollinators on the flowers. Record your observations in a table like this one:

Category	**# Observed**
Bees	
Butterflies	
Moths	
Beetles	
Flies	
Carrioneating flies	
Ants	

Use the Key to Identifying Common Pollinators to help identify some of these major groups. Go through the key before observing insects so you can become familiar with important characteristics necessary to identify groups. To read the key, start with 1a; if the characteristic fits, follow the directions in the right-hand column. If they don't fit, go to 1b. (The insect should fit either 1a or 1b—do not go to 2 unless one of those characters fits.) If you have time, repeat your observations with another patch of flowers that have a different form.

Key to Identifying Common Pollinators

1a.	Wings not visible, or hard wing covers concealing wings	go to 2
1b.	Wings visible	go to 3
2a.	No wings, narrow area between thorax and abdomen	Ants
2b.	Hard wing covers conceal flight wings, form line down middle of back, chewing mouthparts	Beetles
3a.	One set of filamentous wings, eyes large and obvious (careful, some Syrphid flies mimic bees)	Flies
3b.	Two sets of wings	go to 4
4a.	Both sets of wings often colorful, covered with scales	go to 5
4b.	Wings membranous, usually clear	go to 6
5a.	Antennae with knob-like ends, wings usually folded when at rest	Butterflies
5b.	Antennae with feathered ends, no knob, wings often open at rest	Moths
6a.	Thorax and abdomen joined by narrow "waist," abdomen often pointed	Wasps
6b.	"Waist" not as marked, body usually hairy	Bees
7a.	Pollen carried on "belly"	Megachilid leafcutting bee, orchard mason bee
7b.	Pollen carried mainly on leg	go to 8
8a.	Usually small (~5–10 mm), black or metallic green, short tongued	Halictids "sweat bees," Andrenids
8b.	Long tongue, usually over 12 mm	go to 9
9a.	Spur on hind leg, abdomen often appears striped	Anthrophorid digger bees
9b.	No spur, body robust, usually over 20 mm, yellow and black, eyes not hairy	Bumblebees *Bombus*
9c.	No spur, golden brown color, 12–15 mm, hairy, eyeballs	Honeybees *Apis mellifera*

- Describe the flowers you observed (think about shape, size, and color).

- What time of day did you make the observation?

- What was the weather like when you were making your observations?

- Was one group of insects more common on your patch than other groups? Which?

Pollinators are not *trying* to pollinate the flower. As they collect pollen or nectar for food, they brush up against a flower's pistil accidentally and transfer some of the pollen that has collected on their bodies. "Pollination syndromes" are basically the suites of floral traits that attract certain pollinators. Traits include shape, color, size, floral arrangement, and such. Different flowers tend to attract different insect pollinators. Bees tend to prefer yellow, blue, and purple flowers; there are hundreds of types of bees that come in a variety of sizes and have a range of flower preferences. Butterflies tend to prefer red, orange, yellow, pink, blue flowers. They need to land before feeding, so they like flat-topped clusters (e.g., zinnias, calendulas, butterfly weeds) in a sunny location. Moths, on the other hand, pollinate white or dull-colored fragrant flowers since they can't see colors (e.g., potatoes, roses). Flies tend to prefer simple bowl-shaped flowers or clusters in green, white, and cream, and carrion-eating flies pollinate maroon and brown flowers with foul odors (e.g., wild ginger). Although ants like pollen and nectar, they aren't good pollinators, and many flowers have sticky hairs or other mechanisms that keep them out of nectar receptacles.

Some other pollinators to consider are hummingbirds and bats. Hummingbirds use red, orange, purple/red tubular flowers with lots of nectar, since they live exclusively on flowers, but they don't need a landing pad because they hover while feeding. They commonly pollinate sages, fuchsias, honeysuckles, nasturtiums, columbines, jewelweeds, and bee balms. Bats are nocturnal, and they don't have the greatest eyesight. They do have a keen sense of smell, however, so flowers pollinated by bats must be large, light-colored, and night-blooming with a strong fruity odor (e.g., many cactus flowers).

You can also use a dichotomous key to determine the likely pollinators of flowers of different shape, size, and color:

Pollination Syndromes Dichotomous Key

1a.	Flowers small, inconspicuous and usually green or dull in color, petals reduced or absent	Wind
1b.	Flowers conspicuous, usually with white or colored petals	go to 2
2a.	Flowers regular in shape, radially symmetrical	go to 3
2b.	Flowers irregular in shape, bilaterally symmetrical	go to 9
3a.	Flowers purple brown or greenish in color, often with strong odor of rotting fruit or meat, little floral depth	go to 4
3b.	Flowers with little odor, or sweet odor	go to 5
4a.	Flowers purple brown, sometimes with a "light window"	Flies
4b.	Odor day or night, dull color	Beetles
5a.	Flowers with deep corolla tube	go to 6
5b.	Flowers more dish shaped, reward accessible, yellow, or with abundant pollen	Bees, flies, small moths
6a.	Flowers red, open in day, little or no odor, no nectar guide, nectar plentiful	Hummingbirds
6b.	Flowers not pure red, usually sweet odor	go to 7
7a.	Flowers yellow, blue, or purple, corolla tube not narrow, but sometimes needing forced opening, often with nectar guides	Long-tongued bees
7b.	Flowers red, purple, or white, corolla tube or spur narrow, usually lack nectar guide	go to 8
8a.	Flowers purple or pink, diurnal, upright, with landing area	Butterflies
8b.	Flowers white or pale, pendant, open or producing odor at night	Moths
9a.	Flowers red, little or no odor	Hummingbirds
9b.	Flowers with odor, usually with nectar guides	Bees

- What are some of the characteristics of flowers that may affect access to pollen and nectar?

- Do the characteristics of pollinators affect the survival and reproduction of flowers?

- What happens if a pollinator lands on the "wrong" plant?

- Are all pollinators limited to certain kinds of plants? Why or why not?

Adapted from *Pollination Ecology: Field Studies of Insect Visitation and Pollen Transfer Rates* by Judy Parrish (http://tiee.ecoed.net/vol/v2/experiments/pollinate/pdf/pollinate.pdf) and "A Plethora of Pollinators" by Alison Perkins (www.bioed.org/ECOS/inquiries/inquiries/PlethoraofPollinators.pdf).

Explore!

Orchid Bees—Euglossa

Orchids and their pollinators are an incredible example of the extreme adaptations that can evolve in two intimately linked species. This video from Hila Science Video illustrates the remarkable adaptations in both orchids and bees that evolved and the fitness benefits that drove that evolution. Are these "perfect" adaptations though?

http://www.youtube.com/watch?v=gEcv3dBuOe4

Evolution: Toxic Newts

This clip from PBS *Evolution,* "Evolutionary Arms Race," produced by WGBH, takes you out in the field with the Brodie's as they study the evolutionary arms race between newts and garter snakes. The interaction of these two species is driving their evolution; the toxin produced by a single newt can kill 12 people, and although snakes are resistant, that resistance comes with a cost.

http://www.pbs.org/wgbh/evolution/library/01/3/l_013_07.html

Evolution: Ancient Farmers of the Amazon

Another clip from PBS *Evolution,* "Evolutionary Arms Race," produced by WGBH, illustrates a classic example of coevolution of species based on a mutualistic relationship. Cameron R. Currie explains that the relationship between ants and the fungus they "farm" is not so simple; an evolutionary arms race is shaping the species' interactions as well.

http://www.pbs.org/wgbh/evolution/library/01/3/l_013_01.html

Goby-Shrimp Mutualism

This short video shows a unique mutualistic relationship between two marine species and the research that helped scientists understand this relationship (see the "About" tab under the video).

https://sites.google.com/site/coralreefsystems/videos/short-movies/synb

Don't Eat Me! I'm with Those Guys

In 2001, scientists found convincing evidence for the survival advantage of Müllerian mimics.

http://news.sciencemag.org/2001/01/dont-eat-me-im-those-guys

Endosymbiosis

In this YouTube video, Paul Andersen explains the evidence for symbiosis and the evolution of eukaryotes.

http://www.youtube.com/watch?v=-FQmAnmLZtE

Overcoming Misconceptions

Coevolution and Convergent Evolution

Convergent evolution occurs when similar traits evolve in unrelated lineages. Dolphins and fish both have streamlined bodies for swimming, fins, and fanned tails. These organisms converged on similar forms even though they are not closely related. Coevolution is the reciprocal evolutionary change between interacting species, driven by natural selection. When dolphins eat fish, both predators and prey can coevolve. But the important difference to think about is pattern versus process. Convergent evolution *describes* the similarities in the traits of organisms, and examining these traits can help scientists understand the genetic and developmental underpinnings of similar adaptations. Coevolution *explains* why interacting species change.

Coevolution Does NOT Create Mutual Harmony in Nature

Evolution and coevolution cannot create harmony in nature—harmony is a human construct. As Darwin noted, life is a struggle for existence. Mutualistic relationships occur when an increase in fitness of one species can potentially increase the fitness of a partner species. If circumstances were to change, however, such as the arrival of a new species, there is no evolutionary guarantee that the mutualism(s) that originally evolved will be sustained. In addition, coevolution can result from antagonistic relationships—predators eating prey, parasites eating individuals from the inside out, nest parasites pushing less developed offspring out of their parent's nest.

Coevolution Does NOT Promote Stable Coexistence of Species

Although scientists have shown that coevolution can make the dynamics within broad communities of species more stable, coevolution does not promote the stable coexistence of species. Coevolution requires that partner species be able to respond to changes in each other—one must be exerting selective pressure on the other. The random nature of mutations, however, along with constraints imposed by developmental biology, does not guarantee any response. Costs or constraints may prevent proportional responses, and stable coexistence does not necessarily follow. For example, a new mutation in an antagonistic relationship may lead to such high fitness in one partner that the other cannot survive or reproduce at all. Snakes and newts may currently be in an evolutionary arms race, but a change in a single amino acid can make snakes resistant to newt toxin. Newts may not be able to respond to that new selective pressure because making deadlier toxins requires a change in many pathways. Alternatively, coevolution may actually speed up rates of evolutionary responses, promoting biodiversity and reducing the likelihood of stable coexistence. For example, the evolutionary arms race between milkweed toxins and the caterpillars that eat them may have caused both milkweed plants and butterflies (adult caterpillars) to diversify.

Another important factor to consider is that coevolution can intensify extinctions and the *loss* of biological diversity. Extinctions can be caused by a variety of abiotic and biotic factors. Depending on the coevolutionary relationships among species, the loss of one species may lead to the loss of many other species as well.

Which of the following is a true statement?

a. Convergent evolution and coevolution both explain why organisms have similar traits.

b. Once a mutualism between two species evolves, those two species will always coevolve.

c. Coevolutionary relationships do not impose costs to an individual's fitness.

d. Coevolution can lead to the loss of biodiversity.

Go the Distance: Examine the Primary Literature

Scott Carroll and his colleagues were able to show that evolutionary responses can be rapid. They examined the evolution of beak size in soapberry bugs as this native species encountered an invasive exotic species, the balloon vine.

- What evidence did Carroll and his colleagues use to examine changes in soapberry bug beak size?

- What proportion of the population were they able to evaluate?

- How often could they evaluate changes in beak size?

Carroll, S. P., J. E. Loye, H. Dingle, M. Mathieson, T. R. Famula, and M. P. Zalucki. 2005. And the Beak Shall Inherit—Evolution in Response to Invasion. *Ecology Letters* 8 (9): 944–51. doi:10.1111/j.1461- 0248.2005.00800.x. http://onlinelibrary.wiley.com/doi/10.1111/j.1461-0248.2005.00800.x/abstract.

Delve Deeper

1. How do variation among individuals, differential survival or reproduction, and heredity act to generate the patterns of newt toxicity and snake resistance observed by the Brodies?

Test Yourself

1. What is coevolution?
   a. When one species needs to change because another species that it depended on changed
   b. When two or more species interact, and each acts as an agent of selection causing evolution of the other
   c. Evolutionary changes that are shared among two or more species, such as wings
   d. Evolutionary changes among two or more species that are complementary, so that all species benefit
   e. All of the above
2. What is the difference between positive mutualism and positive commensalism?
   a. Positive mutualism occurs when the fitness outcomes for all interacting species are high; positive commensalism occurs when the fitness outcomes some species are high, but others realize no fitness consequences.
   b. Positive mutualism occurs when some interacting species benefit from the relationship and others do not; positive commensalism occurs when only one of the interacting species benefits.
   c. Positive mutualism leads to coevolution of all the interacting species; positive commensalism benefits some species but it does not lead to coevolution.
   d. Natural selection always favors positive mutualism because all interacting species have higher fitness because of the relationship; it does not favor positive commensalism because only one of the interacting species has higher fitness.
   e. Positive mutualism does not lead to coevolution because all the interacting species benefit from the current relationship; positive commensalism leads to coevolution because some of the interacting species benefit, so the others need to evolve to benefit as well.

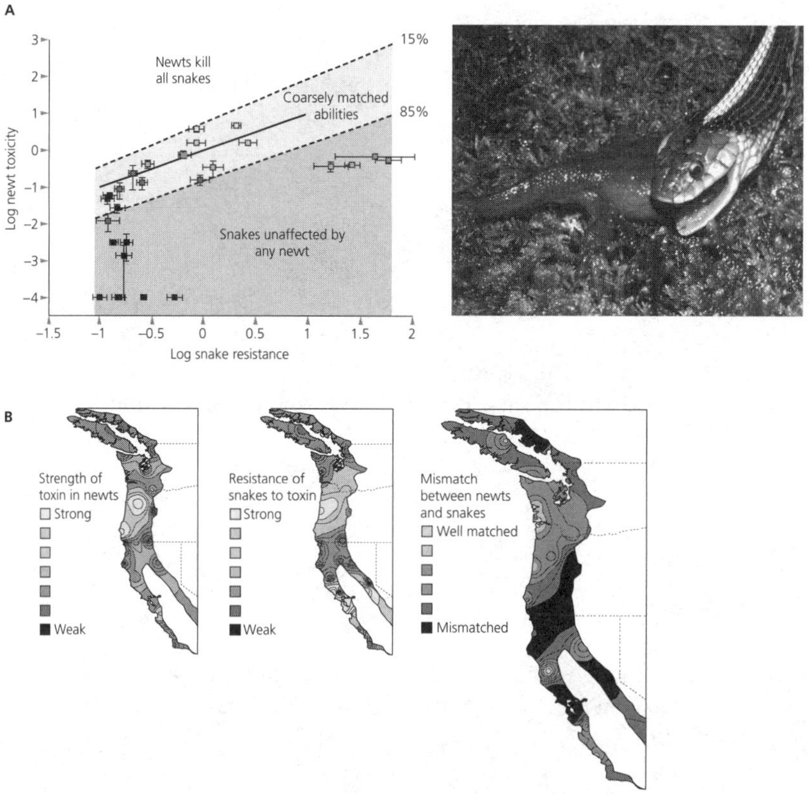

3. According to Figure 15.11, which of the following statements about newt toxicity and snake resistance is/are true?
   a. Among snake populations with low resistance to the toxins, newts display a wide range of toxicity.
   b. Snakes with extremely high resistance do not affect the evolution of toxicity in newts.
   c. Within locations, snakes vary in their ability to resist the toxin.
   d. Within locations, newts do not vary in their ability to produce toxin.
   e. a and c only
   f. All of the above

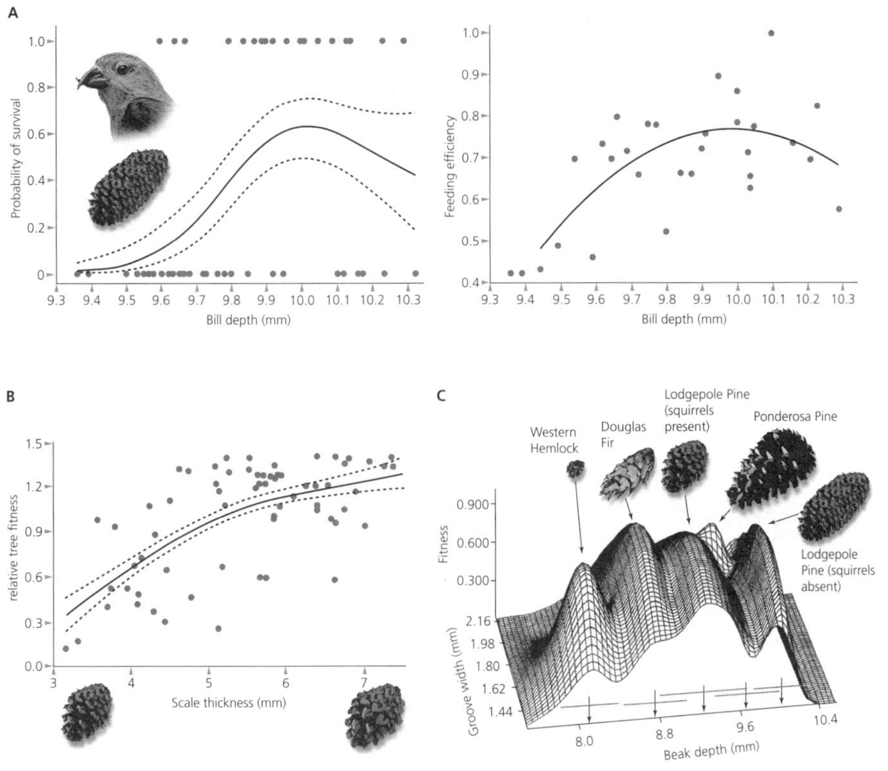

4. How do the data presented in Figure 15.18 support the idea that red crossbills may be diversifying as a result of their coevolutionary arms race with pine trees?
   a. Figure 15.18A shows the trajectories of beak depth over time and the probability of survival (left) associated with feeding efficiency (right), and therefore that speciation will occur as populations with shallow bills diverge from populations with deep bills.
   b. Figure 15.18A (right) and 15.18B show that fitness is greatest in both species when bill depth is slightly larger (4mm) than scale thickness, so that birds with shallower or deeper bills than that optimum size will have to specialize on smaller or larger scales, and the populations will diverge.
   c. Figure 15.18 A (left) and 15.18C show that variation in bill depth within populations of red crossbills is less than variation within the species, and these differences may lead to cladogenesis.
   d. Figure 15.18C shows that within an area, scale thickness varies among pine species, and natural selection will favor different bill depths, maintaining the variation within crossbills.
   e. All of the above.

5. What did Sandra Anderson and her colleagues demonstrate by manually pollinating the flowers of the native *Rhabdothamnus solandri* on mainland New Zealand?
   a. That mainland flowers could produce more fruit than island flowers
   b. That mainland flowers were not naturally producing much fruit, even though they were capable of producing much more
   c. That within island populations, *Rhabdothamnus solandri* flowers were producing fruit at near maximum levels
   d. That the diversity of pollinators was higher on the mainland
   e. All of the above
6. If an evolutionary biologist hypothesized that a lineage of bacteria evolved because of a coevolutionary arms race with its hosts, what prediction(s) might s/he make about their evolutionary history?
   a. That the lineage of bacteria was older than the lineage of the host
   b. That the lineage of the host was older than the lineage of bacteria
   c. That natural selection was negative frequency dependent
   d. That the patterns of speciation events of the bacteria lineage would match closely with the patterns of speciation events of the host lineage
   e. Both a and d
   f. Both c and d
7. How certain are scientists about the origin of mitochondria?
   a. Not very certain. Scientists are still gathering evidence for the beginning of the coevolutionary relationship.
   b. Not very certain. Scientists have not been able to compare the phylogenies of all eukaryotes with the lineage of SAR11, so they cannot draw any conclusions about the beginning of the coevolutionary relationship
   c. Fairly certain. Scientists use the weight of evidence to support or refute hypotheses, and the current evidence supports an origin early in the evolution of eukaryotes.
   d. Fairly certain. Substantial evidence supports the hypothesis that mitochondria were once free-living, oxygen-consuming bacteria, but they are still gathering evidence for the beginning of the coevolutionary relationship.
   e. Both c and d
8. What is a retrovirus?
   a. A virus that is fairly old
   b. An RNA virus that uses an enzyme to become part of the host cells' DNA
   c. A virus that has been resurrected from extinction
   d. A virus that only replicates using RNA
9. Which of the following statements about natural selection is FALSE?
   a. The strength and the direction of natural selection can vary across space.
   b. The strength and the direction of natural selection can change over time.
   c. The strength and the direction of natural selection can vary among species within a coevolutionary relationship.
   d. Natural selection does not vary in strength or direction.
   e. None of the above is a false statement about natural selection.

10. Müllerian mimicry can be considered:
    a. Convergent evolution.
    b. Coevolution.
    c. Speciation.
    d. Both a and b
    e. None of the above
11. Rapid extinction of a group of species can result when:
    a. Species are in antagonistic coevolutionary relationships.
    b. Alleles are swept to fixation.
    c. Species are in mutualistic coevolutionary relationships.
    d. Selection favors highly virulent strains of a virus.
    e. None of the above

Contemplate

Could you use the principles of geographic mosaic theory to examine how the uniforms of sports teams change over time? How? (Think about cities with multiple sports teams—especially the same sport—versus other cities.)	Do horticulturalists (people who cultivate plants for human use) affect coevolutionary partnerships in the natural world? Why or why not?

16 Brains and Behavior

Check Your Understanding

1. What kind of variation is necessary for natural selection to occur?
   a. Variation in the expression of a trait
   b. Phenotypic variation that has a genetic component
   c. Variation in additive alleles that contribute to a phenotypic trait
   d. a and b only
   e. All of the above
   f. None of the above
2. What factors influence a population's evolutionary response to selection?
   a. Differential reproductive success of individuals in the population and the strength of selection
   b. The strength of selection and the amount of phenotypic variation among individuals
   c. The strength of selection and how much of the variation in a phenotypic trait is heritable
   d. The amount of genotypic variation among individuals and how much of that variation is heritable
   e. The reproductive success of some individuals in the population versus the reproductive success of other individuals in the population
3. Why is an understanding of the genetic toolkit important to understanding the evolution of traits?
   a. Because the same underlying networks of genes govern the development of all animals, and new traits evolve as a result of mutations to genes within that network
   b. Because the genetic toolkit consists of all of the genes scientists have identified so far
   c. Because gene networks within the toolkit act like "modules" that can be deployed in new developmental contexts, yielding novel traits
   d. a and c only
   e. All of the above
   f. None of the above

Learning Objectives for Chapter 16

Add important definitions and notes next to each learning objective for this chapter to help guide your understanding.

Learning Objective	Important Definitions	Notes
Explain what a behavioral phenotype is and give three examples.		
Compare and contrast proximate and ultimate questions and how these questions influence studies of behavior.		
Explain how organisms without brains can "behave."		
Propose an ultimate cause for the evolution of a behavior and an experiment that could test that hypothesis.		
Explain the different levels of selection at which behavior can potentially evolve.		
Compare the benefits and costs of living in a group.		
Explain how inclusive fitness can lead to kin selection and how kin selection can lead to differences in male and female behavior.		
Discuss the trade-offs imposed as a result of the evolution of complex cognition.		

Identify Key Terms

Match terms and definitions by filling in the blank to the right of the term with the appropriate letter.

1. **Altruism** ____	a. The safety in numbers that arises through swamping the foraging capacity of local predators
2. **Behavioral ecology** ____	b. A mathematical approach to studying behavior that solves for the optimal decision in strategic situations (games) where the payoff to a particular choice depends on the choices of others
3. **Dilution effect** ____	c. Selection arising from variation in fitness among groups
4. **Eusociality** ____	d. The science that explores the relationship between behavior, ecology, and evolution to elucidate the adaptive significance of animal actions
5. **Evolutionary stable strategy** ____	e. A type of social organization in which species have complete reproductive division of labor. In this group, many individuals never reproduce, instead helping to rear the offspring of a limited number of dominant individuals.
6. **Game theory** ____	f. Selection arising from variation in fitness among individuals
7. **Group selection** ____	g. Occurs whenever a helping individual behaves in a way that benefits another individual at a cost to its own fitness
8. **Haplodiploidy** ____	h. When the number or strength of synaptic connections between neurons is altered in response to stimuli
9. **Inclusive fitness** ____	i. Selection arising from the indirect fitness benefits of helping relatives
10. **Individual selection** ____	j. An individual's combined fitness, including its own reproduction as well as any increase in the reproduction of its relatives due specifically to its own actions
11. **Kin selection** ____	k. A mechanism of sex determination where the sex is determined by the number of copies of each chromosome that an individual receives. Offspring formed from the fertilization of an egg by a sperm (i.e., diploids) are female, while those formed from unfertilized eggs (i.e., haploids) are male.
12. **Synaptic plasticity** ____	l. A behavior which, if adopted by a population in a given environment, cannot be invaded by any alternative behavioral strategy.

Link Concepts

Fill in the bubbles with the appropriate terms.

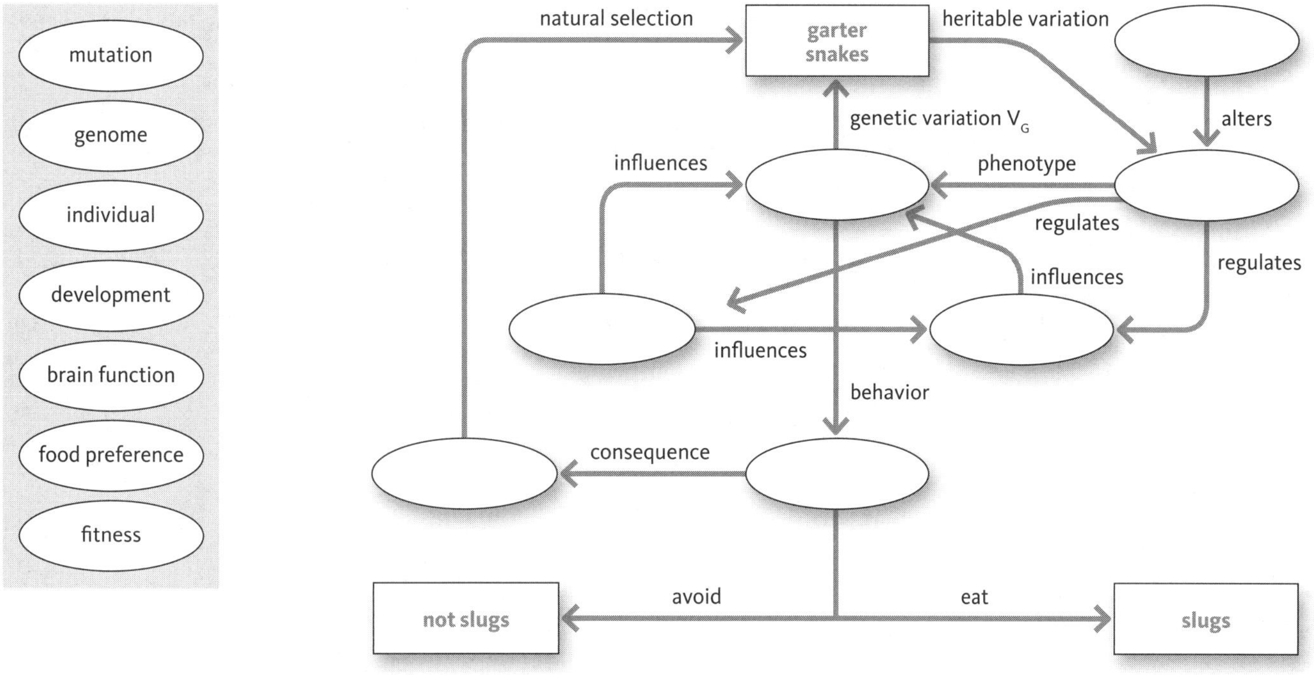

Key Concepts

Fill in the blanks for the key concepts from the chapter.

16.1 Behavior Evolves

Behavior is an internally coordinated ______________ to external ______________ .

Behavioral phenotypes ______________ just like any other phenotypic ______________ .

Proximate studies of behavior focus on how particular behaviors are elicited, including how context-specific ______________ of gene expression, neural signaling, physiology, and anatomical structures ______________ to elicit particular ______________ to stimuli, and how genetic ______________ contributes to population or individual ______________ in the expression of anatomical structures and behavior.

Ultimate studies of behavior (behavioral ecology) focus on ______________ particular behaviors have evolved, including the relationship between ______________ in the expression of a behavior and ______________ in particular ecological settings as well as ______________ patterns of behavioral evolution.

16.2 Behavior without a Brain

Behavior is ____________ limited to animals with nervous systems.

Microbes display a range of ____________ , including aggregating to form spores.

Plants can be ____________ to light and to touch, and they can ____________ by sending and receiving signals. Like animals, these behaviors have a ____________ that can evolve over time.

16.3 Behavior and the Origin of Nervous Systems

Current research suggests the ____________ nervous system ____________ either once or twice.

The animal nervous system enabled ____________ behaviors such as three-dimensional burrowing paths. The fossil record preserves ____________ of the emergence of this complex behavior.

16.4 Innate and Learned Behaviors

Animals use ____________ behaviors to produce adaptive responses to ____________ conditions.

____________ allows the ____________ of behavior to adapt to changes in the environment.

Like other aspects of behavior, learning can ____________ . An evolutionary tradeoff between the ____________ and ____________ of learning influences the ____________ of learning and memory in a given population.

16.5 The Vertebrate Brain

Vertebrates have large, centralized brains that are ____________ into specialized regions. The ____________ size of these regions in different species reflects their evolutionary ____________ and ecological ____________ .

16.6 Individuals, Groups, and the Evolution of Social Behavior

The crucial issue distinguishing the relative importance of ____________ versus ____________ selection is the fate of ____________ that are ____________ to an individual but ____________ to the group in which that individual lives. If individual selection predominates, then these alleles will ____________ ; if group selection predominates, these alleles will ____________ .

Selection is shortsighted. ____________ fitness consequences determine the success or failure of an allele, irrespective of the ultimate ____________ of this process.

16.7 Playing the Evolution Game

Organisms are reproductively ____________ ; they behave in ways that enhance the spread of their ____________ genetic material, even if such behavior is ____________ to their population or species.

An evolutionarily stable strategy is a behavior that, when adopted by a ____________ of players, cannot be ____________ by any ____________ strategy.

16.8 Why Be Social?

There are ____________ and ____________ to group living.

The costs and benefits of a ____________ may not be the same for all ____________ . Factors such as sex, status, or body condition may shift the relative ____________ of specific behaviors.

16.9 The Importance of Kin

The results of personally reproducing and rearing ____________ (direct fitness) and assisting in the rearing of ____________ kin (indirect fitness) are genetically ____________ . Shared ____________ can favor the evolution of behaviors that enhance the reproductive success of close relatives, behaviors often viewed as altruistic.

Kin selection theory predicts that ____________ should help ____________ more than ____________ , and that they should help close relatives more than distant relatives.

Parental care is the most widespread form of ____________ behavior, in part because the physical proximity of parents and offspring keeps the ____________ of relatedness high.

One reason male parental care is ____________ is that certainty of paternity is generally ____________ . Without a high probability of relatedness, parental effort is not evolutionarily ____________ .

16.10 Creativity in the Animal Kingdom

Because learning and memory appear to be represented by interconnected networks of ____________ in the brain, synaptic ____________ is thought to be a neurochemical mechanism responsible for these processes.

The ____________ to learn fast comes at a ____________ , which can offset the benefits of learning in some situations.

Tool use appears to be confined to ____________ and ____________ , and appears to be associated with intelligence. Within each of these lineages, however, tool use evolved ____________ many times.

Interpret the Data

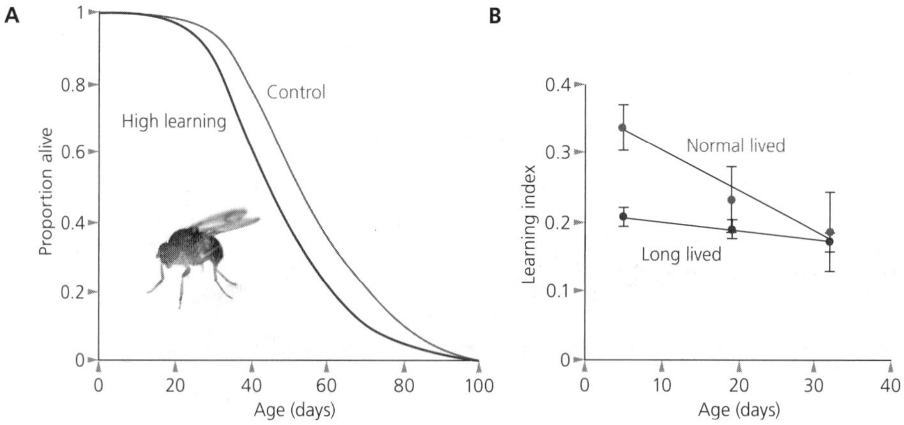

Tadeusz Kawecki and his colleagues at the University of Fribourg in Switzerland examined learning in fruit flies. They conducted an experiment in which they selected for flies that could learn quickly or were long lived, let these flies breed, and over the course of many generations, developed populations of high-learning flies and long-lived flies. A: When the scientists compared the high-learning populations to populations that had not been subjected to their artificial selection (the "controls"), they found that high-learning flies died sooner. B: When the scientists compared the flies selected for longevity, on the other hand, the long-lived flies turned out to be poor learners. (Adapted from Burger et al. 2008.)

- What proportion of high-learning flies were alive at 50 days? What proportion of control flies were alive at 50 days?

- What was the learning index for long-lived flies at 5 days? For normal flies?

- Was the learning advantage of normal flies consistent over the lifetime of the flies?

- Why did the scientists test the effects of learning with both a higher-learning experiment and a longer-lived experiment?

Games and Exercises

Making Observations

Watch how a dragonfly predicts the movements of its prey.

http://news.sciencemag.org/brain-behavior/2014/12/watch-dragonfly-predicts-movements-its-prey?utm_campaign=email-news-latest&utm_source=eloqua

- How did scientists determine that the dragonfly was predicting the prey's movements?

As humans, we are excellent observers. Our capacity for complex cognition allows us to quantify observations, discover patterns, and make predictions as we go through our daily life. Scientists do the same thing, but in a more formalized process. You can study behavior in animals simply by making careful observations within a formal framework.

Start with observations of a focal individual. Pick an organism that interests you. Make sure you can observe it continually for at least a short period of time. Note the date, time, location, weather, and habitat. Identify the individual if you can (a squirrel may have a stubby tail or a white patch) or by location. Record the sequence of behaviors, their duration, and any consequences (such as other individuals' responses). Record your data in a table like this one:

Individual ID	Behavior	Duration (minutes:seconds)

Are all the individuals behaving the same way? Do some spend more time behaving in one way than other ways? Can you tell the difference between males and females, and does gender affect behavior?

What Do Animals Learn?

If you have squirrels or crows in your neighborhood, they can be great tools for learning about learning. Find someplace they can be routinely found foraging or hanging out on the ground, like a park. Set up two experiments: one where you rush toward an individual, and one where you offer food. Think about time of day, presence of other individuals, presence of other people, weather, or any other factors that might influence their behavior. Be sure to use the same methods each time you conduct your experiment. When you rush at the squirrels or birds, do they fly away, move a little, run to a tree? How far away do they react? If you were to conduct the same experiment in the same place (and thus likely to sample the same squirrels, and maybe the same birds) over and over, do you think the reactions would change with more and more experience? What about if you approach slowly and offer the animals food (bird seed, for example)?

Here's what your data might look like:

Date	Sample	Reaction	Distance to Reaction
Day 1	Focal Individual 1	Ran to tree	30 yards
Day 1	Focal Individual 2	Ran to tree	27 yards
Day 1	Focal Individual 3	Ran 20 yards	40 yards
Day 2	Focal Individual 1	Ran to tree	37 yards
Day 2	Focal Individual 2	Ran to tree	35 yards
Day 2	Focal Individual 3	Ran to tree	50 yards

Crows can recognize human faces. Scientists were able to discover this amazing feat with a simple experiment with surprising results. Check out the effects of masks on crow behavior:

http://www.livescience.com/14819-crows-learn-dangerous-faces.html.

Rock, Paper, Scissors, Lizard, Spock

The guys from *The Big Bang Theory* took the Rock-Paper-Scissors game to a new level. Play a couple rounds with your friends and see how it works. Get creative and see if you can come up with two additional phenotypes in side-blotched lizards that would fit the game.

http://bigbangtheory.wikia.com/wiki/Rock_Paper_Scissors_Lizard_Spock

Explore!

Video: Swimming Apes Caught on Tape

Swimming seems instinctive to some mammals, but not to the uninitiated human—or ape. But humans and apes can learn to swim, and this video shows that an orangutan named Suryia has some serious skill.

http://news.sciencemag.org/evolution/2013/08/video-swimming-apes-caught-tape

Slimy, but Not Stupid

Slime molds—eww. Ok, so not so much disgusting as interesting—they can actually navigate through a maze and figure out the shortest path to get to their food. Not bad for not having a brain.

http://news.sciencemag.org/2000/09/slimy-not-stupid

Heritable Differences in Schooling Behavior

Abigail Wark uses video to show heritable differences in schooling in sticklebacks. They used little fake fish on wires, all arranged in tandem, pulled through a tank, and the test fish either lined right up and swam with them, or didn't.

Wark, A. R., A. K. Greenwood, E. M. Taylor, K. Yoshida, and C. L. Peichel. 2011. *Heritable Differences in Schooling Behavior among Threespine Stickleback Populations Revealed by a Novel Assay*. Plos One 6(3): e18316. DOI: 10.1371/journal.pone.0018316.

http://journals.plos.org/plosone/article?id=10.1371/journal.pone.0018316#s5

Dolphin Memories Span at Least 20 Years

Dolphins use whistles to communicate—they have their own unique whistles, and they can remember and repeat the whistles of other dolphins, apparently for 20 years or more!

http://news.sciencemag.org/plants-animals/2013/08/dolphin-memories-span-least-20-years

Kin Recognition in an Annual Plant

Kin recognition is well known in the animal kingdom, but plants behave, too. The amount of roots a plant grows should be a predictor of its competitive ability, at least below ground. This research shows that plants differ in the amount of roots they grow depending on whether they are planted with kin or not.

http://rsbl.royalsocietypublishing.org/content/3/4/435.full

Animal Behavior—Crash Course Biology #25

Hank Green explains animal behavior, including how and why behaviors evolve.

http://www.youtube.com/watch?v=EyyDq19Mi3A

Animal Behavior

Paul Andersen explains the range of animal behaviors, from innate to learned, and the science behind understanding different types of behavior.

http://www.youtube.com/watch?v=6hREwakXmAo

Think Tank

The Smithsonian National Zoological Park developed this online exhibit about how to think about thinking. The site explores tools, language, and society with macaques, hermit crabs, orangutans, mangabeys, and rats.

http://nationalzoo.si.edu/Animals/ThinkTank/default.cfm

Frans de Waal: Moral Behavior in Animals

Frans de Waal shows that morality is not restricted to humans. Using video of research on animal behavior, he shows that two of the pillars of morality, reciprocity and empathy, can be found in many animals, from apes to elephants.

http://www.ted.com/talks/frans_de_waal_do_animals_have_morals.html

Overcoming Misconceptions

What Is "For the Good of the Species"?

As humans, we cannot completely remove ourselves from the wiring in our brains. We have a cognitive capacity that allows us to make judgments about what is "good" and what is "bad"—that is our behavioral legacy. For example, we want to believe that selection has favored altruistic behavior for the "good" of the species. Evolution cannot recognize what is best for a group of organisms, however. Indeed, altruistic behavior is beneficial to the individual—it is "selfish." Individuals may benefit from altruistic behavior (directly or indirectly), and more of their alleles will be represented in the next generation as a result. Similarly, whether an allele spreads or disappears from a population is influenced more by how that allele affects an individual's fitness than how that same allele affects the fitness of the group. Only in rare circumstances can natural selection favor traits that are more beneficial to the group than to the individuals that carry them. Most of the time, these situations are unstable because any mutation that increases the fitness of an individual will spread quickly through the population. The immediate fitness consequences determine the success or failure of an allele, regardless of the ultimate outcome of the process.

Learning and the Capacity to Learn

Although humans, and other animals, clearly transmit learned behaviors through something akin to "culture," the behaviors themselves are not transmitted genetically—the ability to learn is. Teasing those two ideas apart is difficult because we humans are so adept at learning. But just like Lamarckian evolution, pierced ears, tattoos, and knowing all the words to the Best Song Ever are acquired during our lifetimes. Our offspring will not have pierced ears when they are born, but they may like piercings more because we raise them within that culture. The *capacity* to understand, mimic, and prefer that culture has a genetic component, however, and that capacity can be inherited. So some offspring may be more musically inclined than others, and they may be able to memorize words more easily, depending on the suites of genes inherited from their parents. The capacity for complex cognition can, and has, evolved over time.

Morality and Evolution

Humans are animals, and we share behaviors with animals. Those shared behaviors provide amazing evidence for the evolution of our species. Immorality, selfishness, and cruelty are our concepts of "wrong" and "right"—concepts that can be contemplated, because over time, natural selection favored genes that gave humans the capacity to think in abstract terms. We can decide what is wrong or what is right because of our evolutionary history. But natural selection and evolution have no inherent morality. Natural selection favors behaviors that increase survival and reproduction, and ultimately the spread of genes from one generation to the next. Depending on the conditions, cooperation and altruism may be winning strategies.

Which of the following is a true statement?

a. Individuals living in groups often behave for the good of all members.

b. Humans are the only species with culturally transmitted traits.

c. Some mutations can result in alleles that enhance the fitness of a group at the expense of the fitness of the individual with those alleles.

d. Evolution is immoral because it selects for selfish behavior.

Go the Distance: Examine the Primary Literature

In the wild, primates and birds can be pretty creative with tools, showing high degrees of innovation. Apes, corvids, and parrots are particularly innovative, but why these groups are exceptional is unclear. Alex Taylor and colleagues set out to examine cognition in New Caledonian crows (a species of corvid)—specifically, what capacity these birds had to problem solve, and what kind of cognitive mechanism can account for that capacity.

- How did their experiment address problem solving in New Caledonian crows?
- Was this a proximate study of behavior or an ultimate study?
- What was their sample size?

Taylor, A. H., D. Elliffe, G. R. Hunt, and R. D. Gray. 2010. Complex Cognition and Behavioural Innovation in New Caledonian Crows. *Proceedings of the Royal Society Series B: Biological Sciences* 277 (1694): 2637–43. doi:10.1098/rspb.2010.0285.

http://rspb.royalsocietypublishing.org/content/277/1694/2637.abstract.

Delve Deeper

1. What's the difference between learning and the evolution of learning behavior?
2. If the pack size of 20 animals was thrown out of the analysis presented in Figure 16.25C, do you think the relationship would hold? Why or why not?

Test Yourself

1. Which of the following statements about the three graphs in Figure 16.6 is FALSE?
   a. Graph A includes replicates of the experiment; graphs B and C do not.
   b. The three graphs show that the heritability of behavior varies among species.
   c. The three graphs define behavior differently and use different measures (revolutions per day as a measure of activity), number of attacks (as a measure of aggression), and by scoring exploration (as a measure of a willingness or a reluctance to explore).
   d. All three graphs show that scientists conducting experiments can function as strong agents of selection.
   e. Only graphs A and C include the variation within each generation associated with the behavior.
2. Which of the following is an example of behavior in plants?
   a. The trap of a Venus flytrap snaps shut in less than a second.
   b. Tobacco plants that produce nicotine and shunt it to their leaves when neighboring sagebrush are attacked by herbivores.
   c. Plants that release pheromones to attract natural enemies of the herbivores attacking them.
   d. All are examples of behavior in plants.
   e. Plants do not behave; seeing behavior in plants is anthropomorphic.
3. How does group selection differ from individual selection in the effect on alleles within a population?
   a. Group selection favors alleles that the group and the individual need to survive and do well.
   b. Group selection favors cooperation and individual selection does not.
   c. Group selection favors alleles that contribute to the performance of the group, not necessarily the individual.
   d. Group selection does not affect the alleles within a population; individual selection does affect the frequency of alleles because some individuals leave more surviving offspring that are more likely to carry those successful alleles than others.
4. Why is the statement that "evolution happens for the good of the species" a misconception about evolution?
   a. Because evolution occurs when the individuals within the species need to change; the change may not necessarily be good for the species.
   b. Because evolution promotes selfish behavior of individuals, which is not necessarily good for the species.
   c. Because evolution cannot recognize what is best for a group of organisms; it can only act on the best alleles within a population.
   d. Because evolution cannot recognize what is best for a group of organisms; immediate fitness consequences determine the success or failure of an allele, irrespective of the ultimate outcome of the process.

5. What is game theory?
    a. A mathematical approach to studying behavior that solves for the optimal decision in strategic situations
    b. A mathematical model that is used to predict the outcomes of games in certain situations
    c. A strategy for guessing what will happen in different behavioral situations
    d. A mathematical approach to studying the statistical outcomes of games
6. Do all members of a group share equally in the costs and benefits of living in a group?
    a. No. Some individuals can exploit dynamics of other groups, in essence "cheat," to benefit themselves.
    b. No. The costs and benefits of living in a group may be different for different individuals depending on social relationships, such as dominance.
    c. Yes. All individuals in a group benefit when all individuals share equally in the costs.
    d. On average, all individuals share equally; natural selection may favor some individuals at one time, but other individuals will benefit at other times.
    e. None of the above
7. Which of the following examples of altruism has no obvious benefit to the helping individual?
    a. A daughter helping her mother
    b. A friend helping a friend
    c. A stranger helping someone out of harm's way
    d. A team member helping another team member
    e. None of the above
8. If the benefits of helping to raise a sibling's offspring are relatively low, according to Hamilton's rule, what do you predict would happen to helping behavior if predators began cuing in on adult activity?
    a. Helping behavior would cease because the costs of helping would increase relative to the benefits of helping.
    b. Helping behavior would become more common because helping would increase the survival of off-spring.
    c. Helping behavior would stay the same because adult survival should not affect the benefits of helping.
    d. All of the above are viable predictions.
    e. None of the above is a viable prediction.
9. According to Table 16.2, what is the coefficient of relatedness between two first cousins?
    a. 0.05
    b. 0.125
    c. 0.25
    d. 0.5
    e. It depends on which alleles are shared.

10. Which of the following statements about fruit flies and learning in Figure 16.34 is FALSE?
    a. Flies with a higher learning index live longer than flies with a lower learning index.
    b. Both high learning and control flies had similar survival at day 0 and at 100 days.
    c. At 50 days, 20 percent fewer high learning flies were alive than control flies.
    d. Normal flies were better learners early in their lives, but that advantage disappeared as they aged.
    e. The learning index was not that different between normal flies and long-lived flies after flies reached 19 days old.
11. Why might complex social cognition be related to tool use?
    a. Because complex social cognition may lead to extended interaction between individuals and cultural learning about tools
    b. Because animals with complex social cognition skills live in large groups where tool use is common
    c. Because animals with complex social cognition have larger brains than animals that live in small social groups
    d. Because individuals with more social connections are more likely to use tools
    e. Because species that live in more socially connected groups are more likely to use tools

Contemplate

What's your family like at the holidays? What mechanisms do you have for kin recognition? Can you observe altruism? How might inclusive fitness be influencing your behaviors? Your siblings'? Your parents'?	**Why might humans have evolved to be social?**

17 Human Evolution

A New Kind of Ape

Check Your Understanding

1. Why are phylogenies such important tools in evolutionary biology?
   a. Because the relationships described by phylogenies are based on the best available evidence
   b. Because the relationships described by phylogenies are based on different lines of evidence, including morphology, DNA, and fossils
   c. Because the relationships described by phylogenies are hypothetical relationships that can be tested with additional evidence
   d. Because the relationships described by phylogenies are developed with advanced statistical tools that can clarify complex relationships and generate additional hypotheses
   e. All of the above
   f. a, b, and c only
2. Which of the following statements about molecular clocks is FALSE?
   a. Molecular clocks cannot be used to measure divergence in species separated by more than a few hundred million years.
   b. Molecular clocks must be calibrated because different types of DNA segments evolve at different rates.
   c. Molecular clocks result because within DNA, base-pair substitutions accumulate at a roughly clock-like rate, although substitution rates may differ between lineages.
   d. Molecular clocks can be used to predict and test the ages of unknown samples of DNA.
   e. None of the above is a false statement about molecular clocks.
3. Which is NOT a benefit of living in a social group?
   a. Better defense capabilities
   b. Large brain size
   c. Cooperative hunting
   d. Smaller risk to individuals of being killed by a predator
   e. Enhanced opportunities for learning

Learning Objectives for Chapter 17

Add important definitions and notes next to each learning objective for this chapter to help guide your understanding.

Learning Objective	Important Definitions	Notes
Describe the early evidence for classifying humans as primates.		
Explain the kinds of evidence scientists have used to discern major splits in the primate lineage and estimate the timing of their occurrences.		
Describe two hypotheses for the evolution of bipedalism.		
Propose a study examining living primates that could be used to understand fossil primates.		
Discuss the importance of tool making in the evolution of humans.		
Compare and contrast the anatomy of *Australopithecus* sediba and *Homo ergaster*.		
Analyze the scientific debate about the placement of *Homo floresiensis* in the human lineage.		
Differentiate between *Homo heidelbergensis*, *Homo neanderthalensis*, and *Homo sapiens*.		
Explain how DNA can be used to examine the relationships among *Homo heidelbergensis*, *Homo neanderthalensis*, and *Homo sapiens*.		
Describe some of the selective pressures that led to the evolution of the human brain.		
List two adaptations and explain their significance for our understanding of the evolution of language.		
Explain how scientists use genetic signatures to develop and test hypotheses about the geographic distributions of humans over evolutionary time.		
Explain how selection is currently acting on human maternity.		
Predict how variation in the sensitivity to oxytocin and vasopressin can affect human bonding.		
Explain the relationship between the major histocompatibility complex (MHC) in mate choice.		

Identify Key Terms

Match terms and definitions by filling in the blank to the right of the term with the appropriate letter.

1. **Acheulean technology** ____ 2. **Australopithecines** ____ 3. **Hominins** ____ 4. **Levallois tools** ____	a. Members of the "human" branch of the hominid clade, including the genus Homo and its close relatives such as *Australopithecus*, but not *Pan*, *Gorilla*, or *Pongo*. b. Refers to tools associated with hominins between 1.6 million years ago and 100,000 years ago. These tools are found across Africa, much of western Asia, and Europe. They are often found in association with *Homo erectus* remains. Acheulean tools, which include oval hand axes, display more sophistication in construction than Oldowan tools. c. Tools formed by a distinctive method of stone knapping involving the striking of flakes from a prepared core. This technique was much more sophisticated than earlier toolmaking styles, and flakes could be shaped into sharp scrapers, knives, and projectile points. d. Hominins classified in the genus *Australopithecus*. These species, which lived between 4.2 and 1.8 million years ago. They were short, small-brained hominins that were bipedal but still retain adaptations for tree-climbing.

Link Concepts

Map the Evidence

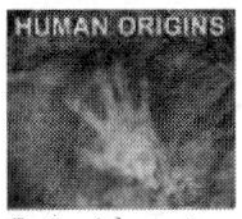

© Smithsonian

Literally hundreds of hominin fossils have been discovered since Darwin, some actually during his lifetime. New fossils are being discovered regularly. As a result, scientists are able to piece together a more and more detailed phylogeny for our species, but they still have questions. Some fossils are perplexing, either because not enough of the specimen was found, or because the bones don't fall neatly into our classification system. Some issues may never be resolved, but scientists are OK with that uncertainty. More importantly, the weight of *all* the fossil evidence, and evidence from other sources, is adding up—it's giving scientists a clearer picture of human evolution.

Whether issues exist about which fossil belongs to which species are completely resolved or not, you can use the fossil record as evidence for where hominins were and when they were there. Mapping out the locations will give you an understanding of how hominins diversified over time. The list of fossils below is a summary of actual fossils available for you to examine on the Smithsonian National Museum of Natural History's website on human origins (http://humanorigins.si.edu/). The list includes the fossil's approximate age, the site where the fossil was discovered, and the year of discovery. The list also includes the specimen name, so you can look up the fossil on the Human Origins website to examine images and descriptions.

Use the map following the table on p. 254 and some colored pens or pencils to track locations, ages, and species. For each species, pick a color and add a point to the center of the country where it was discovered. Make a notation about the age of the fossil.

Age	Species	Site	Year	Specimen
7–6 million	*Sahelanthropus tchadensis*	Toros-Menalla, Chad	2001	TM 266-01-060-1
6 million	*Orrorin tugenensis*	Tugen Hills, Kenya	2001	BAR 1002'00
4.4 million	*Ardipithecus ramidus*	Aramis, Middle Awash, Ethiopia	1994	ARA-VP-6/500
4.1 million	*Australopithecus anamensis*	Kanapoi, Kenya	1994	KNM-KP 29285
3.5 million	*Australopithecus afarensis*	West Turkana, Kenya	1999	KNM-WT 40000
3.3 million	*Australopithecus afarensis*	Dikika, Ethiopia	2000	DIK-1-1
3.2 million	*Australopithecus afarensis*	Hadar, Ethiopia	1974	AL 288-1
3 million	*Australopithecus afarensis*	Hadar, Ethiopia	1992	AL 444-2
2.8 million	*Australopithecus africanus*	Taung, Republic of South Africa	1924	Taung Child
2.8–2.4 million	*Australopithecus africanus*	Sterkfontein, Republic of South Africa	1947	STS 71
2.8–2.4 million	*Australopithecus africanus*	Sterkfontein, South Africa	1971	STW 13
2.5 million	*Australopithecus africanus*	Sterkfontein, Republic of South Africa	1947	STS 14
2.5 million	*Australopithecus garhi*	Bouri, Middle Awash, Ethiopia	1999	BOU-VP-12/1
2.5 million	*Paranthropus aethiopicus*	West Turkana, Kenya	1985	KNM-WT 17000
2.5–2.1 million	*Australopithecus africanus*	Sterkfontein, Republic of South Africa	1947	STS 5
2.4 million	*Homo habilis*	Chemeron, Kenya	1967	KNM-BC 1
2.0–1.5 million	*Paranthropus robustus*	Drimolen, Republic of South Africa	1994	UW DNH 7
1.95 million	*Homo erectus*	Koobi Fora, Kenya	1984	KNM-ER 3228
1.95–1.78 million	*Australopithecus sediba*	Malapa	2008	MH1
1.95–1.78 million	*Australopithecus sediba*	Malapa	2008	MH2
1.9 million	*Homo habilis*	Koobi Fora, Kenya	1973	KNM-ER 1813
1.9 million	*Homo rudolfensis*	Koobi Fora, Kenya	1972	KNM-ER 1470
1.89 million	*Homo erectus*	Koobi Fora, Kenya	1972	KNM-ER 1481
1.8 million	*Homo erectus*	Koobi Fora, Kenya	1975	KNM-ER 3733
1.8 million	*Homo habilis*	Olduvai Gorge, Tanzania	1968	OH 24
1.8 million	*Homo habilis*	Olduvai Gorge, Tanzania	1960	OH 8

continued

Age	Species	Site	Year	Specimen
1.8 million	*Paranthropus boisei*	Olduvai Gorge, Tanzania	1959	OH 5
1.8–1.6 million	*Homo erectus*	Mojokerto, Java, Indonesia	1936	Mojokerto
1.8–1.5 million	*Homo habilis*	Swartkrans, Republic of South Africa	1969	SK 847
1.8–1.5 million	*Paranthropus robustus*	Swartkrans, Republic of South Africa	1936	SK 46
1.8–1.5 million	*Paranthropus robustus*	Swartkrans, Republic of South Africa	1950	SK 48
1.8–1.5 million	*Paranthropus robustus*	Swartkrans, Republic of South Africa		SK 54
1.77 million	*Homo erectus*	Dmanisi, Republic of Georgia	1999	D2282
1.77 million	*Homo erectus*	Dmanisi, Republic of Georgia		D3444
1.7 million	*Homo erectus*	Koobi Fora, Kenya	1974	KNM-ER 1808
1.7 million	*Homo habilis*	Koobi Fora, Kenya	1973	KNM-ER 1805
1.7 million	*Homo habilis*	Olduvai Gorge, Tanzania	1963	OH 16
1.7 million	*Paranthropus boisei*	Koobi Fora, Kenya	1969	KNM-ER 406
1.7 million	*Paranthropus boisei*	Koobi Fora, Kenya	1970	KNM-ER 732 A
1.6 million	*Homo erectus*	Koobi Fora, Kenya	1976	KNM-ER 3883
1.6 million	*Homo erectus*	Nariokotome, West Turkana, Kenya	1984	KNM-WT 15000
1.6 million	*Homo erectus*	Nariokotome, West Turkana, Kenya	1984	KNM-WT 15000
1.55 million	*Homo erectus*	Koobi Fora, Kenya	2000	KNM-ER 42700
1.4 million	*Homo erectus*	Olduvai Gorge, Tanzania	1960	OH 9
1.4 million	*Paranthropus boisei*	Konso, Ethiopia	1993	Konso KGA10-525
1.3–1.0 million	*Homo erectus*	Sangiran, Java, Indonesia	1969	Sangiran 17
> 1 million	*Homo erectus*	Sangiran, Java, Indonesia	1937	Sangiran 2
1 million	*Homo erectus*	Middle Awash, Ethiopia	1997	Daka BOU-VP-2/66
1 million	*Homo erectus*	Buia, Eritrea		Buia UA 31
1 million–700,000	*Homo erectus*	Trinil, Java, Indonesia	1891	Trinil 2
1 million–700,000	*Homo heidelbergensis*	Ceprano, Italy	1994	Ceprano
900,000	*Homo erectus*	Olorgesailie, Kenya	2003	KNM-OG 45500

continued

Age	Species	Site	Year	Specimen
780,000–400,000	*Homo erectus*	Zhoukoudian, China		Zhoukoudian
780,000–400,000	*Homo erectus*	Zhoukoudian, China		Zhoukoudian III
600,000	*Homo heidelbergensis*	Middle Awash, Ethiopia	1976	Bodo
500,000–200,000	*Homo heidelbergensis*	Elandsfontein, Republic of South Africa	1953	Saldanha
450,000	*Homo heidelbergensis*	Tautavel, France	1971	Arago 21
400,000–300,000	*Homo erectus*	Longtandau Cave, Anhui Province, China	1980	Hexian
350,000	*Homo heidelbergensis*	Steinheim, Germany	1933	Steinheim
350,000	*Homo heidelbergensis*	Lake Ndutu, Tanzania	1973	Ndutu
350,000–150,000	*Homo heidelbergensis*	Petralona, Greece		Petralona 1
300,000	*Homo erectus*	Narmada, India		Narmada
300,000–200,000	*Homo heidelbergensis*	West Turkana, Kenya		Eliye Springs ES11693
300,000–200,000	*Homo neanderthalensis*	Wadi Amud, Israel		Zuttiyeh
300,000–125,000	*Homo heidelbergensis*	Kabwe, Zambia	1921	Kabwe 1
259,000	*Homo heidelbergensis*	Florisbad, Republic of South Africa	1932	Florisbad
250,000–200,000	*Homo erectus*	Salé, Morocco	1971	Salé
250,000–70,000	*Homo erectus*	Solo River, Java, Indonesia		Ngandong 13
250,000–70,000	*Homo erectus*	Solo River, Java, Indonesia		Ngandong 14
250,000–70,000	*Homo erectus*	Solo River, Java, Indonesia		Ngandong 7
195,000	*Homo sapiens*	Omo River, Ethiopia	1967	Omo I
160,000	*Homo sapiens*	Jebel Irhoud, Morocco	1961	Irhoud 3
150,000–120,000	*Homo sapiens*	Singa, Sudan	1924	Singa
130,000	*Homo heidelbergensis*	Guangdong Province, China		Maba
130,000	*Homo neanderthalensis*	Krapina, Croatia	1899	Krapina 3
130,000–100,000	*Homo neanderthalensis*	Rome, Italy	1929	Saccopastore 1
122,000–50,000	*Homo neanderthalensis*	Mount Carmel, Israel		Tabun 1
120,000	*Homo sapiens*	Laetoli, Tanzania	1976	Ngaloba LH 18

continued

Age	Species	Site	Year	Specimen
120,000–80,000	*Homo sapiens*	Mount Carmel, Israel	1932	Skhul V
110,000–80,000	*Homo neanderthalensis*	Subalyuk Cave, Hungary		Subalyuk 2
100,000	*Homo sapiens*	Jebel Qafzeh, Israel	1933	Qafzeh 6
90,000–60,000	*Homo neanderthalensis*	Pech de l'Azé, France	1909	Pech de l'Azé I
70,000	*Homo neanderthalensis*	Bajsuntau, Uzbekistan	1938	Teshik-Tash
70,000–50,000	*Homo neanderthalensis*	Dederiyeh, Syria	1993	Dederiyeh 1
70,000–50,000	*Homo neanderthalensis*	La Ferrassie Cave, France	1909	La Ferrassie
70,000–45,000	*Homo neanderthalensis*	Forbes' Quarry, Gibraltar	1848	Gibraltar 1
68,000	*Homo sapiens*	Liujiang, China		Liujiang
65,000	*Homo neanderthalensis*	La Quina Rock Shelter, France	1915	La Quina 18
65,000	*Homo neanderthalensis*	La Quina Rock Shelter, France	1911	La Quina 5
60,000	*Homo neanderthalensis*	La Chapelle-aux-Saints, France	1908	La Chapelle-aux-Saints
45,000–39,000	*Homo sapiens*	Sarawak, Malaysia	1958	Niah Cave
45,000–35,000	*Homo neanderthalensis*	Shanidar, Iraq		Shanidar 1
41,500–39,500	*Homo sapiens*	Pestera cu Oase, Romania	2003	Oase 2
41,000	*Homo neanderthalensis*	Wadi Amud, Israel	1961	Amud
40,000	*Homo neanderthalensis*	Feldhofer Cave, Neander Valley, Germany	1856	Feldhofer
39,000–33,000	*Homo sapiens*	Hofmeyr, Republic of South Africa	1952	Hofmeyr
30,000	*Homo sapiens*	Cro-Magnon, France	1868	Cro-Magnon 1
36,000	*Homo neanderthalensis*	Saint-Césaire, France	1979	Saint-Césaire
18,000	*Homo floresiensis*	Liang Bua, Flores, Indonesia	2003	LB-1
13,000–9,000	*Homo sapiens*	Kow Swamp, Australia	1967	Kow Swamp
11,500	*Homo sapiens*	Minas Gerais, Brazil	1976	Lapa Vermelha IV Hominid 1
4,700	*Homo sapiens*	Tepexpan, Mexico	1947	Tepexpan 1
Age uncertain	*Homo neanderthalensis*	Awirs Cave, Engis, Belgium	1829	Engis 2

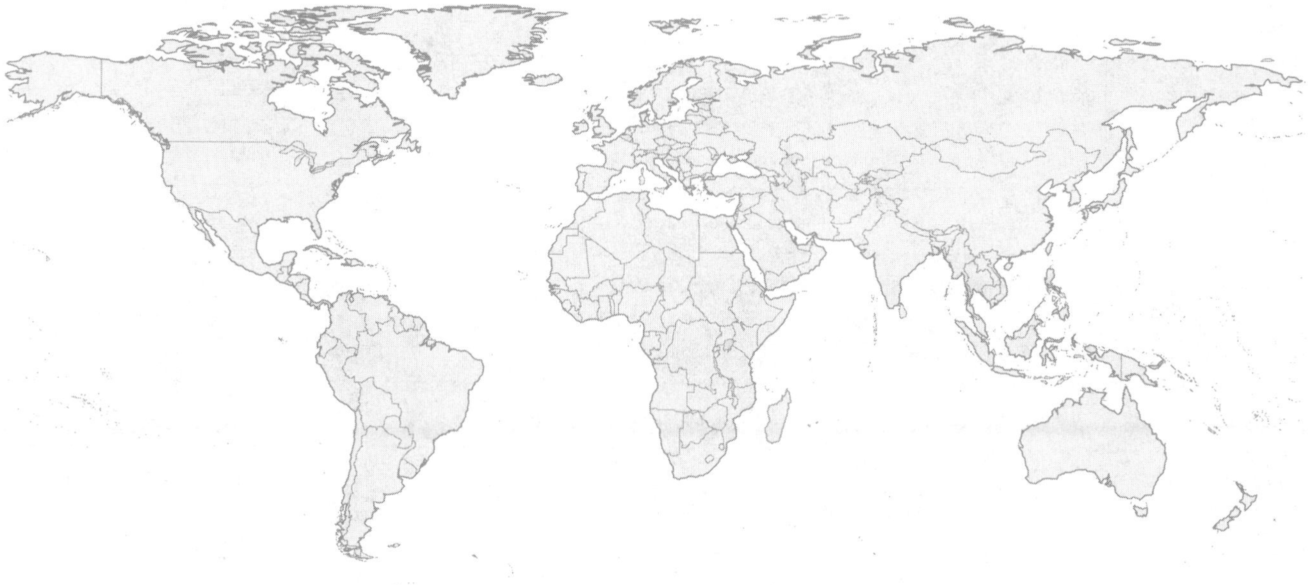

- What can you infer about the diversity of hominins from your map?

- Do you generally agree with the consensus of scientists below?

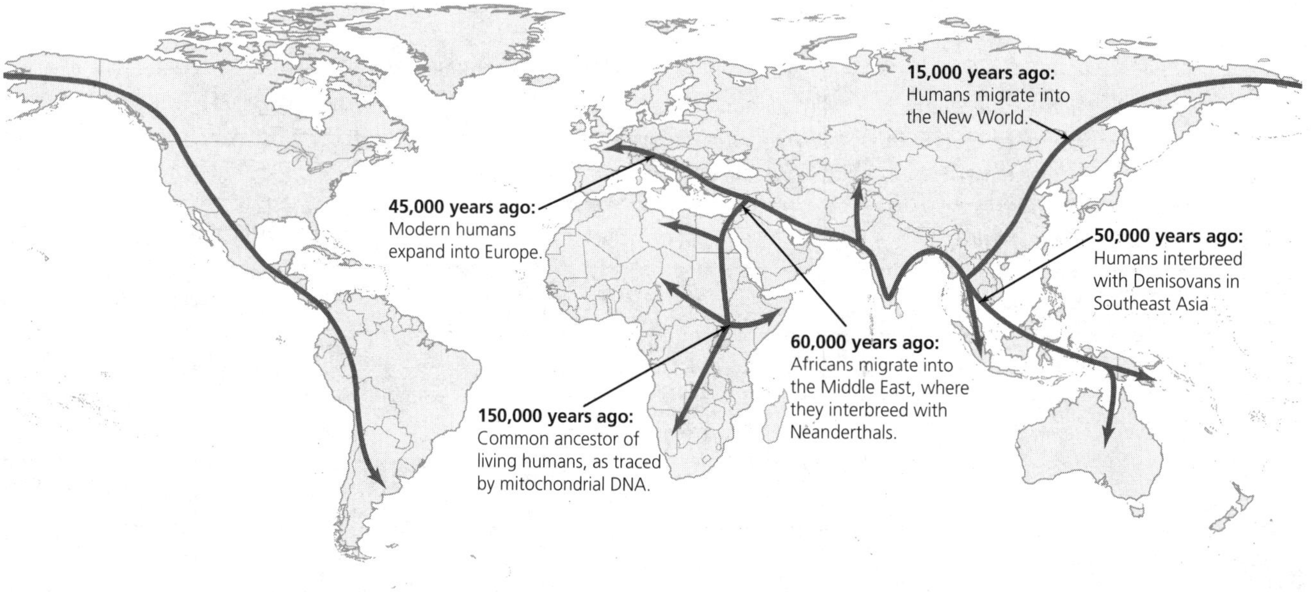

Develop your own concept map that explains the genetic evidence for the evolution of humans. Think about complex cognition, Denisovans, gene duplication, genetic drift, humans, hybridization, language, mutations, Neanderthals, regulatory regions, sociality, and trichromatic vision.

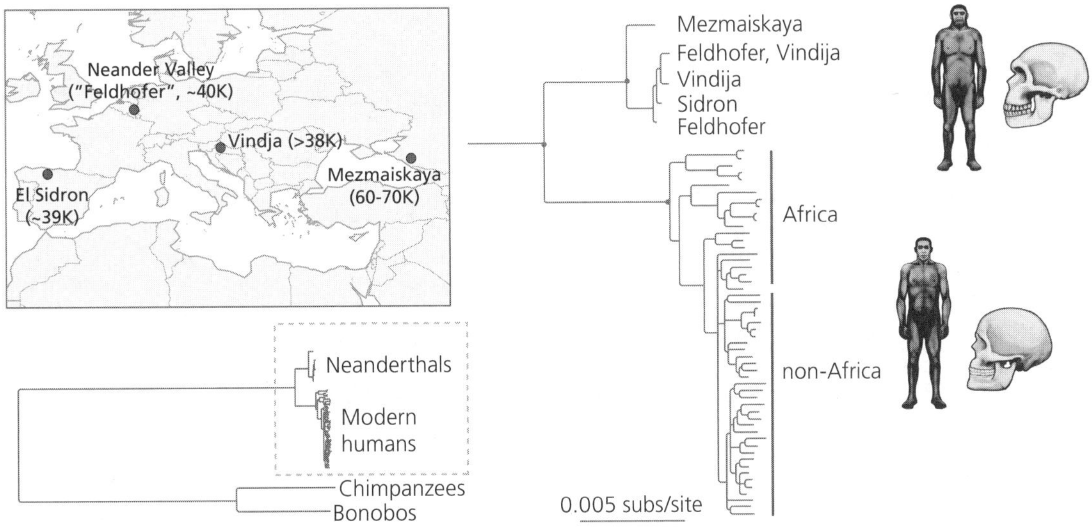

Key Concepts

Fill in the blanks for the key concepts from the chapter.

17.1 Discovering Our Primate Origins

Humans were classified with ____________ based on similar ____________ well before Darwin proposed an ____________ for the relationship.

17.2 Primate Evolution: Molecular and Fossil Evidence

Scientists have used different ____________ to determine the origins of primate lineages, including humans. Although the evidence is as diverse as ____________ and ____________, it has ____________ identified humans as ____________.

17.3 Making Sense of Hominin Evolution

Scientists may not agree about some specific aspects of the human lineage, such as ____________ and exact phylogenetic relationships, but they have come to a strong ____________ about the relationships among most hominins. More importantly, they continue to look for evidence that will ____________ or ____________ our current understanding.

17.4 Walking into a New Kind of Life

Scientists can gain important clues about the evolution of ________ in the ________ from changes in important ________ of fossils, such as size, shape, orientation of bones, and fossil trackways. They also can examine modes of ________ of living primates and ________ their bones to fossils to see how closely they match.

17.5 A Changing Environment

Shifts in climate brought significant changes to the ________ in which Humans evolved.

Early hominins responded to the complicated ________ of open-habitat grasslands and forested areas by evolving into obligate bipeds that retained a number of significant arboreal traits.

17.6 Staying Alive on the Savanna

The availability of food and water may have been a strong selector shaping the evolution of our skulls, jaws, and teeth.

17.7 The Toolmakers

Oldowan tools are the oldest stone tool industry in M- (from 2.6 million years ago up until 1.7 million years ago). They consisted of simple chipped stones used for tasks such as butchering meat.

17.8 The Emergence of *Homo*

Australopithecus sediba may represent a transitional species in the human ________. It has characteristics of both australopithecines and the human genus, *Homo*.

Some early *Homo* species have a taller body, narrower pelvis, and larger brain. *Homo* also emerged out of Africa for the first time and developed new types of tools.

17.9 Parallel Humans

________ hominins emerged within the last 800,000 years.

Several ________ of these large-brained hominins lived as recently as 40,000 years ago.

Anatomically modern humans migrated out of Africa and settled across the entire world.

17.10 New Discoveries from Ancient Genes

Both molecular and fossil evidence suggest that humans, Neanderthals, and Denisovans represent ____________, ____________ lineages descended from a ____________ living in Africa several hundred thousand years ago.

17.11 Evolving a Human Brain

One ____________ for the evolution of the human brain is an ____________ to a complex variety of selective pressures related to social living. Some of the ____________ for this hypothesis may be recorded in our genes.

17.12 The Language Instinct

Many of the adaptations for understanding language likely evolved early in the ____________ lineage, but some specialization is uniquely human.

The brains of our most recent common ancestors had many ____________ for language, including Broca's area, but ____________ to regulatory genes like *POU3F2*, may have occurred after humans branched off from our *Homo heidelbergensis* relatives.

17.13 Bottlenecks in the Origin of Modern Humans

Signatures of ____________ are evident in human populations today. These can be used to reconstruct movements and bottleneck events of ancestral populations from as recently as tens of thousands of years ago.

17.14 Recent Natural Selection

Humans have adapted ____________ to recent changes in their selection environment, such as changes in our diet due to the advent of agriculture and domestication of animals.

17.15 Emotions and Other Evolutionary Legacies

The ____________ of human emotions such as fear, anticipation, nurturing, and love is the result of a history of selection and adaptation. Many ___feelings___ are shared between humans and other mammals.

Evolutionary psychology explores the extent to which human behavioral predispositions arose as psychological ___________ that evolved to solve ___________ problems in ancestral environments.

Interpret the Data

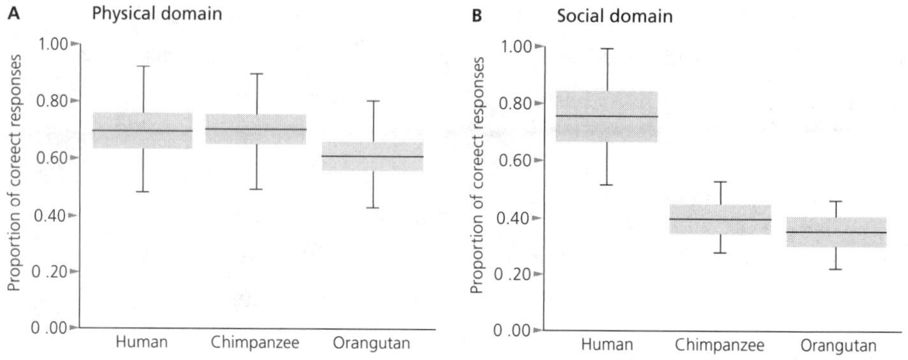

Esther Herrmann and her colleagues (2007) tested the ultra-social hypothesis by examining the development of social skills in human children versus other apes. They performed two kinds of tests—one examining how children, chimpanzees, and orangutans responded to physical properties (such as space, quantity, and physical cause and effect), and the other examining how they responded to social cues (such as observation and learning).

- If ultra-social relationships acted as selective agents on human brain development, would you predict that children, chimpanzees, and orangutans should differ in their understanding of basic math? Why or why not?

- What do the width of the gray bars and the "whiskers" for children, chimpanzees, and orangutans indicate?

- If the whiskers indicate individuals with test results at the ends of the distributions of each species, how would you interpret the differences among children and chimpanzees in social skills test?

Games and Exercises

Toolmaking might not be unique to humans, but the depth and breadth of tools crafted by hominins are exceptional in the animal kingdom. Our hominin ancestors didn't have the mental capacity to craft highly specific tools. The evolution of stone tools reflects the evolution of the hominin brain. Stone technology changed over time as hominin lineages evolved the capacity to control their hands, to plan, and to create elaborate tools. Oldowan tools are very different from Acheulean technology. Our *Homo sapiens* ancestors were skilled craftspeople; their creations indicate a capacity for self-expression and trade.

Go online and search for images of stone tools. There is no shortage of images, and they are often grouped according to similarities (time, shape, function).

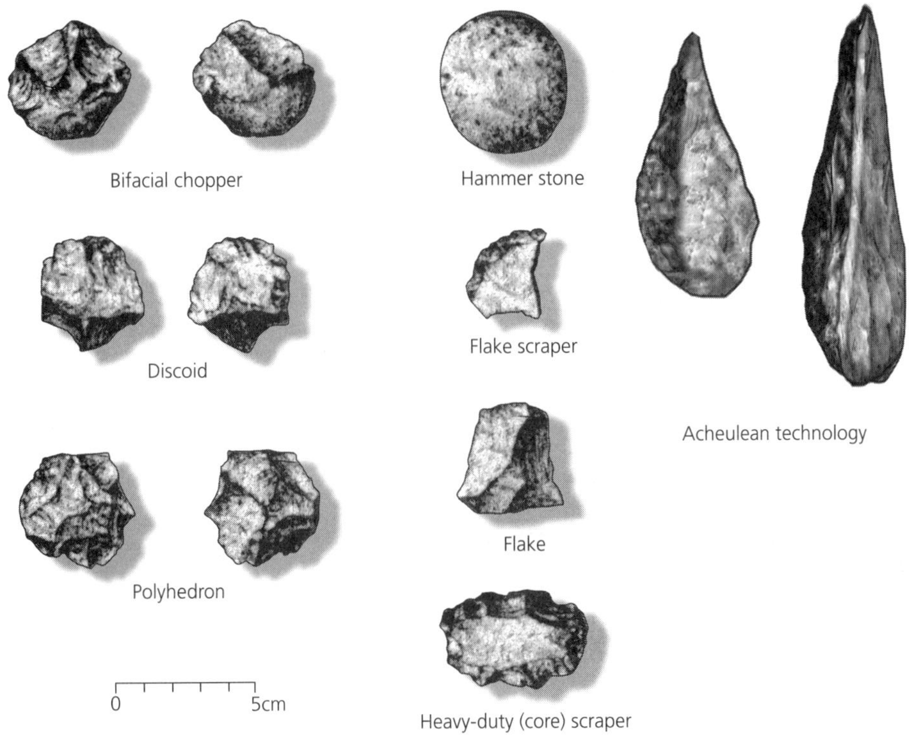

- What characteristics distinguish older and newer technology?

- Can you guess what some of the tools might have been used for?

- Why might images of newer technology on the web be more prevalent than images of older technology?

- How might understanding the evolution of tool making apply to human's ability to fashion technology today?

Explore!

What Does It Mean to Be Human?

This exhibit from the Smithsonian Institution allows users to explore the research and the evidence scientists use to examine important questions about our evolutionary history.

http://humanorigins.si.edu/

Solving the Mystery of the Neandertals

Cold Springs Harbor Laboratory created a website to help students understand their own genetic origins (http://www.geneticorigins.org). You can participate in two different experiments using your own DNA. This animation explains the mitochondrial control region and shows how Svante Pääbo and his colleagues discovered the relationship between Neanderthals and humans.

http://www.geneticorigins.org/mito/media2.html

The Human Journey: Migration Routes

The Human Journey is an extensive project to map genetic markers in our population. Individuals can contribute their own DNA literally to become a part of the evidence. The resulting information can be used to trace the migratory patterns of humans from Africa. They also have some excellent videos.

https://genographic.nationalgeographic.com/human-journey/

Becoming Human

This series explores what it means to be human. Each episode is one hour and available online from PBS *NOVA*.

http://www.pbs.org/wgbh/nova/evolution/becoming-human.html#becoming-human-part-1

http://www.pbs.org/wgbh/nova/evolution/becoming-human.html#becoming-human-part-2

http://www.pbs.org/wgbh/nova/evolution/becoming-human.html#becoming-human-part-3

Origins of Bipedalism

From PBS *NOVA*, this interactive runs through the diverse hypotheses for the evolution of bipedalism. After exploring each hypothesis, you can vote on the hypothesis that you think is the most likely.

http://www.pbs.org/wgbh/nova/evolution/origins-bipedalism.html

PBS *Evolution* and WGBH offer a wealth of videos and interactive websites related to human evolution. Here are a few links.

Finding Lucy

This video relives the excitement of the discovery of the fossil specimen known as Lucy.

http://www.pbs.org/wgbh/evolution/library/07/1/l_071_01.html

Humankind

This interactive explores the hominid family tree. The timeline shows details on species, details of their discovery, and time of existence.

http://www.pbs.org/wgbh/evolution/humans/humankind/

Walking Tall

"Walking Tall" is a short video that explains our evolutionary legacies—the adaptations that permit a graceful gait and the imperfections that lead to some our physical problems.

http://www.pbs.org/wgbh/evolution/library/07/1/l_071_02.html

Riddle of the Bones

Another interactive explains the evidence scientists use to decipher fossil bones, including determining how ancient hominins moved, what they looked like, species relationships, and when they lived.

http://www.pbs.org/wgbh/evolution/humans/riddle/

Overcoming Misconceptions

Humans Are NOT Distinct "Kinds" of Organisms

Outside of science, people often want to separate species as clearly defined entities, but defining a species as a "distinct kind" does not provide any useful way of distinguishing groups of organisms, especially taxonomic levels. A "kind" could be fungi (a kingdom) or mammals (a class), a duck (an order) or a duck-billed platypus (a species). Defining exactly what constitutes a species is difficult, but scientists understand that classification is a human artifact. "Species" doesn't have to be defined precisely—unequivocally—for evolutionary theory to explain the existence of groups of organisms that share characteristics with a common ancestor.

Our species shares many characteristics with other hominins, and as scientists discover more and more fossil specimens, distinguishing each species is becoming more and more challenging. Our species *is unique* in that we, humans, are the only one of an apparently diverse group of hominins to survive. More importantly, however, scientists are uncovering the transitional fossils that together make up the human phylogenetic tree.

Humans Are Still Evolving

We tend to think our species has overcome the pressures of natural selection. It's true that we can modify our environments with technology, such as medicine, agriculture, and education. But natural selection remains powerful. Just as living at high altitudes and lactose tolerance shaped the evolution of mountain people and cattle herders in the not so distant past, humans will continue to face challenges to survival and reproduction. For example, women with a genetic tendency for low cholesterol are reproducing at greater rates in Framingham, Massachusetts, than women with a genetic tendency for high cholesterol. Predicting exactly what will shape us is nearly impossible, but the effects we are having on our climate and our environment may turn out to be strong selective factors in our future.

Phylogenetic Trees and "Advanced" Organisms

As humans, our need to classify also possessed us to categorize organisms as "primitive" and "advanced." Even when we look at phylogenetic trees, it's hard to escape believing that lineages that branch off earlier in the tree are "lower" or "primitive" forms. But branches can swing—like a mobile—and don't necessarily represent a "ladder-like" progression to more advanced species. As humans, we want to consider ourselves as the pinnacle of evolution. Our human brain is exceptional and unique, but other adaptations in other organisms are exceptional as well. For example, the duck-billed platypus may seem "primitive," but many of its extraordinary traits (e.g., their unique bill) are highly advanced, having evolved after the lineage split off from other mammals (see Box 4.1).

Which of the following is a true statement?

a. Scientists agree about how to define species, especially human species.

b. Because of all our advances in medicine and technology, humans are no longer evolving

c. Humans are considered an advanced species because they are found at the end of the phylogenetic tree.

d. As scientists discover more and more hominin fossils, they may be able to define the genes that distinguish humans from other species.

Go the Distance: Examine the Primary Literature

Stephen Stearns and his colleagues tackle a question most of us think about when we contemplate human evolution: Are humans still evolving?

- Besides the Framingham study, what are two other groups of individuals Stearns and his colleagues suggest looking at to measure *fitness*?

- What do those studies currently measure?

- According to Stearns and his colleagues, what are major unresolved issues facing scientists studying evolution in modern humans?

- Do Stearns and his colleagues contend that scientists can *know* how our species will evolve?

Stearns, S. C., S. G. Byars, D. R. Govindaraju, and D. Ewbank. 2010. Measuring Selection in Contemporary Human Populations. *Nature Reviews Genetics* 11 (9): 611–22. doi:10.1038/nrg2831.

Delve Deeper

1. Which molecule(s) related to our emotions likely evolved early in our mammalian history?

2. Why is the discovery of stone tools in the fossil record important in understanding the evolution of our genus, *Homo*?

3. What are some of the important factors influencing gene expression in humans?

Test Yourself

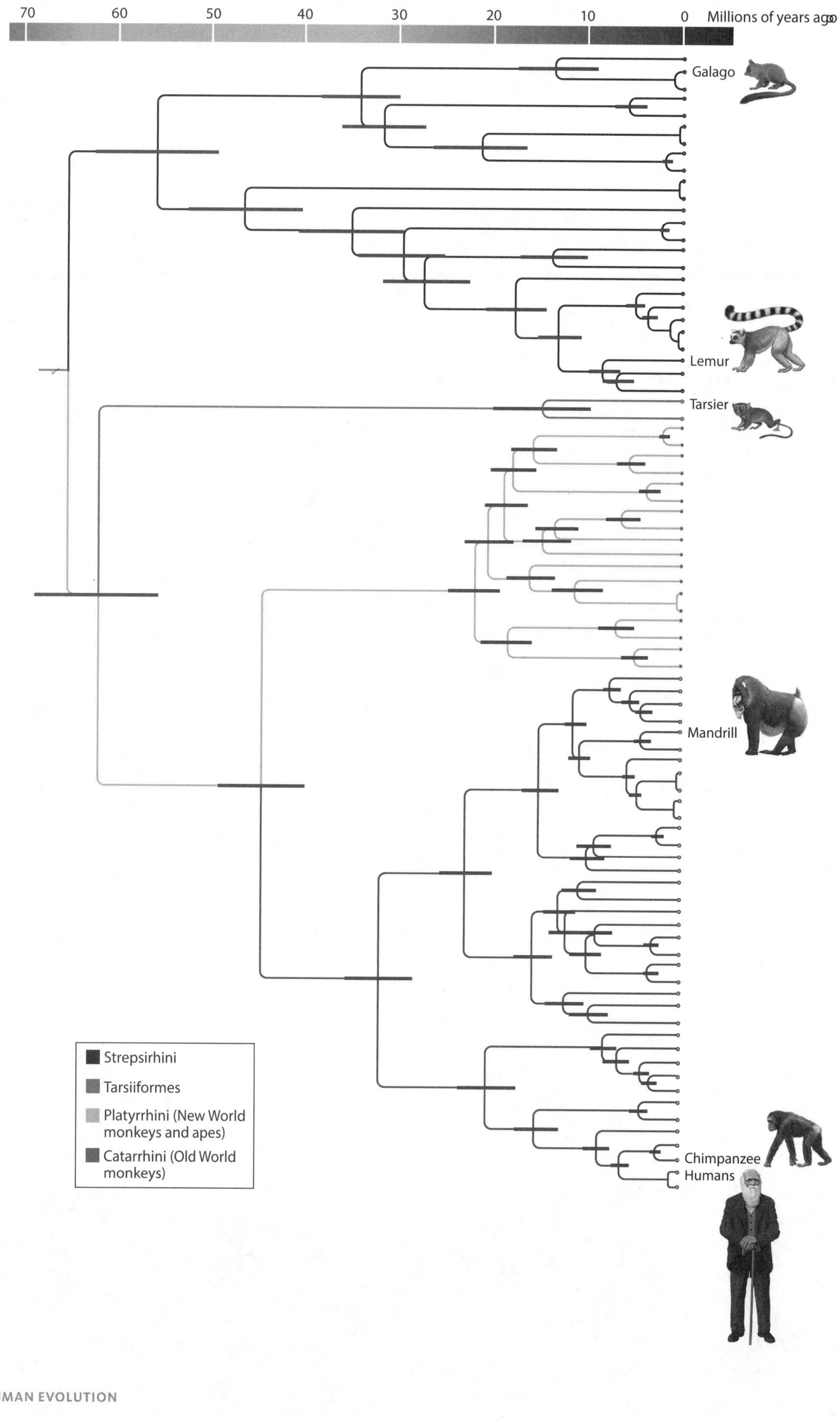

1. Which of the following statements about the phylogeny of primates in Figure 17.3 is TRUE?
   a. Humans share a common ancestor with lemurs (Lemuridea).
   b. The confidence intervals reflect the ages of nodes, not the relationships depicted.
   c. The lineage containing humans split from the lineage containing apes about 8.4 million years ago.
   d. The variation around estimates of the *Homo-Pan* split is smaller than the variation around the split of the Platyrrhini (New World monkeys) and Catarrhini (apes and Old World monkeys).
   e. All of the above
2. When evolutionary biologists refer to the hominin clade, to what are they referring?
   a. The great ape branch of our evolutionary tree that includes the genus *Homo* and all of its ancestors, including *Australopithecus, Paranthropus, Pan, Gorilla,* and *Pongo*.
   b. The human branch our evolutionary tree that includes the genus *Homo* and its close relatives such as *Australopithecus* and *Paranthropus*, but not chimpanzees.
   c. Members of the genus *Homo*, including *ergaster, habilis, erectus, neanderthalensis, heidelbergensis, floresiensis*, and *sapiens*, but excluding *Australopithecus* and *Paranthropus*.
   d. Members of the genus *Homo*, including *ergaster, habilis, erectus, neanderthalensis, heidelbergensis*, and *floresiensis*, but excluding humans.
   e. None of the above
3. What was the tool technology used by *Homo heidelbergensis* called?
   a. Acheulean
   b. Oldowan
   c. Levallois
   d. Stone
4. What did Esther Herrmann and her colleagues predict would happen when they tested children, chimpanzees, and orangutans in math and social skills (Figure 17.32)?
   a. Children would do better than chimpanzees and orangutans with math tests because humans, chimpanzees, and orangutans are all social animals and the ultra-social hypothesis predicts that all should develop social skills early in development.
   b. Children would do better than chimpanzees and orangutans with social skills tests because the ultra-social hypothesis predicts that humans develop social skills early in their development.
   c. Children would not do better than chimpanzees and orangutans with physics tests because the ultra-social hypothesis makes no prediction about humans, math skills, and development.
   d. All of the above
   e. b and c only

5. Which of the following statements is TRUE according to Figure 17.35?
   a. The peoples of Europe separate out perfectly according to the genetic structure of their populations.
   b. The genetic structure of European populations is too complex to observe any patterns.
   c. Isolated countries tend to have greater differences in their genetic structure than countries close together.
   d. The high levels of clumping that align with the geography of Europe are likely a product of genetic drift and founder events.
   e. The genetic structure of European populations indicates a high likelihood for sympatric speciation.
6. How did Larry Young test whether vasopressin activity affected mate-attachment behavior (Figure 17.38)?
   a. He artificially increased the expression of a vasopressin receptor in some prairie voles and found they tended to be even more monogamous than control prairie voles.
   b. He artificially increased the expression of a vasopressin receptor in some meadow voles and found they tended to be monogamous more than other meadow voles.
   c. He artificially increased the expression of a vasopressin receptor in some meadow voles and compared social behavior of control meadow voles and experimental meadow voles with monogamous prairie voles.
   d. All of the above
   e. b and c only

Contemplate

Could you define different populations of humans today as different species? Why or why not?	How might our changing climate affect our evolution?

18 Evolutionary Medicine

Check Your Understanding

1. Which is NOT one of the three conditions necessary for evolution by natural selection to occur?
   a. Individuals must differ in the characteristics of a trait.
   b. The differences among individuals in a trait must be at least partially heritable.
   c. Some individuals survive and reproduce more successfully than others because of differences in a trait.
   d. The more an individual needs a trait, the more quickly it will adapt to its environment.
   e. All are necessary for natural selection to occur.
2. Which of the following is NOT an important factor influencing gene expression in humans?
   a. The external environment a gene is exposed to during development
   b. Pleiotropy
   c. Epigenetic effects
   d. Genomic imprinting
   e. All of the above are important factors influencing gene expression.
   f. Only b and c are important factors influencing gene expression.
3. What would you predict would be the outcome of natural selection on a mutation that increases fertility early in life but increases susceptibility to cancerous growths later in life?
   a. Natural selection would favor individuals with the mutation because the fitness effects early in life would be bigger than the harm the mutation causes in old age.
   b. Natural selection would not favor individuals with the mutation because individuals susceptible to cancerous growths would die sooner and have lower lifetime reproductive success than individuals without the mutation.
   c. Natural selection would favor a reaction norm that balanced reproductive success and susceptibility to cancerous growths.
   d. Natural selection would remove individuals with the mutation from the population because mutations are detrimental.
   e. A single mutation cannot be both beneficial and detrimental.

Learning Objectives for Chapter 18

Add important definitions and notes next to each learning objective for this chapter to help guide your understanding.

Learning Objective	Important Definitions	Notes
Analyze the role of the selective environment within a host and its effect on pathogen evolution.		
Describe the conditions necessary for the evolution of antibiotic resistance.		
Explain why phylogenies are effective tools for understanding the origins of infectious diseases.		
Examine the costs and benefits of generating variation within viruses to understand disease.		
Demonstrate the adaptive significance of the *HbS* allele.		
Explain why an understanding of human genetic variation is important when considering drug treatment options.		
Review the pleiotropic effects of genes and their role in senescence.		
Apply models used to understand the rise of antibiotic-resistant bacteria to anticancer drug resistance.		
Explain the hygiene hypothesis.		
Compare and contrast the thrifty genotype hypothesis, the thrifty phenotype hypothesis, and the thrifty epigenotype hypothesis.		
Explain how understanding our evolutionary history may help in the search for medically relevant genes.		

Identify Key Terms

Match terms and definitions by filling in the blank to the right of the term with the appropriate letter.

1. **Epigenetics** ____	a. The study of modifications to the state of DNA along the genome, including methylation, coiling of DNA around histones, and binding of noncoding RNAs. Modifications to the conformational state of DNA are independent of the nucleotide sequence (epi means "over" or "upon"), but they can alter the expression of genes in ways that are heritable across mitotic (and occasionally meiotic) cell divisions. Epigenetic mechanisms are an important way that cell types diverge in patterns of gene expression during development, and environmental circumstances (e.g., stress) alter the expression of genes over an individual's lifetime.
2. **Evolutionary medicine** ____	b. The capacity of pathogens to defend against antibiotics or other drugs.
3. **Hygiene hypothesis** ____	c. Genes that suppress cell growth and proliferation. Many are transcription factors activated by stress or DNA damage that arrest mitosis until DNA can be repaired. Mutations interfering with their expression can lead to excessive proliferation and cancer.
4. **Oncogenes** ____	d. Proposes that epigenetic mechanisms are responsible for coupling fetal nutritional conditions with the establishment of a particular physiology (e.g., a "starvation physiology") that persists for life and can be at least partially inherited by offspring.
5. **Proto-oncogenes** ____	e. Occurs when genetic material from different strains gets mixed into new combinations within a single individual.
6. **Resistance** ____	f. Proposes that a lack of early childhood exposure to infectious agents, symbiotic microorganisms (e.g., gut flora or probiotics), and parasites increases a person's susceptibility to allergic and autoimmune diseases.
7. **Thrifty epigenotype hypothesis** ____	g. The integrated study of evolution and medicine to improve scientific understanding of the reasons for disease and actions that can be taken to improve health.
8. **Thrifty genotype hypothesis** ____	h. Mutated versions of proto-oncogenes. Increased expression of oncogenes can lead to cancer.
9. **Thrifty phenotype hypothesis** ____	i. Proposes that alleles that were advantageous in the past (e.g., because they were "thrifty" and stored nutrients well) may have become detrimental in the modern world, contributing to metabolic syndrome, obesity, and type 2 diabetes.
10. **Tumor suppressor genes** ____	j. Proposes that the conditions a fetus experiences during pregnancy can affect physiology throughout an individual's life. Type 2 diabetes may be the result of a "starvation physiology" resulting from exposure to nutrient-poor conditions during development coupled with a "Westernized" lifestyle that is nutrient rich and low in exercise.
11. **Viral reassortment** ____	k. Normal genes whose functions, when altered by mutation, have the potential to cause cancer.
12. **Virulence** ____	l. Describes the ability of a pathogen to cause disease

Link Concepts

Fill in the bubbles with the appropriate terms.

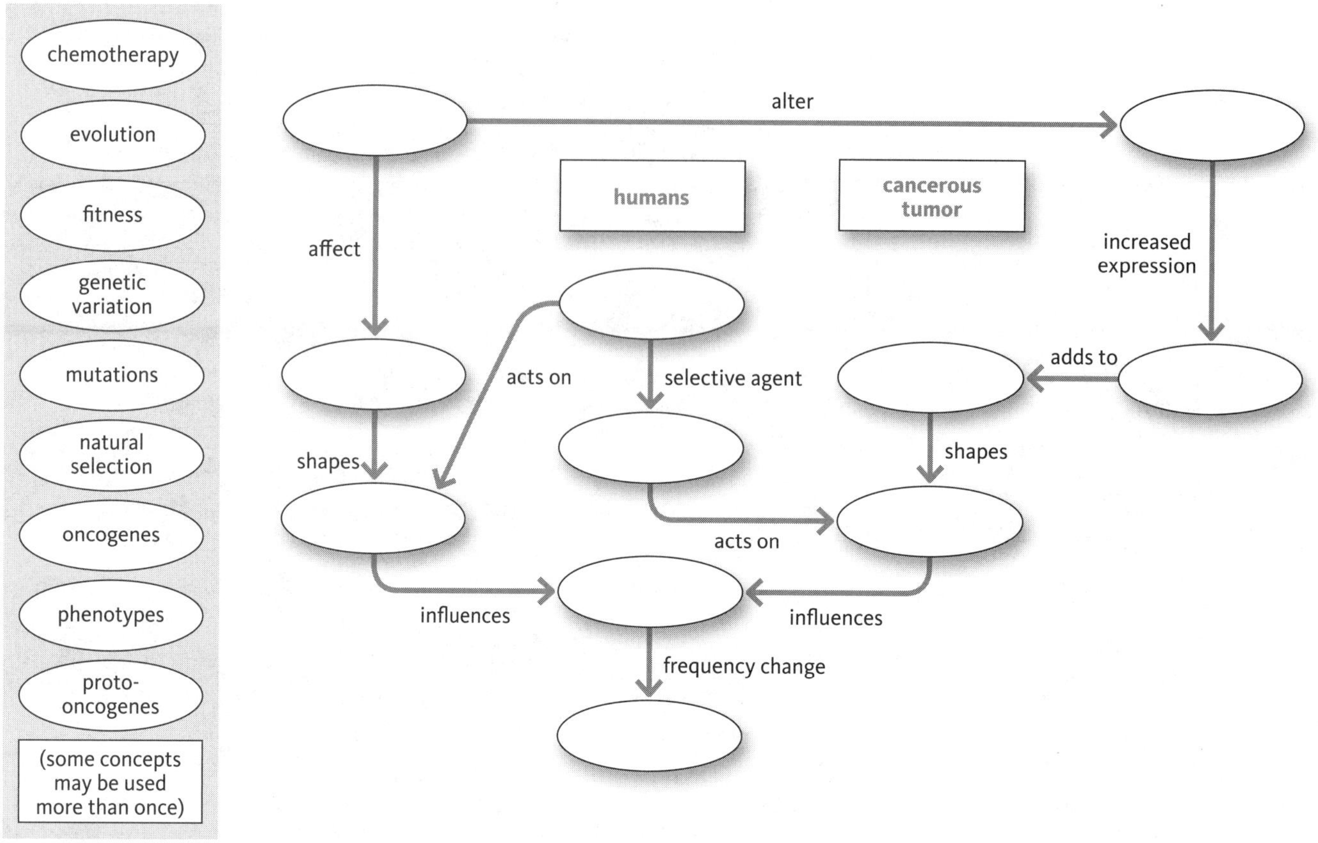

Key Concepts

Fill in the blanks for the key concepts from the chapter.

Many genes participate in ____________ and ____________ regulatory networks. Knowledge of these gene interactions can help scientists identify new candidate genes for ____________ .

18.1 Maladaptation and Medicine

Six evolutionary explanations for why we are vulnerable to disease include:

1. Pathogens evolve ____________ than their hosts.
2. Natural selection often ____________ environmental change.
3. ____________ make it impossible for natural selection to "solve" certain biological problems, irrespective of time.
4. A species' evolutionary history creates ____________ on the potential changes natural selection can bring about.
5. Some traits increase reproductive fitness at the cost of increasing ____________ to disease.
6. What may appear to be disease may actually be an ____________ .

18.2 Evolving Pathogens

Pathogens have the potential for ____________ evolution because high rates of ____________ are coupled with high ____________ that lead to extraordinary genetic variation.

Virulence can evolve in pathogens when (1) individuals within the disease population ____________ in their replication rate, and (2) the genetic basis for that replication can be ____________ to other individuals (through either vertical or horizontal gene transfer).

____________ can influence the evolution of pathogen virulence. By making it harder (or easier) for pathogens to be ____________ from one person to the next, these practices select for lower (or higher) ____________ .

18.3 Defeating Antibiotics

Because bacteria are capable of acquiring new mutations through both ____________ and ____________ gene transfer, antibiotic resistance can evolve especially ____________ .

Understanding the evolutionary biology of these important pathogens could influence ____________ options and slow the pace of the evolution of ____________ .

18.4 The Origin of New Diseases

____________ studies of pathogen evolution can provide important clues to the native hosts—and therefore ____________ —of newly emerging infectious diseases.

Understanding the biology of pathogens as they evolve in their native hosts can help us ____________ how and where future human ____________ will erupt, and can suggest ____________ measures to keep outbreaks from occurring.

18.5 Ever-Evolving Flu

The influenza virus can evolve rapidly to evade ____________ by our immune system and our ____________ .

Wartime conditions changed the nature of ____________ acting on the influenza virus, permitting the evolution and spread of an extraordinarily ____________ strain in 1918.

Public health efforts can select for less ____________ strains of pathogens by making it more difficult for them to be ____________ from person to person. But breakdown of these efforts—as occurs after natural disasters or in populations with high density and poor sanitation, like refugee camps—can have the reverse effect. They can select for ____________ deadly pathogen strains.

Many scientists suspect that world travel has created a ____________ similar to the one that facilitated the 1918 pandemic: mass movement and crowding relax the selective constraint on ____________ .

Experiments can shed light on the potential paths of evolution that ____________ and other ____________ may take in the future.

18.6 Molded by Pathogens

Disease-causing organisms, like parasites, exert strong ____________ on humans, especially on genes related to our immune systems. Adaptations, such as the *HbS* allele, are not ____________ ; they reflect the diversity of ____________ that can evolve when they confer fitness ____________ in some contexts but not in others.

18.7 Human Variation and Medicine

Understanding the role of ____________ in human diseases can be critical not just for treatment of genetic disorders but also for diagnosing the ____________ that can accompany treatment.

18.8 Old Age and Cancer: Evolution's Tradeoffs

Aging is a ____________ by-product of ____________ acting at other stages of life.

18.9 The Natural Selection of Cancer

Under normal conditions, the ____________ between proliferation and programmed cell death is tightly regulated. ____________ mutations that break down these regulatory processes can lead to uncontrolled cell ____________ and cancer.

Once cells start to divide uncontrollably, they begin to evolve by ____________ inside our body. Their fitness is no longer aligned with ours, and selection acts at the level of ____________ cell lines within the tumor.

Somatic cell-line evolution within a cancerous tumor exemplifies how natural selection can be "shortsighted." Immediate fitness ____________ drive the evolution of increased cell division and metastasis even though this activity ultimately ____________ the host and, with it, all of the cancer cells in the tumor.

18.10 Mismatched with Modern Life

One legacy of the ____________ between parasites and humans may be that signals from parasites and bacteria are needed to keep the immune system from becoming ____________ . People living in clean environments free of these parasites are more likely to develop ____________ ranging from eczema and allergies to Crohn's disease and type 1 diabetes.

Exposure to nutrient-poor conditions during early pr ____________ silencing of genes) to a "starvation-mode" physiology for life. This physiology is ____________ to resource-poor environments, but it can be dangerous for individuals who later adopt a "Westernized" lifestyle rich in nutrients and low in exercise.

18.11 Evolutionary Medicine: Limits and Clues

Clues to the functions of human genes lie in the genomes of some ____________ related species because of ____________ . These deep homologies can guide researchers in their search for the ____________ of diseases and also allow scientists to identify new drugs that could be beneficial to humans.

Interpret the Data

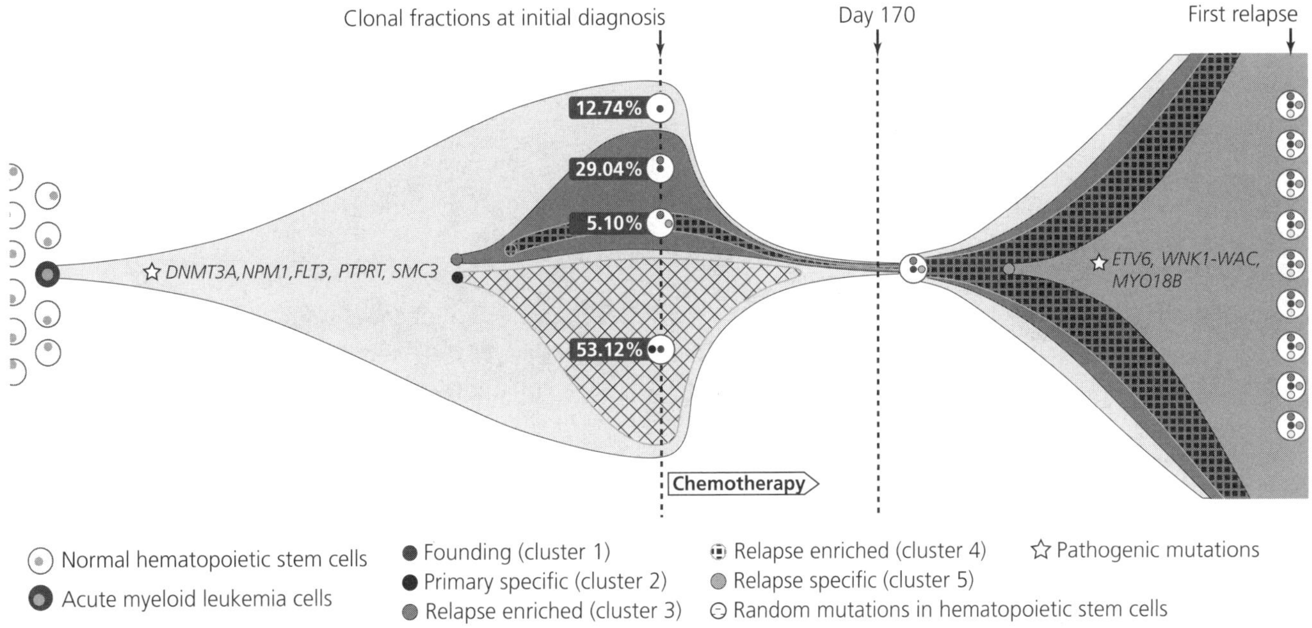

Cancer is subject to the same evolutionary processes as other organisms. Like other organisms, three conditions are necessary for evolution by natural selection: (1) Individual cancer cells must differ in their expression of a trait; (2) the differences must be at least partially heritable; and (3) some individuals must survive and reproduce more effectively than others because of these differences. Figure 15.15 illustrates the operation of these conditions in a woman who died of acute myeloid leukemia, a cancer of immune cells. A lineage of hematopoietic stem cell (progenitors of immune cells) acquired mutations: DNMT3A, NPM1, FLT3, PTPRT, and SMC3 (indicated on the left side of the diagram). Heritable mutations arose in one lineage, leading to faster reproduction of the cancerous cells. (The width of each wedge represents relative population size.) Within this lineage, new mutations arose leading to new lineages (dark gray and hatched areas). Other mutations arose in the cluster 3 lineage, leading to yet another new lineage. Chemotherapy functioned to reduce population sizes until about Day 170, and a new mutation arose (the light gray area on the right side of the diagram, including ETV6, WNK1-WAC,MYO18B) before treatment ended. (Adapted from Ding et al. 2012.)

- What happened to the cluster 3 cancer cell lineage (dark gray hatched area on the left side [marked by 5.10%]) after chemotherapy?
- What happened to the cluster 2 cancer cells after chemotherapy?
- Why did the cluster 4 lineage increase so dramatically after chemotherapy?
- Was the cluster 5 lineage resistant to chemotherapy?

Games and Exercises

Is It Antibiotic Resistance?

Suppose you are infected with a bacterial disease. Your sister had this same illness last week and took a full cycle of antibiotics. She quickly became better. You started taking the same antibiotic, but it had no effect. In fact, you had to return to the doctor after a week because you did not feel better. What has happened? Why did you remain sick after taking antibiotics, while your sister quickly recovered?

Three possible hypotheses explain the events:

1. You developed a tolerance for the antibiotic (i.e., you experienced a nongenetic change that made you less sensitive to the effects of the antibiotic).
2. The bacteria infecting you developed a tolerance for the antibiotic (i.e., individual bacteria experienced a nongenetic change that made them less sensitive to the effects of the antibiotic).
3. The bacteria infecting you evolved to be resistant to the antibiotic (i.e., a genetic mutation for resistance occurred in a bacterial cell, it had a reproductive advantage and increased in the population).

- Which hypothesis do you think is most likely? Why?

Now suppose that when you first visited your doctor, she told you that she is conducting research on antibiotics and asked you to be a part of the study. You agreed. As part of the study, you went to the doctor every day and let her take a new sample from your infection, which she then conducted tests on. She discovered that on the first day, your bacteria were susceptible to the antibiotic—the bacteria were killed by the antibiotic. She prescribed the antibiotic for you, which you immediately began taking. Later in the week, the bacteria from your infection were found to be resistant to the antibiotic—the bacteria were not killed by the antibiotic.

- Which of the three hypotheses does this result rule out (1, 2, or 3)? Why?

Another result from the study is that initially, all the bacteria were susceptible to the antibiotic, but by the third day, some of the bacteria were resistant to the antibiotic. With each passing day, more of the bacteria were resistant, until finally all of the bacteria were resistant.

- Does this result support either (or both) of the remaining hypotheses?

- Does it rule out either of them? Why or why not?

Finally, suppose your doctor performs a DNA analysis of the bacteria causing your infection and discovers that the resistant bacteria differ from the susceptible bacteria by a single gene: the gene that encodes the protein on the bacterial cell that is the "target" of the antibiotic (the target site is the place in the bacterial cell where the antibiotic binds and does its dirty work). The resistant bacteria have an altered form of this target site, with the result that the antibiotic is unable to bind to the target site and thus is unable kill the bacterium.

- How did the resistant bacteria come to be different genetically from their susceptible ancestors?

- Which hypothesis does this result support? Why?

- Which hypothesis (1, 2, or 3) do you think is most likely correct?

Evolving Resistance

Scientists collected data on antibiotic use and the evolution of resistance from a community in Finland from 1978 to 1993. They looked at the annual amount of antibiotics used by members of the community and compared that with samples of bacteria from young children with middle ear infections. They examined the bacterial strains to see if they were susceptible to (killed by) or resistant to (not killed by) these antibiotics. The table below lists the year, the amount of antibiotics used, and the percentage of the bacterial strains that were resistant to the antibiotic (from 0 to 100 percent).

Year	Annual Antibiotic Usage	% Resistant Strains
1978	0.84	0
1979	0.92	2
1980	1.04	29
1981	0.98	46
1982	1.02	45
1983	1.03	58
1984	0.95	61
1985	1.12	60
1986	1.06	49
1987	1.14	59
1988	1.21	58
1989	1.28	71
1990	1.32	84
1991	1.31	79
1992	1.27	78
1993	1.28	91

Use this data to make two different graphs: (1) antibiotic usage versus year (graph year on the x-axis and annual antibiotic usage on the y-axis); and (2) percent resistant strains versus year (graph year on the x-axis and percent resistant strains on the y-axis).

- Did the annual antibiotic usage in this community increase, decrease, or stay the same over the course of the study?

- Did the percentage of bacterial strains resistant to these antibiotics increase, decrease, or stay the same over this same time period?

- From these data, do you think antibiotic usage was related to percentage of strains resistant to the antibiotics?

- Do you think one of these factors was the cause of the other factor? If you do, which one?

- How might you be able to tell for sure?

- In another study, also in Finland, researchers investigated the effect of greatly limiting the use of an antibiotic on a community (kids still got ear infections, but they now used a "wait and see" approach to see if the infection would clear up on its own). After antibiotic usage was greatly decreased, the percentage of bacterial strains that were resistant to this antibiotic was decreased by 50 percent (cut in half). What does this new information tell you?

This activity was developed by Dr. Kerry Bright at the University of Montana. If you use this lesson in your classroom, please send along some feedback (brightk@mso.umt.edu).

Explore!

Intelligent Design: Curing Diseases with Darwinian Medicine

Richard Dawkins interviews Randolph Nesse about the genius of evolutionary medicine. Ness coauthored *Why We Get Sick: The New Science of Darwinian Medicine* with George C. Williams). Ness explains some important research on humans, including life-history trade-offs in men and women.

http://www.youtube.com/watch?v=EWldEn8zQ68

TEDx SantaCruz: Rachel Abrams—(R)evolutionary Medicine

TEDx is a program of local, self-organized events that bring people together to share a TED-like experience. Rachel Abrams shares her thoughts on how our evolutionary history shaped us and the medical issues facing our species today.

http://www.youtube.com/watch?v=vUP0yt-6ba4

How Is Darwinian Medicine Useful?

This short article explains how medical scientists are beginning to look at our bodies and our responses, including obesity, anxiety, symptoms, and disease.

http://www.ncbi.nlm.nih.gov/pmc/articles/PMC1071402/

Palo Alto Talks: Evolutionary Medicine

If you want to go deep into evolutionary medicine, these 17 videos take you into the inagural conference for scientists and entepreneurs sponsored by the Stanford University School of Medicine, Department of Neurology and Neurological Sciences. Sessions include disease, cancer, behavior, and mental disorders.

https://www.youtube.com/watch?v=ZIfrBrX-eeo

Big Food: Health, Culture, and the Evolution of Eating

This online exhibit from the Yale Peabody Museum of Natural History includes a video about our evolution and our food. You can play a game to help understand the balance of food intake and energy expenditure.

http://peabody.yale.edu/exhibits/big-food-health-culture-and-evolution-eating

Overcoming Misconceptions

Eugenics Is NOT about Science

Eugenics is the idea that people with "good" genes (the literal translation of "eugenics") should breed, at the expense of people with "bad" genes. Eugenics is a social philosophy; it's not a component of evolutionary theory.

The theory of evolution *explains* our human-ness; it does NOT guide it. In our early history, human races evolved because of the same principles that lead to speciation. People in the same geographic areas intermarried, local adaptations (like height or skin color) were favored, and lineages evolved. However, humans have never been very prone to isolating barriers, and "races" based on artificial classifications, such as height or skin color, are not well defined at all. Ironically, our different social customs often prevented intermarrying—not our evolutionary biology.

In our recent history, evolutionary theory has been co-opted by those who want to push their own frightening agendas. Concepts like racial "purity" and "good breeding stock" are not components of the theory—they were used by members of society for power and control, by those claiming to be superior in some way. But "superiority" is also in the eyes of the beholder, and science has very little to say about such ideas.

Evolutionary Theory Is Important to Practicing Physicians

Although most physicians don't use evolutionary medicine on a day-to-day basis, understanding evolution is critical to their everyday practice. For example, understanding the evolution of resistance is crucial to any physician prescribing antibiotics. Cancer is similarly subject to natural selection, and resistance to cancer drugs can be a problem for some patients. In addition, treatments based on evolutionary principles for autoimmune diseases, such as type 1 diabetes and asthma, may be just around the corner.

Go the Distance: Examine the Primary Literature

Gabriel G. Perron, Michael Zasloff, and Graham Bell examined the evolution of resistance to an antimicrobial peptide in *Escherichia coli* and *Pseudomonas fluorescens.* They were interested in specific types of antimicrobial peptides, those that interfere with the membrane structure of the bacterial cell. Using a selection experiment, Perron and his colleagues were able to show that both bacteria species independently evolved adaptations that conferred resistance to antimicrobial peptides.

- Why did evolutionary biologists hypothesize that the evolution of resistance to these specific types of antimicrobial peptides might be unlikely?

- Why are Perron's results important?

Perron, G. G., M. Zasloff, and G. Bell. 2006. Experimental Evolution of Resistance to an Antimicrobial Peptide. *Proceedings of the Royal Society Series B: Biological Sciences* 273 (1583): 251–56. doi:10.1098/rspb.2005.3301. http://rspb.royalsocietypublishing.org/content/273/1583/251.full.

Delve Deeper

1. How might repeated infections function to enable host shifting in some pathogens?

2. Why do devastating diseases, such as Huntington's disease, continue to plague humans?

3. What are the three conditions necessary for evolution by natural selection to occur?

Test Yourself

1. What two opposing agents of selection influence the virulence of a pathogen?
    a. Selection for high mutation rates and selection against detrimental mutations
    b. Selection for rapid within-host replication and selection for high survival
    c. Selection for within-host replication and selection for between-host transmission
    d. Selection for between-host transmission and selection for between-host survival
2. Why doesn't natural selection "weed out" the *HbS* allele from the human population if it causes sickle cell anemia?
    a. Because the allele keeps reappearing through mutation
    b. Because heterozygous individuals have a fitness advantage over homozygous individuals
    c. Because natural selection can only remove recessive alleles from a population when they are rare
    d. Because sickle cell anemia only affects an individual's ability to survive; people with the allele can still have offspring that carry the allele
    e. Because natural selection is weeding out the allele; it just hasn't completely disappeared yet.
3. What factor(s) related to evolutionary biology likely contributed to the devastating diseases that resulted when members of different populations of humans met over the course of history?
    a. Localized evolutionary arms races led to resistance to some pathogens.
    b. Barriers to gene flow prevented alleles for resistance from spreading among human populations.
    c. Natural selection within hosts favored a shift toward increased virulence.
    d. All of the above
    e. a and b only
    f. b and c only
4. What does the fact that the first antibiotic (Rifampin) given to JH to treat his *Staphylococcus aureus* infection failed to work indicate?
    a. That the strain of *Staphylococcus aureus* JH had become infected with may have already evolved resistance to some antibiotics.
    b. That natural selection had favored strains of *Staphylococcus aureus* within JH that could mutate rapidly.
    c. That JH was re-infected after treatment by a new strain that doctors failed to detect.
    d. That Rifampin is not a functional antibiotic.
    e. That the strain of *Staphylococcus aureus* JH had become infected would eventually evolve resistance to every antibiotic.
5. In their experiment, Stuart Levy and his colleagues found that in just a few short months bacteria infesting chickens became resistant to drugs, and those bacteria were able to spread to chickens that were not exposed to antibiotic treatments (control chickens). What would you predict would happen in the control chickens if Levy and his colleagues eliminated the antibiotics from all the chickens' diets?
    a. The frequency of drug resistant bacteria would not change in the control chickens.
    b. The frequency of drug resistant bacteria in the control chickens would decline.

c. The frequency of drug resistant bacteria in the control chickens would continue to increase.

d. Resistant bacteria from the control chickens would contaminate the experimental chickens.

e. Natural selection would continue to favor drug resistant bacteria in the control chickens.

6. What happened to the cluster 4 cancer cell lineage (orange line in Figure 18.23) after chemotherapy?

a. The lineage survived chemotherapy.

b. The lineage evolved into the red lineage.

c. The lineage split as a result of new mutations.

d. The lineage went extinct.

e. Both a and b

f. Both a and c

7. Will an individual exposed to a mutagenic environmental stimulus always develop cancer?

a. Yes, because the mutagen will cause somatic mutations within the individual permitting cancerous growth.

b. Yes, but only in individuals with alleles that make them more susceptible to cancerous growth.

c. No, because cancerous growth requires multiple mutations that occur simultaneously to both oncogenes and tumor suppressor genes.

d. No, because multiple factors influence the risk of cancerous growth, including the type of tissue exposed, the number of mutations within a cell, and the amount of exposure to the stimulus.

e. Both a and b

f. Both c and d

8. How did Ed Marcotte and his colleagues investigate gene function and their role in human diseases?

a. They determined the number of different organisms that shared gene sequences with human diseases.

b. They examined the new functions of the genes in different lineages to find clues about the selective pressures the organisms faced.

c. They experimentally manipulated orthologous genes from other species to determine gene function and then predicted the function of genes in humans.

d. They isolated clusters of genes in yeast, nematodes, and mice that might be implicated in human diseases because they shared similar structure.

e. Both a and b

f. Both c and d

9. Why are scientists concerned about the H5N1 flu virus?

a. They aren't because H5N1 is not contagious from one person to the next.

b. Because only a few mutations may be necessary for the H5N1 virus to shift hosts

c. Because H5N1 is especially virulent

d. Because H5N1 contains only 10 genes on eight segments of RNA, and it can replicate quickly

e. They aren't. H5N1 only infects birds.

10. Which of these statements about vaccines is TRUE?
    a. Vaccines against flu viruses are usually ineffective because viruses have high mutation rates.
    b. Vaccines developed using cells from monkeys or chimpanzees rarely work to prevent disease in humans.
    c. Vaccines can be developed by altering the selective environment that a virus is adapted to.
    d. Vaccines rarely function to prevent disease because viruses can adapt so quickly.
11. What factors associated with the evolutionary biology of bacteria facilitate the evolution of resistance to antibiotics?
    a. High mutation rates
    b. Horizontal gene transfer
    c. Sub-lethal doses of antibiotics
    d. All of the above facilitate the evolution of resistance.
    e. None of the above facilitates the evolution of resistance.

Contemplate

What would you predict about the evolution of antibiotic resistance if antibiotic drugs can pass through our bodies and into our urine?	Besides diet and exercise, where might scientists turn to combat obesity? Why?

Answers

1 The Whale and the Virus

How Scientists Study Evolution

Check Your Understanding

1. Because evolution is a theory, that means:
    a. It is a guess or a hunch.

       Incorrect. Many people use the word theory in place of "guess" of "hunch," but to scientists, the word theory carries a lot of weight. Evolutionary theory is well supported by evidence and years and years of examination, experimentation, and testing. *Evolution: Making Sense of Life* will introduce you to some of that evidence.
    b. It has very little evidence to support it.

       Incorrect. There is much evidence to support the theory of evolution from Darwin's own research to the development of the Modern Synthesis incorporating evidence from diverse fields such as geology, genetics, and geography, among others. *Evolution: Making Sense of Life* will introduce you to some of that evidence.
    c. Scientists may or may not believe in it.

       Incorrect. Science is not about "believing" in theories; it moves forward based on the weight of evidence, and evolutionary theory is well supported by evidence and years and years of examination, experimentation, and testing. *Evolution: Making Sense of Life* will introduce you to some of that evidence.
    d. It has never been observed.

       Incorrect. Scientists have been able to observe evolution directly, both in the lab and in nature. Theories in science do not require direct observations—they are built on evidence that comes from myriad sources. Evolutionary theory is well supported by evidence from the natural world and years and years of examination, experimentation, and testing. *Evolution: Making Sense of Life* will introduce you to the evidence, as well as the historical development of the theory.
    e. None of the above

       Correct. Because evolution is a theory, like Einstein's General Theory of Relativity, it functions as a set of overarching mechanisms and principles that explain a major aspect of the natural world utilizing natural rather than supernatural phenomenon to explain the process. Evolutionary theory is well supported by evidence and years and years of examination, experimentation, and testing; it explains the diversity of life on Earth. *Evolution: Making Sense of Life* will introduce you to some of that evidence.
2. Why is understanding evolution important?
    a. Understanding evolution can help us understand biodiversity issues associated with deforestation and global warming.

       Correct, but so are other answers. The world's biodiversity faces many threats. *Evolution: Making Sense of Life* will show how, for example, studying past mass extinctions can help us understand the effects our actions may have on the world's biodiversity.

b. Understanding evolution can help us understand the evolution of antibiotic resistance and cancer.

   Correct, but so are other answers. Through the process of evolution, some dangerous bacteria have become resistant to even the most powerful antibiotics. *Evolution: Making Sense of Life* will explain how through observations of the evolution of antibiotic resistance in laboratory experiments, scientists may be able to offer solutions to help slow resistance in nature.

c. Understanding evolution can help us understand our own genetic makeup and how it affects our lives.

   Correct, but so are other answers. Our species, and our DNA is a product of the evolutionary process. *Evolution: Making Sense of Life* will clarify how understanding evolution may provide answers to what it means to be human, how our DNA interacts with our environment, and what ancient segments lurk in our genomes.

d. All of the above

   Correct. Understanding evolution can help us these issues and many others understand biodiversity issues facing our society. It can help us understand the evolution of resistance to antibiotics, and it can help us understand how our DNA makes us human. *Evolution: Making Sense of Life* will introduce you to the evolutionary theory, the evidence, experiments, and critical examination that have led to our current understanding, and the importance of understanding evolution to our future on this planet.

3. Which of the following statements about evolution is true?

   a. Once biologists find all the missing links, they will be able to understand evolution.

      Incorrect. Biologists do not need to find all the "missing links" before they can understand evolution. Missing "links" is a bit of a misnomer. It implies that the fossil record is a complete record that scientists have yet to uncover—once they find all the fossils, they will have a record of every direct ancestor of every species. The fossil record will *always* be incomplete, but scientists have discovered some key fossils that can be broadly considered as missing links. *Tiktaalik* (Chapter 4) is a fossil that "links" fish and amphibians—not through direct ancestry, but by providing important evidence about their common ancestry. The fossil record is only one source of evidence evolutionary biologists use to examine the historical relationships of organisms over time and the processes that affect genetic lineages. This evidence comes from disciplines as diverse as molecular biology and biogeography. *Evolution: Making Sense of Life* will explain some of the difficulties with the fossil record and the volumes of evidence scientists use to draw conclusions about evolutionary relationships and our historical past.

   b. Evolution is a process that leads to more and more complex organisms.

      Incorrect. Complex traits can certainly evolve from simpler traits, but evolution does not necessarily progress from simple to complex. Evolution results because populations change over time, and that change often results from natural selection acting to favor traits that are more effective for the survival and reproduction of individuals relative to other traits in that population. The traits natural selection favors, adaptations, can just as easily be less "complex" as more "complex." *Evolution: Making Sense of Life* will introduce you to adaptations and the evidence scientists use to unravel their evolutionary history, as well as some of the more elegant examples of lineages that lost their complexity over time.

c. Evolution is entirely random.

Incorrect. Mutations may occur at random, but evolution involves much more than just mutations. A mutation that arises by chance is subject to both random and non-random processes. If that mutation affects the reproductive success of the individuals that have it, non-random processes, such as natural selection and sexual selection, can have striking effects. In fact, the non-random nature of evolution is quite apparent in animals that occupy the same ecological niche in the same physical environment. Because these animals face similar selection pressures, populations often converge on amazingly similar forms—think about whales and fishes. *Evolution: Making Sense of Life* presents some of the random and very non-random processes that affect the likelihood that a mutation will spread or disappear from a population.

d. Some forms of life are higher on the ladder than other forms of life.

Incorrect. It's tempting to think about the evolution of organisms as a ladder from lower to higher forms, but in fact the relationships among organisms are like a highly branched bush. Every organism can boast of adaptations to its environment. As humans, we tend to think of ourselves as the pinnacle of evolution with our complex brains capable of abstract thought and language, but other organisms have complex adaptations that humans lack. Dolphins communicate with other dolphins, insects see complex patterns in flowers, and plants produce their own food through photosynthesis. These adaptations could just as easily be considered "higher" forms of life than humans. *Evolution: Making Sense of Life* explores the diversity of life and some of the remarkable adaptations found in that diversity.

e. None of the above is a true statement.

Correct. Evolutionary biologists understand evolution even though the fossil record is incomplete because the theory of evolution is based on volumes of evidence from a diversity of sources. Evolutionary biologists are interested in the historical relationships that result from the random and non-random processes that affect genetic lineages. *Evolution: Making Sense of Life* will introduce you to the theory of evolution, the evidence, experiments, and critical examination that have led to our current understanding, and the importance of understanding evolution to our future on this planet.

Identify Key Terms

1. c; 2. f; 3. e; 4. g; 5. d; 6. b; 7. j; 8. h; 9. i; 10. a

Link Concepts

- evolution
- genetic variation
- genetic drift
- mutations
- natural selection
- phenotypes
- scientific theory
- viral reassortment

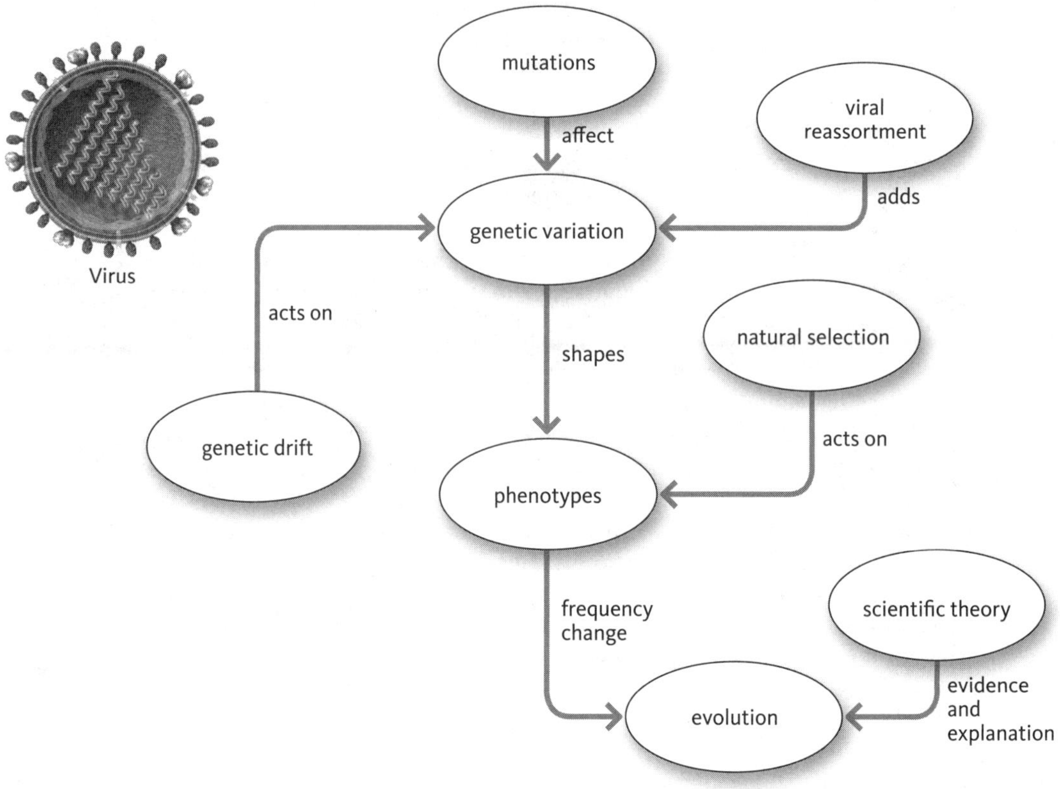

- evolution
- genetic drift
- genetic variation
- mutations
- natural selection
- phenotypes
- scientific theory

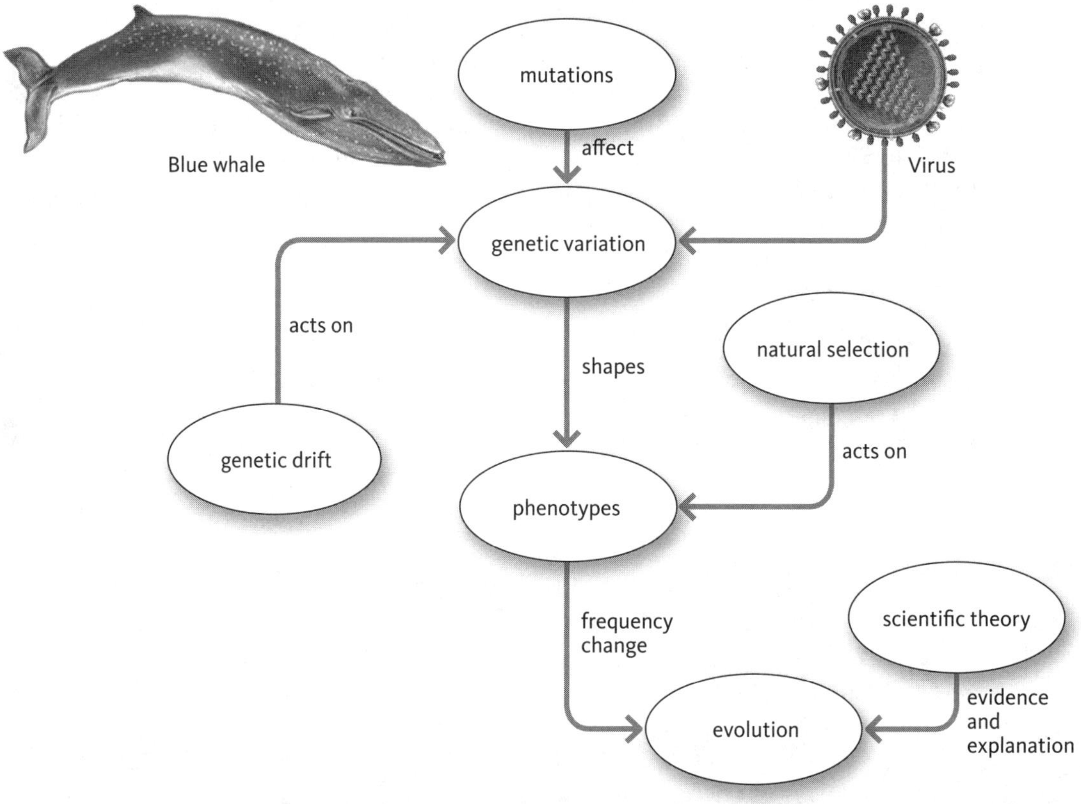

Interpret the Data

- When did the diversity of baleen whales (Mysticetes) peak?

  During the Serravallian about 12 million years ago (see the systems line) and then again in the Zanclean around 4 million years ago
- When did the diversity of toothed whales (Odontocetes) peak?

  During the Tortonian around 10 million years ago and then again in the Zanclean around 4 million years ago
- When was the diversity of diatoms high?

  During the Tortonian around 10 million years ago, again in the Zanclean around 4 million years ago, and in the Holocene/Pleistocene (present stage)
- From this graph, what might you predict about the diversity of toothed whales (Odontocetes) in the future? What is your prediction based on?

  The diversity of dolphins (odontocetes) will increase. The prediction is based on the correlation between odontocetes and diatom diversity in the past and the current increase in both odontocetes and diatom diversity in the Holocene/Pleistocene.
- Can you predict the same about the diversity of baleen whales (Mysticetes)? Why or why not?

  You could, but the graph indicates a continuous decline in Mysticetes diversity despite the increase in diatom diversity in the Holocene/Pleistocene (present stage). So even though Mysticetes diversity may increase in the future, the evidence from fossil diversity does not support the prediction.

Overcoming Misconceptions

Which of the following is a true statement?

b. Having bigger brains gave both advantages and disadvantages to human ancestors.

Delve Deeper

1. How do scientists use evidence to explain events that happened in the past?

   Scientists can use observations of the present as indirect evidence of events that occurred in the past. For example, they can examine skeletons of living organisms and how those animals move in their environments and infer that ancient animals must have used structural elements similarly.
2. Does evolution produce a peaceful balance in the natural world?

   No. Evolution does not produce a peaceful balance, nor does it strive to. The influenza virus evolves quickly, and with the right combination of mutation and reassortment, it can wipe out huge numbers of individuals. Even more benign strains attack the young, the elderly, and the weak. In addition, predators hunt and kill prey, and parasites devour their hosts from the inside out. Survival and reproduction are part of the natural process of evolution.

Test Yourself

1. a; 2. c; 3. b; 4. c; 5. b; 6. c; 7. b; 8. e; 9. a

2 From Natural Philosophy to Darwin

A Brief History of Evolutionary Ideas

Check Your Understanding

1. How would you define biological evolution?

   a. A gradual process in which something changes into a different and usually more complex or better form

      Incorrect. Evolution does not have a purpose; it does not necessarily move from simple to complex or from worse to better forms. Misconceptions about what evolution is or how it acts are pretty common, so each chapter of the study guide includes an Overcoming Misconceptions section designed to help you think understand evolutionary theory.

   b. Any change in the frequency of heritable traits within a population from one generation to the next

      Correct. Chapter 1 introduces the concept of biological evolution and its effects on organisms as small as viruses and as big as blue whales.

   c. A process of slow, progressive change

      Incorrect. Biological evolution is not necessarily slow, let alone progressive. Evolution can occur quite rapidly (within months in viruses and other short-lived organisms; see Section 1.2), and it does not have a purpose. Misconceptions about what evolution is or how it acts are pretty common, so each chapter of the study guide includes an Overcoming Misconceptions section that explains the way that scientists understand evolutionary theory.

   d. Any kind of change over time

      Incorrect. Change over time does not necessarily reflect the heritable changes that are necessary for biological evolution. A whale can grow and develop, changing over time from a small, young calf to a large adult, but this change occurs within an individual's lifespan. Biological evolution is the change in the frequency of heritable traits within a population from one generation to the next, so that a trait, such as peg-shaped teeth, may become more or less frequent within the population depending on the selective pressures affecting that population (Chapter 1).

   e. A theory that humans have their origin in other types of animals, such as apes, and that the distinguishable differences between humans and apes are due to modifications in successive generations

      Incorrect. Biological evolution certainly can explain the historical relationship between humans and apes, but that is not its only implication. Biological evolution affects every living organism. In an essay entitled "Nothing in Biology Makes Sense Except in the Light of Evolution" (Dobzhansky 1973), Theodosius Dobzhansky emphasized the important role evolution plays in understanding everything from fossils to the diversity of life. See Chapter 1.

2. Why are mutations important in evolution?

   a. Because mutations are always deleterious, and organisms with these deleterious mutations do not survive and reproduce

      Incorrect. Mutations can often be deleterious, but they can also be beneficial, or even neutral. Mutations provide the variation among individuals on which the mechanisms of evolution can act. Beneficial mutations may lead to greater survival or reproduction through their effects on the characteristics, or phenotype, of an organism. See Chapter 1.

b. Because mutations only occur in viruses

   Incorrect. Mutations can occur in any organisms during the process of DNA (or RNA) replication or when these molecules are damaged. Because viruses reproduce at such a rapid rate, lineages can evolve rapidly, and this evolution can be observed both in natural settings (e.g., with flu viruses) and in the lab. So, the mutations that occur in viruses do offer important support for evolutionary theory, especially as they relate to human health and welfare. See Chapter 1.

c. Because mutations are random, and evolution is a random process

   Incorrect. Mutations are random, but evolution is not necessarily a random process. In fact, natural selection is decidedly *not* random. When a random mutation does arise in a particular individual, that mutation may have beneficial, harmful, or even neutral effects. Natural selection acts on mutations that affect the success or failure of the individual—those individuals with mutations that increase survival or reproduction will do better than those individuals without those mutations. Likewise, those individuals with mutations that decrease survival or reproduction will do worse than those individuals without those mutations. This non-randomness is clear in adaptations that have converged on the same form, such as the streamlined shape of whales and fishes. Because these animals occupy similar ecological niches in the same physical environment, they face some of the same pressures. The lineages evolved similar forms as a result of natural selection operating on different random mutations. See Chapter 1.

d. Because mutations create the variation among individuals on which other mechanisms of evolution can act

   Correct. Mutations affect the genetic sequences that generate an organism's phenotype. So, even when the effect on the phenotype may not be readily apparent to evolutionary biologists, mutations can lead to genetic variation among phenotypes—the raw material for evolution through both natural selection and genetic drift (see Chapter 1).

3. What kinds of evidence have scientists used to study the evolution of whales?

   a. Evidence from DNA

      Correct, but so are other answers. In fact, DNA from living mammals provided evidence that whales were related to artiodactyls. Scientists compared snippets of DNA to determine patterns and understand historical relationships. These patterns led to the hypothesis that whales were most closely related to hippos—a hypothesis that could be tested with new fossil evidence. See Chapter 1.

   b. Evidence from fossils, such as their anatomical traits

      Correct, but so are other answers. Although Darwin proposed that whales descended from land mammals, very few fossil whales had been discovered to test that hypothesis. Darwin based his hypothesis on evidence from living animals—their similarities (e.g., milk production) and differences (e.g., gills). As more and more fossil whales were discovered, scientists tested Darwin's hypothesis about whale evolution, and began understanding how whales evolved from land mammals. See Chapter 1.

   c. Evidence from geology

      Correct, but so are other answers. Geology provides important information about the origin and ages of different rocks. Evolutionary biologists can use this information to understand what kind of habitats the fossils formed in. For example, *Pakicetus* fossils were found in rocks that formed from the sediments of shallow streams that flowed seasonally through hot, dry landscapes—not in rocks formed in ocean habitats. *Pakicetus* probably lived on land, even though it has traits shared by modern whales. See Chapter 1.

d. Evidence from living species

Correct, but so are other answers. Evolutionary biologists often use evidence from living species to understand the past. Clues can come from many sources, from how organisms behave to their DNA. Patterns of DNA from cows, goats, camels, and hippos strongly suggested that cetaceans were indeed related to artiodactyls—they shared a common ancestor. And that common ancestor most closely linked whales and hippos. So using living species, scientists could develop a hypothesis about similarities between fossil whales and hippos, and they were able to go out and test that prediction with new evidence from fossil whales. See Chapter 1.

e. All of the above

Correct. Scientists have used DNA, fossils, geology, and living species to examine the evolution of whale lineages (see Chapter 1). This evidence for evolution is presented throughout *Making Sense of Life.*

Identify Key Terms

1. c; 2. e; 3. g; 4. k; 5. h; 6. f; 7. d; 8. j; 9. b; 10. a; 11. i; 12. l

Link Concepts

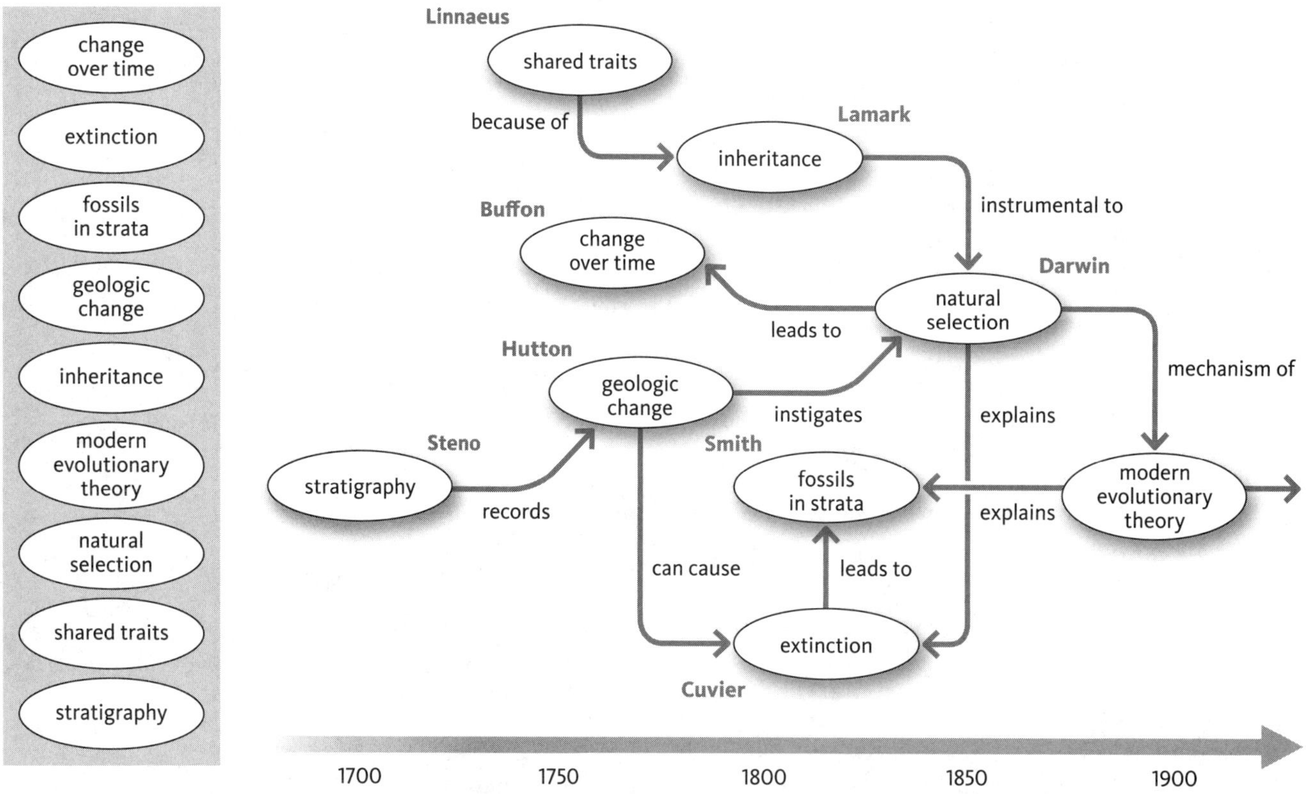

Interpret the Data

- At which of the stage of human development pictured does the pattern of blood vessels look most similar to the pattern found in sharks?

  29 days

- If scientists were only to examine human embryos at 56 days, would the homologies be apparent?

  No. By 56 days, the course of human development has altered the arrangement of blood vessels so that the looping pattern apparent early in development is no longer evident.

- What does this homology indicate about the relationship between fishes and humans?

  The fact that fishes and humans share this important homology indicates that we share a common ancestor.

Overcoming Misconceptions

Which of the following is a misconception about evolution?

b. Evolution is entirely random.

Delve Deeper

1. What is the difference between how scientists use the word theory and how it is used in everyday language?

   Scientists use the word "theory" to describe an overarching set of principles that explain how the natural world works. The concept is very powerful and at the root of most scientific work. Scientists use theories to develop hypotheses and make predictions. Theories organize the way scientists think.

   In everyday language, the concept is much less powerful—the word generally refers to a guess or a hunch. Television crime dramas may use the word to summarize evidence that supports a crime—the theory of the crime—however, that use is still very different from its use in science.

2. Consider the following discussion question and fill in the boxes.

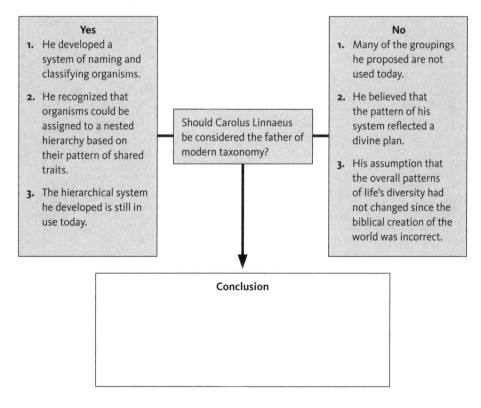

Should Carolus Linnaeus be considered the father of modern taxonomy?

Yes

1. He developed a system of naming and classifying organisms.
2. He recognized that organisms could be assigned to a nested hierarchy based on their pattern of shared traits.
3. The hierarchical system he developed is still in use today.

No

1. Many of the groupings he proposed are not used today.
2. He believed that the pattern of his system reflected a divine plan.
3. His assumption that the overall patterns of life's diversity had not changed since the biblical creation of the world was incorrect.

Conclusion

Test Yourself

1. e; 2. b; 3. a; 4. c; 5. d; 6. d; 7. d

3 What the Rocks Say

How Geology and Paleontology Reveal the History of Life

Check Your Understanding

1. Why is the study of stratigraphy important to understanding evolutionary theory?

   a. Because stratigraphy provides evidence for the relative age of fossils

   Correct. Stratigraphy is the study of layering in rock (stratification) as a method for reconstructing the past. It was first developed by Nicolaus Steno in the mid-1600s, before the development of evolutionary theory. Stratigraphy was instrumental in helping scientists examine change over time (see Chapter 2).

   b. Because Nicolaus Steno recognized that rocks occurred in layers even though he didn't believe in evolution

   Incorrect. Nicolaus Steno did develop his ideas about stratigraphy very early on in the development of evolutionary theory, but his beliefs about evolution did not affect his conceptual understanding of stratification. He is considered the father of stratigraphy because he recognized that layers of rock were laid down in succession with the oldest layers at the bottom and the youngest ones at the top (see Section 2.1). He also recognized that fossils were the remains of living organisms.

   c. Because stratigraphy characterizes the process of fossilization

   Incorrect. Stratigraphy characterizes layers of rock (see Section 2.1), what the layers are and how they are temporally related to each other, not how fossils appear within those layers. However, fossils can occur within layers of rock, trapped by sedimentation, for example, and stratigraphy can provide evidence for the ages of those fossils (you will learn about fossilization in Section 3.4).

   d. All of the above

   Incorrect. Nicolaus Steno developed his ideas about stratigraphy in the mid-1600s, but his beliefs about evolution did not affect his conceptual understanding of stratification. Nor does stratigraphy characterize the process of fossilization; it can provide evidence for fossil ages, however. Stratigraphy is the study of layering in rock (stratification) as a method for reconstructing the past (see Section 2.1).

   e. None of the above

   Incorrect. Stratigraphy is the study of layering in rock (stratification) as a method for reconstructing the past. It was first developed by Nicolaus Steno in the mid-1600s, before the development of evolutionary theory. Stratigraphy was instrumental in helping scientists examine change over time (see Chapter 2).

2. What was Georges Cuvier's contribution to evolutionary theory?

   a. He strongly objected to Jean-Baptiste Lamarck's ideas about life evolving from simple to complex, drumming Lamarck out of the scientific establishment.

   Incorrect. Cuvier certainly disagreed with Lamarck's theory, but he did not destroy the career of a scientist with whom he disagreed. Scientists regularly disagree, often until the balance of evidence weighs in favor of one theory or another. Cuvier's contributions to evolutionary theory come from his work on comparative anatomy and the concept of extinction (see Section 2.2).

b. He invented paleontology.

Incorrect. The science of paleontology was not invented but developed over years as scientists began to uncover and study fossils. Cuvier contributed significantly to the science of paleontology through his exhaustive anatomical studies of fossils. However, many individuals were contributing to the science of paleontology prior to Cuvier's work.

c. He recognized that some fossils were both similar to and distinct from living species, and many fossil animals no longer existed.

Correct. Cuvier studied the fossil remains of elephants and compared them to living elephants. He discovered that some characters of the fossils, such as the shapes of teeth, were distinct from living elephants, and provided some of the first compelling evidence for extinction (see Chapter 2).

d. He was the first to organize and map strata according to a geological history.

Incorrect. William Smith discovered that layers of rocks contained distinctive groups of fossils, and he was able to organize strata into a geological history as a result. Cuvier used the system developed by Smith to map strata in other parts of the world (see Section 2.2).

e. He rejected the idea that species evolved, instead believing that life's history was a series of appearances and extinctions of species.

Incorrect. Cuvier believed that life's history was a series of appearances and extinctions of species, rather than accepting the concept of evolution. His contributions were significant to the developing field of paleontology, however (see Section 2.2).

3. What is a scientific theory?

a. A belief that scientists try to prove as fact

Incorrect. Scientists try to avoid letting their beliefs interfere with their work. More importantly, science is not about proving anything as fact. Science is about testing theories, weight of evidence, and ultimately understanding and explanation. Facts are simply consistent observations. Evolution is a testable theory that provides an overarching set of mechanisms or principles that explain a major aspect of the natural world.

b. A set of laws that define the natural world

Incorrect. Laws can be important in science, but laws simply define relationships, such as $E = mc^2$. In fact, laws can be essential components of scientific theories. Theories are an overarching set of mechanisms or principles that explain major aspects of the natural world. Darwin contributed the first overarching set of mechanisms for the theory of evolution as natural selection and sexual selection (see Section 2.3).

c. A set of mechanisms or principles that explain a major aspect of the natural world

Correct. A scientific theory is different from the word theory used in everyday language. Scientific theories are not just guesses; they are overarching sets of mechanisms or principles that explain and provide testable predictions. Darwin contributed the first overarching set of mechanisms for the theory of evolution as natural selection and sexual selection (see Section 2.3).

d. A guess based on a few facts that scientists try to prove as correct

Incorrect. Although in everyday language, a theory is often considered a guess, in science a theory is not a guess. A theory, such as evolution, is an overarching set of mechanisms or principles that explain a major aspect of the natural world. More importantly, science is not about facts or proving ideas as correct. Science is about testing theories, weight of evidence, and ultimately understanding and explanation.

e. An educated guess based on some experience that allows scientists to test evidence

Incorrect. The idea of an educated guess is closer to the concept of a hypothesis—a tentative explanation grounded in some evidence. Many theories can develop from hypotheses as the overarching mechanisms and principles that guide the explanation and offer predictions are developed and tested. Evolutionary theory initially developed from early hypotheses about fossils and their geologic relationships; Darwin contributed the first overarching set of mechanisms as natural selection and sexual selection (see Section 2.3).

Identify Key Terms

1. q; 2. e; 3. a; 4. g; 5. k; 6. m; 7. f; 8. o; 9. l; 10. n; 11. d; 12. i; 13. c; 14. h; 15. p; 16. b; 17. j

Interpret the Data

- Which species had the lowest ratio of ^{13}C to ^{12}C, and when did it live?

  Australopithecus anamensis. 4 million years ago
- What is the range of $\delta^{13}C$ values for the genus *Homo* (our genus)?

  –9 to –5
- If the $\delta^{13}C$ values generally range between –24‰ and –32‰ for plants using the C3 pathway, and between –10‰ and –14‰ for plants using the C4 pathway, what can you surmise about the diets of species with low and high ratios?

  Species with low $\delta^{13}C$ values likely ate more plants that used the C3 pathway, and species with high $\delta^{13}C$ values likely ate more plants that used the C4 pathway.
- What can you determine about the diet of our genus?

  Our genus, *Homo*, likely ate more plants that used the C4 pathway.

Overcoming Misconceptions

Which of the following is a true statement?

c. Scientists use independent lines of evidence to validate results.

Delve Deeper

1. Why do scientists consider radiometric dating a valid way to measure the age of rocks?

   Scientists often spend a lot of time critically examining other scientists' evidence, and radiometric dating has been used as a tool to estimate age for a very long time. They've tested and retested radiometric dating and compared ages derived from radiometric dating with other methods for dating and found very consistent results.
2. The beginning of the Cambrian period, 542 million years ago, is often referred to as the Cambrian Explosion. The name is appropriate in some ways, and inappropriate in others. Why is this a fitting name for this time? Why isn't the term *explosion* a good descriptor?

   The Cambrian Explosion was often called that because of the first appearance of so many living groups, seemingly from nowhere. The "explosion" actually lasted millions and millions of years—about 13 million. The lineage that gave rise to vertebrates, including humans, first appeared in the fossil record during the Cambrian period. Also, as more evidence was collected about life in the pre-Cambrian, such as the Ediacaran fauna, the rise of new groups did not seem quite as explosive as previously thought, however.

3. In the space provided, indicate whether you think each statement is a Hypothesis (H), a Prediction (P), or an Observation (O).

C_4 plants have higher levels of carbon-13 than both living and extinct C_3 plants. O

The ratio of carbon isotopes in extant animals reflects the kinds of animals and plants that they eat. O or H

Fossil animals will show the same relationship between the ratios of carbon isotopes and diet as living animals. P

Hominin fossils in East Africa dating from 4.2 million years ago have relatively low ratio of carbon-13 to carbon-12. O

The relatively low ratio of carbon-13 to carbon-12 in 4.2 million-year-old hominin fossils indicates a diet rich in C_3 plants. H

The relatively low ratio of carbon-13 to carbon-12 in hominin fossils is similar to the ratio found today in the teeth of chimpanzees. O

Chimpanzees feed on fruits and leaves. O

Plant fossils from the same sites where the teeth of 4.2 million-year-old hominin fossils were discovered will contain C_4 plants, consistent with early hominins living in grassy woodlands where they easily could have found C_4 plants. P

4.2 million-year-old hominin fossils were actively selecting C_3 plants for their diet. H

Test Yourself

1. a; 2. a; 3. c; 4. b; 5. d; 6. d; 7. a; 8. d

4 The Tree of Life

How Biologists Use Phylogeny to Reconstruct the Deep Past

Check Your Understanding

1. How does homology relate to the theory of evolution?

   a. Homology refers to traits that only superficially look alike, like bat wings and human arms; the theory of evolution cannot explain these similarities.

   Incorrect. Homology refers to traits that are similar in two or more species because the traits were inherited from a common ancestor with those traits. Similarities may be "beneath the surface," like bat wings and human arms—the underlying structure is similar in the different species despite overt differences in appearance. Darwin, Cuvier, and many scientists before them, recognized homology as a basis for organization for comparative biology—that related organisms often had similar morphologies. The theory of evolution provides a mechanism for those observations—descent with modification (see Chapter 2).

   b. Homology refers to traits that look alike but have entirely different origins, like the fins of dolphins and of fish; the theory of evolution cannot explain the origins.

   Incorrect. Homology refers to traits that are similar in two or more species because they were inherited from a common ancestor with those traits. Traits that look alike but have different origins are said to be analogous. Darwin, Cuvier, and many scientists before them (see Chapter 2), recognized homology as a basis for organization for comparative biology—that related organisms often had similar morphologies. The theory of evolution provides a mechanism for those observations—descent with modification.

c. Homology refers to traits that are structurally similar in different organisms, like bat wings and human arms, because they each were inherited from a shared common ancestor with those traits; the theory of evolution provides a mechanism for those observations.

   Correct. Darwin, and many scientists before him, recognized homology as a basis for organization for comparative biology—that related organisms often had similar morphologies. The fact that many homologies are found together in the same groups of species strengthens support for Darwin's idea of descent with modification (see Chapter 2).

d. Homology refers to traits that have converged on a shared form; the theory of evolution provides a mechanism for those observations.

   Incorrect. Homology refers to traits that are similar in two or more species because they were inherited from a common ancestor with those traits—like bat wings and human arms—where the underlying structure is similar in the different species. Analogous traits are traits that have converged on a shared form, like the fins of dolphins and fish. Darwin, Cuvier, and many scientists before them, recognized homology as a basis for organization for comparative biology—that related organisms often had similar morphologies, and the theory of evolution provides a mechanism for those observations—descent with modification (see Chapter 2).

2. How accurate is radiometric dating?

   a. Not very accurate because scientists cannot know how much of a particular isotope was originally present in a rock, and thus, they cannot know how much of it has decayed

      Incorrect. It's true that scientists were not around when a rock formed and cannot know how much of a particular isotope was originally present, but they can examine different minerals within a rock and compare the ratios of the different isotopes using fairly straightforward mathematical formulas to determine the rock's age (see Chapter 3).

   b. Very accurate because scientists can determine the exact age of rocks and the fossils found within them

      Incorrect. Although radiometric dating is a valuable tool, it cannot be used to determine the *exact* ages of rocks. Radiometric dating provides estimates of age based on the half-life of the isotopes within the rocks. Different isotopes can be used to measure different time scales, and the time scale being addressed can affect the precision of the estimate. Isotopes with very long half-lives can be used to determine the ages of very old rocks with relative accuracy—say a 3 million year range for stromatolite fossils dating back 3.4 *billion* years. Whereas isotopes with shorter half-lives can provide more precise estimates for younger rocks—for example, a margin of error of 7000 years (see Chapter 3).

   c. Accurate because radiometric dating can determine estimates for the ages of rocks and fossils, often with relatively small margins of error

      Correct. Radiometric dating provides estimates of age based on the half-life of the isotopes within the rocks. Different isotopes can be used to measure different time scales, and the time scale being addressed can affect the precision of the estimate. Isotopes with very long half-lives, can be used to determine the ages of very old rocks with relative accuracy—say a 3 million year range for stromatolite fossils dating back 3.4 *billion* years. Whereas isotopes with shorter half-lives can provide more precise estimates for younger rocks—for example, a margin of error of 7000 years. Scientists can also use different elements to check and hone their estimates (see Chapter 3).

d. Not very accurate because scientists use probabilities to determine decay rates for each isotope they use in radiometric dating

Incorrect. Scientists definitely use probabilities to determine decay rates, but the use of statistical techniques does not negate the accuracy of radiometric dating as a methodology. Statistics validate and can provide insight to the accuracy and precision of measures, such as the age of rocks. Scientists accept that *no measure* is completely accurate, and often use several lines of evidence to improve their estimates. See Chapter 3 for a discussion of the role of different isotopes in estimated the ages of rocks.

3. Besides radiometric dating, what other lines of evidence can be used to determine the ages of fossils?

a. Stratigraphy

Correct. Stratigraphy can be used to determine the relative ages of fossils from strata that are widespread and can be correlated in time. With the advent of radiometric dating, more precise estimates of age can be applied to fossils (see Chapter 3).

b. Homology

Incorrect. Homology can be used to identify fossils that share a common ancestor, but it cannot be used to determine the ages of fossils. See Chapter 1 for an introduction to homology, and Chapter 2 to learn how homology is important to common descent.

c. Complexity

Incorrect. Although some early scientists suggested that life developed from simple to complex (see Chapter 2), complexity of fossils is not appropriate to determine the age of fossils—even the relative age of fossils. Evolution has not been a "march of progress" (see Chapter 2 Common Misconceptions), and the development of complexity is not necessarily related to age.

d. a and b only

Incorrect. Stratigraphy can be used to determine the relative ages of fossils from strata that are widespread and can be correlated in time, but homology can only identify fossils that share a common ancestor (see Chapter 1 and Chapter 2).

e. All of the above

Incorrect. Stratigraphy can be used to determine the relative ages of fossils from strata that are widespread and can be correlated in time, but homology can only identify fossils that share a common ancestor (see Chapter 2). Evolution has not been a "march of progress" (see Chapter 2 Common Misconceptions), and the development of complexity is not necessarily related to age.

Identify Key Terms

1. k; 2. g; 3. a; 4. d; 5. e; 6. s; 7. n; 8. q; 9. i; 10. b; 11. j; 12. f; 13. l; 14. h; 15. o; 16. t; 17. p; 18. m; 19. r; 20. c

Link Concepts

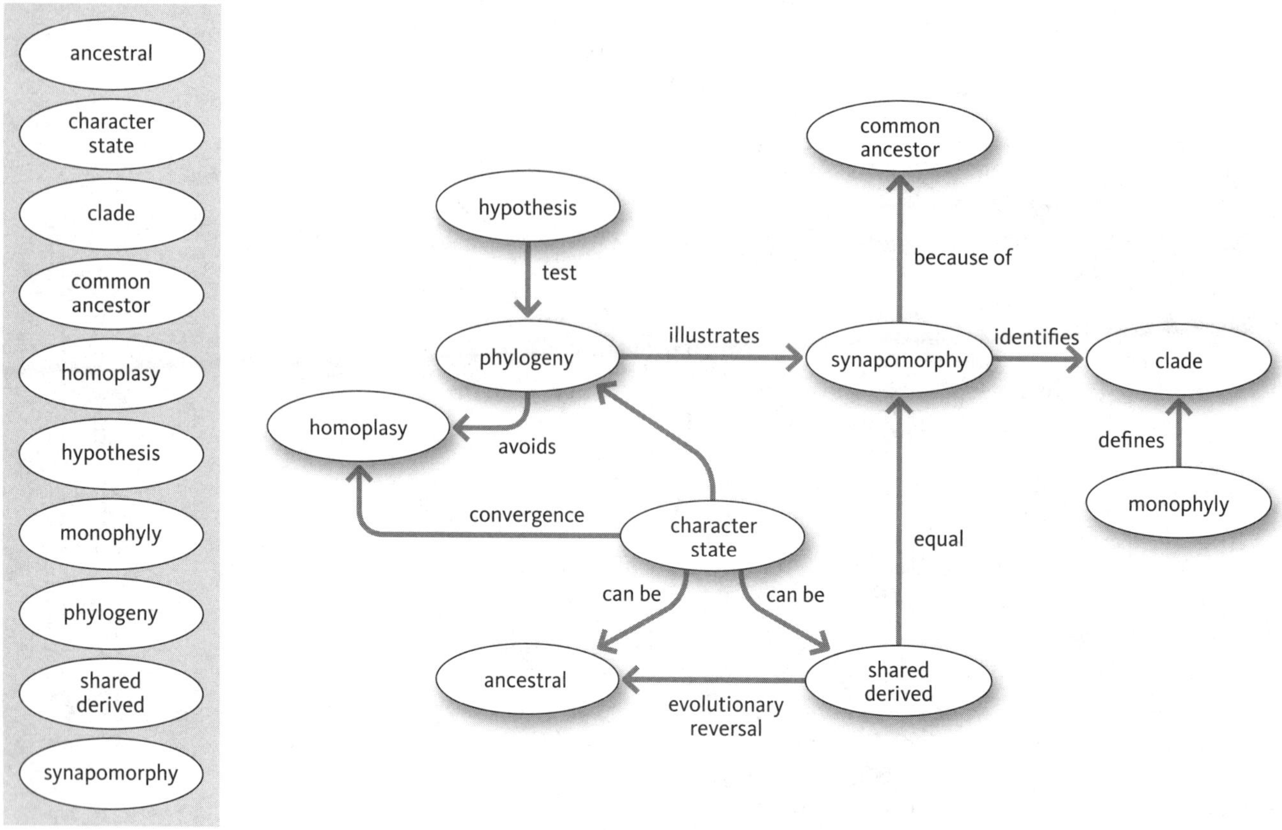

Interpret the Data

- Explain why living birds are considered dinosaurs.

  Because they are members of the theropod clade and share numerous synapomorphies, such as feathers
- What trait(s) distinguish living birds from theropod dinosaurs?

  Toothless beaks, fused wing digits, and short feathered tails
- Which branches of theropods have feathers?

  Compsognathids, Tyrannosauroids, Oviraptorosaurs, Dromeosaurids, Archaeopteryx, living birds
- Which branches may have been able to fly?

  Dromeosaurids, Archaeopteryx

Games and Exercises

- Are the Euphasmatodea a monophyletic clade?

  Yes, the Euphasmatodea are depicted as a monophyletic clade—the branch consists of the common ancestor and all the descendent taxa. However, the cladogram indicates that there is no clear resolution of the extant taxa within the Euphasmatodea.

• Did the addition of *Eophyllium* clarify the resolution of the Euphasmatodea?

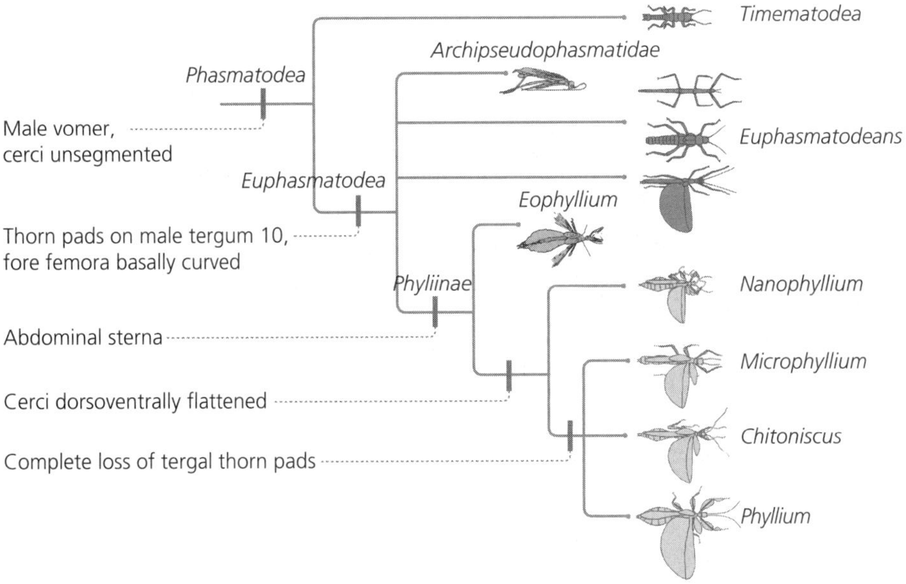

No. *Eophyllium* has abdominal sterna, so it is considered a member of the Phylliinae, leaf insects that exhibit an extreme form of morphological and behavioral leaf mimicry. Since it doesn't have dorsoventrally flattened cerci though, *Eophyllium* must have arisen before the *Nanophyllium, Microphyllium, Chitoniscus,* and *Phyllium* clade split off. *Archipseudophasmatidae,* however, must be a member of the Euphasmatodea because it lacks abdominal sterna. However, this extinct group provides little resolution for the extant Euphasmatodea.

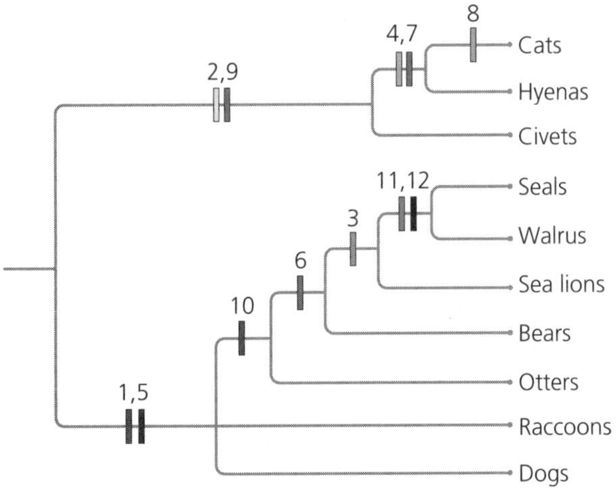

Overcoming Misconceptions

Which of the following is a true statement?

d Scientists are not looking for missing links.

Delve Deeper

1. Why are phylogenies based on shared derived character? What does including an outgroup do for understanding those relationships?

   Shared derived characters, or synapomorphies, represent characters that evolved in the immediate common ancestor, so they should be present in all the descendants. Phylogenies are constructed from shared characters, and classification depends on the derived nature of those character states. Outgroups serve to help identify shared derived characters versus ancestral characters.

2. Do you agree with how time is portrayed in this phylogeny? Why or why not?

   No. The arrows incorrectly indicate that time is going from left to right, rather than from bottom to top. On this tree, the oldest nodes are at the bottom, so the branching event occurred earlier at those nodes than nodes further up the tree. So the lizard lineage branched off earlier than the bird lineage, but that doesn't mean that within these taxonomic groupings, *all* lizards are older than *all* birds.

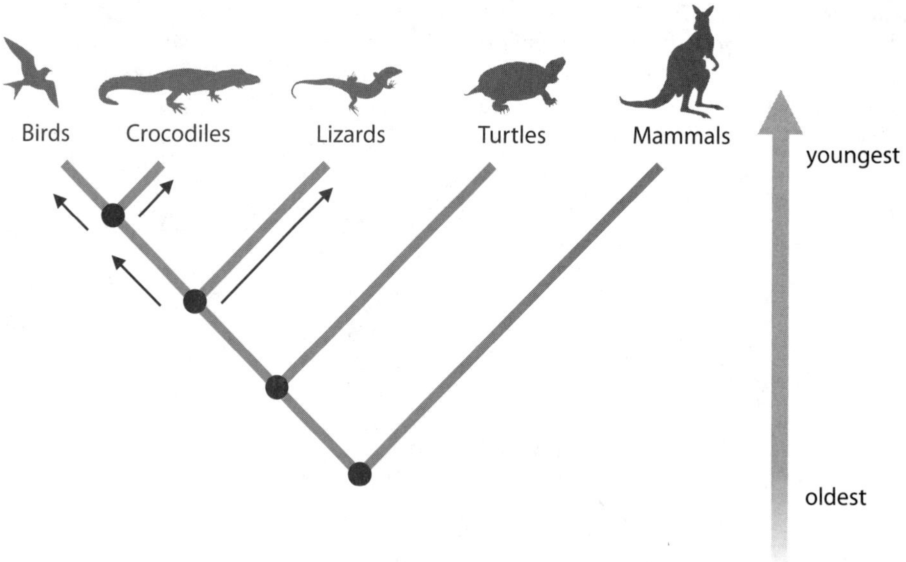

3. Scientists have been examining the correlation between brain size and basal metabolic rate. Some scientists have suggested that basal metabolic rate influences brain size because the brain tissue is metabolically active and costly to run. Maternal investment, either through gestation or lactation, can influence the transfer of metabolic energy during growth and development of offspring. Therians provide a unique opportunity to test hypotheses about basal metabolic rate because eutherians invest in gestation (by providing a placenta to feed developing offspring), and marsupials invest in lactation. How might tests of these hypotheses be influenced by phylogenetic relationships? What might be the effect on the relationship between basal metabolic rate and brain size if one group, the primates (eutherians), have inordinately large brain sizes?

   Read Weisbecker, V. and A. Goswami. 2012. Brain size, life history, and metabolism at the marsupial/placental dichotomy. Proceedings of the National Academy of Sciences (http://www.pnas.org/content/early/2010/08/31/0906486107.abstract). What did the authors find?

Small brain size may be an ancestral trait common to both eutherians and marsupials. Therefore, data points within the phylogenetic contrasts might not be independent—the data might be influenced by these historical relationships. The effect of one group with exceptionally large brain sizes may influence the results even more. Because primates are a diverse group of eutherians and they have exceptionally large brains relative to their basal metabolic rates, they may influence the statistical outcome of the comparison of brain size among eutherians relative to marsupials. Indeed, the specific relationship between brain size and basal metabolic rate found in primates may have evolved only in that lineage, and without independent contrasts could lead unsuspecting researchers to believe that marsupials have smaller brains than eutherians for any given basal metabolic rate.

Test Yourself

1. b; 2. d; 3. e; 4. e; 5. a

5 Raw Material

Heritable Variation among Individuals

Check Your Understanding

1. Who incorporated the idea that the theory of evolution required the capacity for one generation to pass on its traits to the next?
   a. Charles Darwin

      Correct, but so are other answers. Darwin conducted exhaustive studies that indicated a capacity for one generation to pass on its traits to the next, but he had no explanation for how that transference took place. Darwin's evidence was substantial and significant; scientists have discovered much about the mechanisms for passing on traits. Chapter 5 will review the important elements of these discoveries and tie them into evolutionary theory as it is understood today.

   b. Jean-Baptiste Lamarck

      Correct, but so are other answers. Lamarck incorporated the capacity for one generation to pass on its traits to the next in his ideas about evolution, but he believed these traits could be acquired in an individual's lifetime. Although Lamarck's ideas were intuitive at the time, scientists later rejected his ideas about inheritance of acquired characteristics. Since the early 1900s, scientists have discovered much about the mechanisms for inheriting traits. Chapter 5 will review the important elements of these discoveries and tie them into evolutionary theory as it is understood today.

   c. Alfred Russel Wallace

      Correct, but so are other answers. Wallace developed a mechanism for evolution very similar to Darwin's idea of natural selection (see Section 2.3). That mechanism requires that traits be heritable—passed down from one generation to the next, and that possession of those traits confers some benefit in terms of reproductive success or survival.

d. All of the above

Correct. Each of these naturalists incorporated some capacity for one generation to pass on traits to another (see Chapter 2), although none of them understood the mechanism by which that transference took place. Since the early development of the theory of evolution, scientists have discovered much about the mechanisms for passing on traits. Chapter 5 will review the important elements of these discoveries and tie them into evolutionary theory as it is understood today.

e. None of the above

Incorrect. Each of these naturalists incorporated some capacity for one generation to pass on traits to another (see Chapter 2), although none of them understood the mechanism by which that transference took place. Since the early development of the theory of evolution, scientists have discovered much about the mechanisms for passing on traits. Chapter 5 will review the important elements of these discoveries and tie them into evolutionary theory as it is understood today.

2. Why is heritable variation among individuals an important factor for natural selection?

a. Because variation has to be heritable for a species to survive

Incorrect. Theoretically, variation does not *have* to be heritable for a species to continue in existence or for an individual to survive for that matter. In fact, horizontal gene transfer is an important mechanism that does not result from genes shared between parents and offspring (see Box 4.3). However, heritable variation is integral to natural selection for most organisms because it is an effective mechanism by which beneficial mutations can increase in frequency in a population. Natural selection acts to alter the abundances of those mutations in the population based on the relative reproductive success of individuals possessing those mutations.

b. Because variation has to be heritable for individuals to pass down their beneficial mutations

Correct, but so are other answers. Heritable variation is integral to natural selection for most organisms because it is an effective mechanism by which beneficial mutations may be transmitted. Natural selection acts to alter the abundances of those mutations in the population based on the relative reproductive success of individuals possessing those mutations.

c. Because when individuals respond to the environment, they can pass the traits they acquired on to offspring

Incorrect. Jean-Baptiste Lamarck suggested that individuals that responded to the environment could pass those acquired characteristics to their offspring as a mechanism of evolution. In fact, except in special situations (modern-day epigenetics, for example, is the study of how experiences that parents have over their lifetimes can influence how genes are expressed in offspring) variation arises as a result of new mutations. In any given environment, some individuals do better than other individuals thanks to differences in their phenotypes. When those phenotypes are heritable, natural selection acts to alter their abundances in the population based on the relative reproductive success of individuals possessing those traits.

d. Because natural selection cannot act when all individuals are absolutely identical

Correct, but so are other answers. Individuals of a species or group must vary in some characteristic, and some variants of the characteristic must confer an advantage on those who bear them. Natural selection acts to alter the abundances of those characteristics in the population based on the relative reproductive success of individuals possessing them. So, heritable variation is integral to natural selection for most organisms because it is an effective mechanism by which beneficial mutations may be transmitted.

e. Both b and d

Correct. Individuals of a species or group must vary in some characteristic (e.g., via beneficial mutations), and some of those variants must be advantageous. Natural selection acts to alter the abundances of those variants in the population based on the relative reproductive success of individuals possessing them. So, heritable variation is integral to natural selection for most organisms because it is an effective mechanism by which beneficial mutations may be transmitted.

3. How can the phylogeny on the left be used to understand the phylogeny on the right?

a. The phylogeny on the left shows that you and your sister are more closely related to each other than you are to your cousins, just as humans and frogs are more closely related to each other than they are to goldfish or trout.

Correct. Although the phylogeny on the left shows the relationships among individual family members and not groups of organisms, the idea of common ancestry is essentially the same. Humans are more closely related to frogs than they are to fish because they share a more recent common ancestor with frogs (P)—just like you and your sister share a more recent common ancestor (P) than you and your cousins (G). Humans share a more distant common ancestor with fish (G) (see Chapter 4).

b. The phylogeny on the left shows that the relationship between humans and frogs cannot be compared with the relationship between siblings.

Incorrect. Although the phylogeny on the left shows the relationships among individual family members and not groups of organisms, the idea of common ancestry is essentially the same. You and your sister share a more recent common ancestor (P) than you and your cousins (G). Similarly, humans and frogs share a more recent common ancestor than humans and trout (see Chapter 4).

c. The phylogeny on the left shows that your cousins must be more closely related to each other than you and your sister because trout and goldfish are more closely related to each other than frogs and humans.

Incorrect. You and your sister share a common ancestor (P), and each of your cousins also shares a common ancestor (A). You share a more distant common ancestor with your cousins (G). The tree on the left indicates that the relationship between you and your sister is similar to the relationship between your cousins. The phylogeny on the right shows that humans and frogs share a common ancestor (P) and, similarly, trout and goldfish share a common ancestor (A). So humans and frogs are more closely related to each other because they share a more recent common ancestor than with goldfish, and trout and goldfish are more closely related to each other because they share a more recent common ancestor than with humans (see Chapter 4).

d. The phylogeny on the right shows that humans are more closely related to goldfish than to trout.

Incorrect. The branches on a phylogeny can swing freely, just like the different parts of a mobile. So the branch with you and your sister can rotate freely on its axis, and the branch with humans and frogs can rotate freely on its axis. If you and your sister were flipped, that wouldn't make her any more related to your cousins than you are. You and she share a more recent common ancestor than you and your cousins. Likewise, if humans and frogs were flipped, that doesn't make frogs any more related to goldfish (or trout, if those two were flipped). Humans and frogs share a more recent common ancestor than they do with fish (see Chapter 4).

Identify Key Terms

1. m; 2. o; 3. e; 4. ag; 5. c; 6. z; 7. al; 8. j; 9. w; 10. as; 11. aj; 12. k; 13. am; 14. f; 15. q; 16. x; 17. p; 18. af; 19. n; 20. i; 21. y; 22. b; 23. v; 24. at; 25. ak; 26. l; 27. t; 28. ae; 29. g; 30. ab; 31. aa; 32. ao; 33. s; 34. ad; 35. aq; 36. ar; 37. ap; 38. d; 39. ac; 40. an; 41. a; 42. u; 43. r; 44. ai; 45. h; 46. ah

Link Concepts

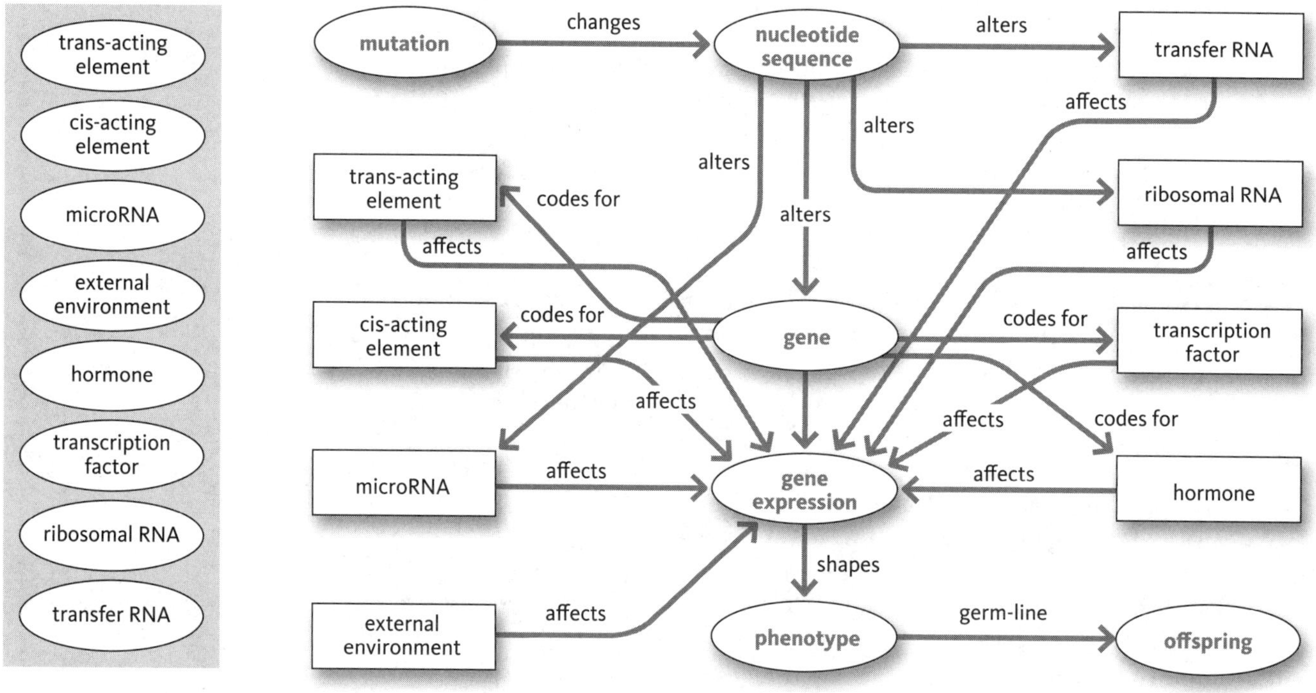

Interpret the Data

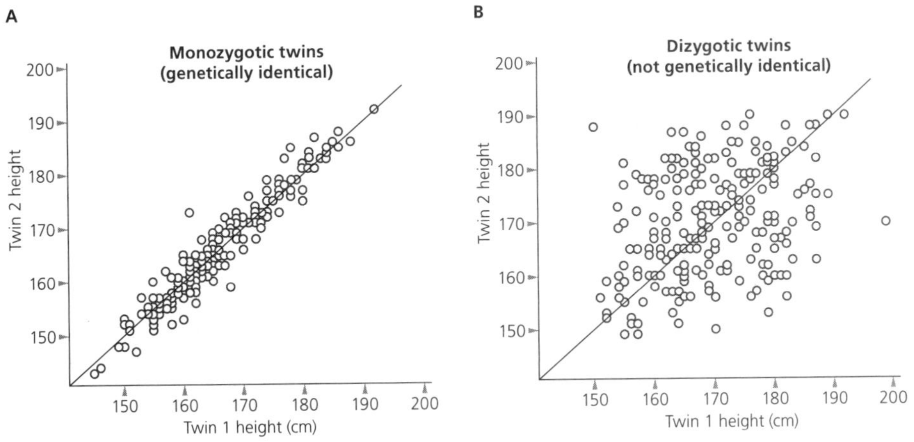

- Can you predict the height of Twin 2 when twins are identical? For example, what would you expect the height of Twin 2 to be if the height of Twin 1 is 160 cm?

  Yes. You can predict the height of Twin 2 with some certainty when twins are identical. If Twin 1 is 160 cm, Twin 2 is likely to be about 160 cm. In part A, the height of Twin 2 ranged from approximately 152 to 165 cm when Twin 1 was exactly 160 cm.
- Is there variation in the heights of identical twins?

  Yes. Heights of identical twins vary—their heights do not match up exactly. Even though the twins share identical genes, other factors, such as subtle differences in the environment within which the individuals develop, affect the height of each individual.
- Why don't the heights of identical twins match exactly?

  Identical twins inherit identical sets of genes, but environmental factors can affect gene expression differently in different individuals. So, while identical twins have a strong tendency to grow to the same height, their heights still include some variation.
- Do you see any correlation in height between fraternal twins?

  Although predicting the heights of fraternal twins is not as straightforward as predicting the heights of identical twins, the heights of fraternal twins could still be correlated because the individuals are related and they likely experienced generally similar environments as they developed.
- What might be contributing to the variation between fraternal twins?

  Fraternal twins develop from separate eggs; the alleles they carry will thus differ because of genetic recombination and independent assortment. As a result, fraternal twins are less likely to grow to the same height than identical twins.

Games and Exercises

Strand 1: CACGTGGACTGAGGACTC

mRNA: GUGCACCUGACUCCUGAG

Amino acids: Val His Leu Thr Pro Glu

Strand 2: CACGTGGACTGAGGACAC

mRNA: GUGCACCUGACUCCUGUG

Amino acids: Val His Leu Thr Pro Val

- What's the difference between the two strands?

  The last thymine is changed to an adenine (so from CTC to CAC).

- What's the difference between the amino acid sequences that result?

  The last amino acid is different—the sequence now transcribes Val (valine) instead of Glu (glutamic acid).

- If the mutation in Strand 2 was at the third base in that codon instead of the second, would the hemoglobin protein form normally?

  It would depend on the mutation. If the mutation changed the third base in the DNA from a cytosine to an adenine (or to a guanine), the sequence would read CTA (or CTG), the RNA strand would read GAU (or GAC), and the codon would read for the amino acid Asp (aspartic acid). If the mutation changed the third base in the codon from a cytosine to either a thymine or a cytosine, however, the codon will still produce Glu (glutamic acid), and the protein would remain the same.

Overcoming Misconceptions

Which of the following is a true statement?

d. A low concentration of transcription factors can influence the expression of a gene.

Delve Deeper

1. Why aren't the mutations that occur in skin cells or in other organs, such as the heart or brain, heritable?

   Because the cells dividing in the skin, heart, and brain are not involved in sexual reproduction. All these cells die when the individual dies. A mutation that occurs in these cells can't be passed down to offspring. Only eggs and sperm transmit information stored in DNA to the next generation, so only mutations that occur in these germ cells are heritable.

2. Why don't all phenotypic traits occur as discrete, alternative states like Mendel's peas?

   Phenotypic traits are rarely determined by single Mendelian loci. Besides the fact that variation in some traits can be attributed to the cumulative action of many genes, the environment can also influence phenotypic traits. Variation in the environment can lead to variation in the phenotypes that arise from a single genotype, as does the complex interactions between many different genes and the environment.

3. What sources of genotypic variation among individuals are random?

   Mutations are one source of random variation—their effects can vary depending on where they occur within the genome and when. Mutations are the ultimate source of genetic variation. Independent assortment is another random element. Pairs of chromosomes separate during meiosis and segregate randomly, so that any particular gamete may have any combination of maternal and paternal chromosomes. In addition, genetic recombination is a random event that affects variation every time a germ cell divides. So gametes can end up with very different combinations of alleles, even though they are produced by the same individual.

   Evolution itself is not entirely random, however. In fact, natural selection is decidedly NOT random. Natural selection is the process that acts on these random sources of variation, favoring the variation manifested in phenotypes that increases survival or reproduction.

Test Yourself

1. a; 2. c; 3. c; 4. d; 5. e; 6. a; 7. e; 8. b; 9. e; 10. c

6 The Ways of Change
Drift and Selection

Check Your Understanding

1. According to the text, what is an organism's phenotype?

   a. The interaction of an organism's genes with the environment to produce characteristics such as how the amount of light a plant is exposed to influences its height

   Incorrect. An organism's genes do interact with the environment (both external and internal environments) during development to produce phenotypic characteristics, but the phenotype is the characteristic itself. The phenotype is any aspect of an organism that can be measured such as morphology, physiology, and behavior (see Chapter 5).

   b. Characteristics of an organism that can be classified into discrete categories, such as gender or eye color

   Incorrect. Some phenotypic characteristics can be classified into discrete categories, such as gender (female or male) or eye color (blue, green, brown), but many other characteristics cannot, such as height, length of the femur, antler size (these traits vary in a range around a mean). The phenotype is any aspect of an organism that can be measured such as morphology, physiology, and behavior (see Chapter 5).

   c. Any aspect of an organism that can be measured, such as how it looks, how it behaves, how it's structured

   Correct. The phenotype is linked to the genotype—genes interact with other genes and with the environment during the development of the phenotype. Some of those aspects can be classified into discrete categories, such as gender (female or male) or eye color (blue, green, brown), but many other characteristics cannot, such as height, length of the femur, antler size (these traits vary in a range around a mean). See Chapter 5 for a discussion of the complexities of the links between phenotypes and genotypes.

   d. The genetic makeup of an individual

   Incorrect. An organism's phenotype is any measurable aspect, such as morphology, physiology, or behavior. The genetic makeup of an individual is that individual's genotype. See Chapter 5 for a discussion of the complexities of the links between genotypes and phenotypes.

   e. Both b and d

   Incorrect. Some phenotypic characteristics can be classified into discrete categories, such as gender (female or male) or eye color (blue, green, brown), but many other characteristics cannot, such as height, length of the femur, antler size (these traits vary in a range around a mean). The phenotype is any aspect of an organism that can be measured such as morphology, physiology, and behavior, and the genetic makeup of an individual is that individual's genotype (see Chapter 5).

2. How many alleles can a genetic locus in a diploid individual have?

   a. One

   Incorrect. Fixed alleles, by definition, are the only allele at a particular genetic locus within a population; all members of that population are homozygous for that allele at that locus. Fixed loci are not necessarily typical, however; there are no hard and fast rules about the number of alleles at any particular genetic locus. Mutations can create new alleles at any genetic locus, and the number of alleles and their frequency within a population can change as a result of natural selection and genetic drift.

b. Two

Incorrect. Simple models of loci often represent alleles in upper- and lowercase letters, such as AA, Aa, and aa, implying that a locus only has two alleles. The number of alleles at any locus is not limited to two, however, and more accurate models assign allele names that reflect the option for greater diversity. There are no hard and fast rules about the number of alleles at any particular genetic locus, however. Mutations can create new alleles at any genetic locus, and the number of alleles and their frequency within a population persist or disappear as a result of natural selection and genetic drift.

c. More than two

Incorrect. There are no hard and fast rules about the number of alleles at any particular genetic locus. Mutations can create new alleles at any genetic locus, and the number of alleles and their frequency within a population persist or disappear as a result of natural selection and genetic drift. Although the simple models of loci imply that a locus only has two alleles because they use upper- and lowercase letters, such as AA, Aa, and aa, that doesn't necessarily mean that genetic locus has only two alleles. More accurate models assign allele names that reflect the option for greater diversity of alleles at any particular locus—whether that be one, two, or more than two.

d. It depends on the locus.

Correct. There are no hard and fast rules about the number of alleles at any particular genetic locus. Mutations can create new alleles at any genetic locus, and the number of alleles and their frequency within a population can change as a result of natural selection and genetic drift. A more appropriate way of representing alleles incorporates subscripts (A1, A2, A3, A4) or superscripts (Ester1, Ester4), reflecting the diverse number of alleles that may or may not be known (yet) for any particular locus.

3. Why is the variation of phenotypic traits often continuous, distributed around a mean in a bell-shaped curve?

a. Because phenotypic traits are a result of dominance

Incorrect. Dominance can affect phenotypic traits, but dominance refers to alleles at a single genetic locus. Although Mendel was able to predict some phenotypes, most phenotypic traits cannot be reduced to the simple patterns of inheritance Mendel studied. Most phenotypic traits involve more than one gene (i.e., they are polygenic), and the different alleles from these different genes contribute to the trait differently. The result—a continuous distribution (like height, from 5'5" to 5'5 ¼" to 5'5 ½" and all the possible measures in between). Most individuals will have measurements near the mean, with fewer and fewer in the "tails," so the curve appears bell shaped.

b. Because phenotypic traits are not related to genotypes

Incorrect. Although the relationship between genotype and phenotype is not necessarily straightforward, the genotype and phenotype are in fact intimately related. Most phenotypic traits involve more than one gene (i.e., they are polygenic), and the different alleles from these different genes contribute to the trait differently. The result—a continuous distribution (like height, from 5'5" to 5'5 ¼" to 5'5 ½" and all the possible measures in between). Most individuals will have measurements near the mean, with fewer and fewer in the "tails", so the curve appears bell shaped.

c. Because phenotypic traits are only influenced by the environment

Incorrect. The environment is indeed an important component of the expression of phenotypic traits, but a distinct genetic architecture underlies that influence. Indeed, most phenotypic traits involve more than one gene (i.e., they are polygenic), and the different alleles from these different genes contribute to the trait differently. The result—a continuous distribution (like height, from 5'5" to 5'5 ¼" to 5'5 ½" and all the possible measures in between). Most individuals will have measurements near the mean, with fewer and fewer in the "tails," so the curve appears bell shaped.

d. Because phenotypic traits are often polygenic

Correct. Most phenotypic traits involve more than one gene (i.e., they are polygenic), and the different alleles from these different genes contribute to the trait differently. The result—a continuous distribution (like height, from 5'5" to 5'5 ¼" to 5'5 ½" and all the possible measures in between). Most individuals will have measurements near the mean, with fewer and fewer in the "tails," so the curve appears bell shaped.

Identify Key Terms

1. a; 2. x; 3. u; 4. f; 5. r; 6. q; 7. h; 8. k; 9. j; 10. w; 11. y; 12. t; 13. o; 14. a; 15. l; 16. i; 17. v; 18. n; 19. b; 20. z; 21. s; 22. m; 23. p; 24. e; 25. c; 26. g

Link Concepts

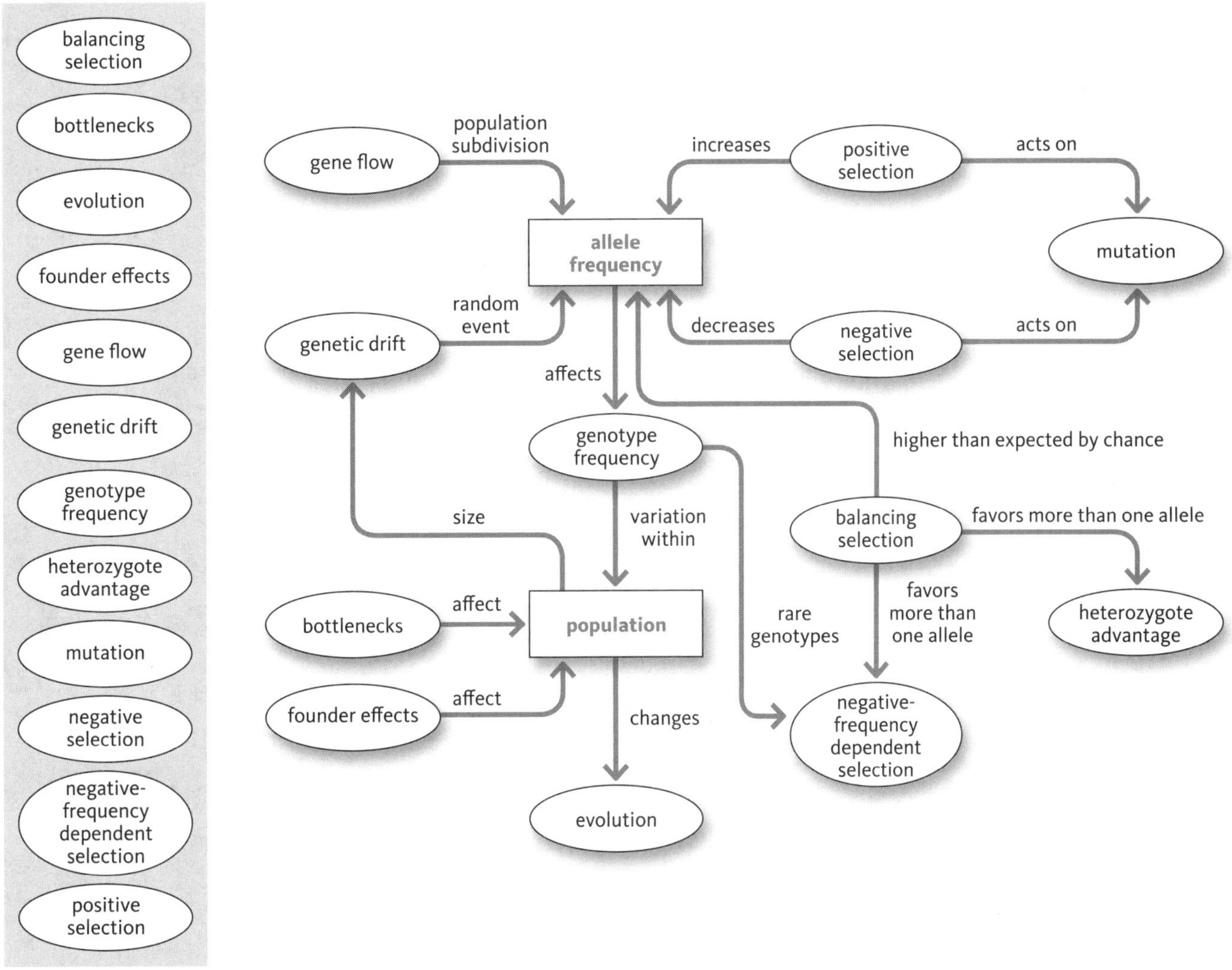

Interpret the Data

- At the rarest frequency of the yellow morph, what was the relative reproductive success of males? Of females?

  Roughly 1.4 for males (range 1.35 to 1.5); 1.45 for females (range 1.3 to 1.6)
- At the most common frequency of the yellow morph, what was the relative reproductive success of males? Of females?

  Roughly 0.9 for males (range 0.8 to 0.95); 0.85 for females (range 0.7 to 0.9)
- Why might the foraging experiences of bumblebees drive this relationship?

  Because bumblebees may learn that flowers of a specific color do not offer a nectar reward. The more common one color is, the more quickly bees may learn. So, when yellow orchids are rare, they may be visited by naïve bees expecting to find nectar because the bees have learned to avoid the purple orchids. As a result, the yellow orchids will produce more seeds, and have higher fitness than the purple orchids. Their relative frequency will increase in the population. As the yellow orchids become more and more common, naïve bees will visit them less often because they have learned they don't produce nectar, instead opting to try the rarer purple orchids. At that point, the rarer purple orchids will have higher fitness than the yellow orchids.

Games and Exercises

- What does q^2 equal?

  We know the frequency of white bears on the islands = 20%, and *W* is recessive. So *WW* must be 0.20 (the decimal version of 20% = 20/100 = 0.20). Therefore, $q^2 = 0.20$.

  BTW, q = the square root of 0.20 = 0.45
- What does p^2 equal?

  $p = 1 - q = 0.55$, so $p^2 = 0.31$
- What does $2pq$ equal?

  2 x 0.55 x 0.45 = 0.49

  BB genotypes should occur at a frequency of $p^2 = 0.31$, or 31%
  BW genotypes should occur at a frequency of $2pq = 0.49$, or 49%
  and *WW* genotypes should occur at a frequency of $q^2 = 0.20$, or 20%
- If you had a population of 37 bears on the two islands, how many would you expect to be black?

  (37 x 0.31) + (37 x 0.49) = 29.6, so about 30 should be black.
- How many would you expect to be white?

  37 x 0.20 = 7.4, so about 7 should be white.
- How many bears would you expect to have black fur? How many Spirit Bears?

  $q = 0.08$, so $q^2 = 0.01$

  $p = 1 - q = 0.92$, so $p^2 = 0.85$

  $2pq = 0.15$

  (112 x 0.92) + (112 x 0.15) = 111.3, so about 111 should be black.

  112 x 0.8 = 0.7, so 1 bear may be white.

White Bears and Selection

- Complete this table:

  What is p? $p = 1 - 0.369 = 0.631$

	Observed	Expected	(Observed – Expected)2
BB (black)	42	34.64	54.169
BW (black)	24	40.51	272.712
WW (white)	21	11.85	83.796
Total	87		410.677

- What are the degrees of freedom?

  2

- What is the chi square value?

  Far less than 0.0005

- How would you interpret this result?

  Kermode bears are not in Hardy–Weinberg equilibrium, so one or more assumptions may be violated. The population may be too small and experiencing drift, all of the genotypes may not be equally likely to survive and reproduce, the population may be experiencing immigration or emigration, or mutations may have arisen. Non-random mating could also lead to this pattern (see Chapter 11).

- Which genotypes are overrepresented in the population? Which are underrepresented?

  Both homozygous black and homozygous white bears are overrepresented. Heterozygotes are underrepresented.

  If black bears are at a disadvantage with a selection coefficient of $s = 0.2$, then what should the genotype frequencies be after selection in time$_{t+1}$?

Genotype:	*BB*	*BW*	*WW*
f_{t+1}	$p^2 \times w_{11}$	$2pq \times w_{12}$	$q^2 \times w_{22}$
	$0.631^2 \times 0.8$	$2(0.631)(0.369) \times 0.8$	$0.369^2 \times 1.0$
	0.319	0.373	0.136

$= p^2 \times w_{11} + 2pq \times w_{12} + q^2 \times w_{22} = 0.\ 0.319 + 0.373 + 0.136 = 0.827$

The relative frequencies of individuals with each genotype after selection using the average fitness of the population:

Genotype:	BB	BW	WW
f_{t+1}	$(p^2 \times w_{11}) / \overline{w}$	$(2pq \times w_{12}) / \overline{w}$	$(q^2 \times w_{22}) / \overline{w}$
	0.319 / 0.827	0.373 / 0.827	0.136 / 0.827
	0.385	0.450	0.165

$p_{t+1} = [(p^2 \times w_{11}) / \overline{w}] + (pq \times w_{12}) / \overline{w})$

$= (p^2 \times w_{11} + pq \times w_{12}) / \overline{w}$

$= (0.631^2 \times 0.8 + 0.631 \times 0.369 \times 0.8) / 0.827$

$= 0.610$

and

$q_{t+1} = [(q^2 \times w_{22}) / \overline{w}] + (pq \times w_{12}) / \overline{w})$

$= (q^2 \times w_{22} + pq \times w_{12}) / \overline{w}$

$= (0.369^2 \times 1.0 + 0.631 \times 0.369 \times 0.8) / 0.827$

$= 0.390$

- Is there another way to calculate q_{t+1} if p_{t+1} is known?

  Yes. $1 - p_{t+1} = q_{t+1}$

  $1 - 0.610 = 0.390$

- What would be the proportion of each genotype found in time$_{t+1}$?

  $p^2 = 0.610^2 = 0.372$ or 37.2% homozygous black bears

  $2pq = 2(0.610)(0.390) = 0.476$ or 47.6% heterozygous black bears

  $q^2 = 0.390^2 = 0.152$ or 15.2% white bears

- What does the selection coefficient mean in terms of persistence of the *B* allele within the population?

  The *B* allele is selected against, and its frequency will continue to decline if the selection coefficient stays the same. It may persist in the population at very low frequencies unless genetic drift drives the locus to fixation.

- Determine the change in allele frequencies resulting from selection in time$_{t+2}$ as a function of the average excess of fitness. Assume the fitnesses have stayed the same.

Genotype:	*BB*	*BW*	*WW*
f_{t+1}	$p_{t+1}^{\;2} \times w_{11}$	$2p_{t+1}q_{t+1} \times w_{12}$	$q_{t+1}^{\;2} \times w_{22}$
	$0.610^2 \times 0.8$	$2(0.610)(0.390) \times 0.8$	$0.390^2 \times 1.0$
	0.298	0.381	0.152

$\overline{w} = 0.830$

$a_B = [0610. \times (0.8 - 0.830)] + [0.390 \times (0.8 - 0.830)]$

$= -0.030$

$\Delta p = (0.610 / 0.830) \times -0.030$

$= -0.022$

- What does the average excess of fitness of the *B* allele indicate?

  Because it's a negative number, the average excess of fitness shows that the *B* allele is decreasing from generation to generation.

- Calculate the average excess of fitness for the *W* allele:

  $a_W = [p \times (w_{12} - \overline{w})] + [q \times (w_{22} - \overline{w})]$

  $a_W = 0.048$

  and the predicted change in frequency of the *W* allele as a result of selection:

  $\Delta q = (q / \overline{w}) \times a_W$

  $\Delta q = 0.022$

- Record the allele frequencies in a table:

	p	*q*
t	0.631	0.369
t +1	0.610	0.390
t + 2	0.588	0.412

- What would be the proportion of each genotype in $time_{t+2}$?

  $p^2 = 0.588^2 = 0.346$ or 34.6% homozygous black bears

  $2pq = 2(0.588)(0.412) = 0.485$ or 48.5 % heterozygous black bears

  $q^2 = 0.412^2 = 0.170$ or 17.0% white bears
- Based on your understanding of Hardy–Weinberg and the distribution of observed phenotypes, what can you say about this population?

  The expected numbers of heterozygotes are consistently higher than the observed. Also, the expected proportion of heterozygotes is increasing. So other violations of the model may be affecting this population. For example, not all of the genotypes may be equally likely to survive and reproduce. Black homozygotes and black heterozygotes may not have the same fitness—the heterozygotes may be at a greater disadvantage than the homozygotes. Similarly, the population on the islands may be influenced by migration of either black homozygotes or black heterozygotes.

Overcoming Misconceptions

Which of the following is a true statement?

a. A dominant allele is an allele that produces the same phenotype whether it is homozygous or heterozygous.

Delve Deeper

1. If a mutation that produces a new allele that is deleterious arises in a population, what will most likely happen to the frequency of that allele?

   Not all deleterious mutations disappear from populations; the outcome depends on the allele's effect on the phenotypes. If the mutant allele is recessive, it may be rare enough that it is almost never expressed in a homozygous state. It can even remain at a low frequency within the population for a very long time. Of course, drift may also determine whether the allele persists in the population.
2. Why has the evolution of resistance to insecticides in mosquitoes been so rapid?

   Insecticides, and other pesticides, can impose extremely strong selection. Pest populations often are highly genetically variable. They don't necessarily have mutation rates any higher than other organisms; the variation comes from their large effective population size and rapid reproduction. However, large amounts of standing genetic variation mean that populations are poised for rapid evolutionary responses to selection (h^2 is very large, and $R = h^{2*} s$), and vast population sizes mean that effects of drift are likely to be weak compared with effects of selection.

Test Yourself

1. e; 2. e; 3. b; 4. a; 5. c; 6. b; 7. b; 8. a; 9. d; 10. b; 11. e; 12. e

7 Beyond Alleles

Quantitative Genetics and the Evolution of Phenotypes

Check Your Understanding

1. What is the difference between polyphenic traits and polygenic traits?

   a. Polyphenic traits are the different traits that arise because different alleles lead to different phenotypes; polygenic traits are traits influenced by many genes leading to a continuous distribution of phenotypes over a given range.

   Incorrect. Polyphenic traits are multiple, discrete phenotypes that can arise from a *single* genotype; individuals carrying different alleles would be considered different genotypes (see Section 5.4). Polygenic traits are traits influenced by many genes, resulting in a continuous distribution of phenotypic variation rather than discrete categories such as blue or brown eyes (see Section 5.4). They are also known as quantitative traits—the topic of this chapter.

   b. Polyphenic traits are the multiple, discrete phenotypes that can arise from different alleles within a population; polygenic traits vary continuously within a population because of heritable variation.

   Incorrect. Different phenotypic traits can exist within a population, but those traits can result from different genotypes or from polyphenisms. Polyphenic traits, specifically, are the multiple, discrete phenotypes that can arise from a *single* genotype (see Section 5.4). Polygenic traits are phenotypic traits that are influenced by many genes, leading to a continuous distribution of phenotypic variation rather than discrete categories such as blue or brown eyes (see Section 5.4). They are also known as quantitative traits—the topic of this chapter.

   c. Polyphenic traits are the multiple, discrete phenotypes that can arise from a single genotype depending on environmental circumstances; polygenic traits are traits influenced by many genes leading to a continuous distribution of phenotypic variation over a given range.

   Correct. Section 5.4 introduces polyphenic and polygenic traits. Polyphenic traits often arise because of a threshold of sensitivity to the environment (e.g., an individual male beetle that would produce horns if food was plentiful but not produce horns if food was scarce). Polygenic traits are also known as quantitative traits—the topic of this chapter.

   d. Polyphenic traits are traits that result when natural selection favors rare genotypes leading to multiple phenotypes within a population; polygenic traits are traits that arise because a single gene affects the expression of many different phenotypic traits.

   Incorrect. Negative frequency-dependent selection favors rare genotypes, but the genotypes differ. Polyphenic traits are the multiple, discrete phenotypes that can arise from a single genotype (see Section 5.4). Pleiotropy is the condition when a single gene affects the expression of many different phenotypic traits. Polygenic traits are traits influenced by many genes, resulting in a continuous distribution of phenotypic variation rather than discrete categories such as blue or brown eyes (see Section 5.4). They are also known as quantitative traits—the topic of this chapter.

2. Which of the following were important facts in Charles Darwin's development of theory of evolution by natural selection?

   a. No two individuals are exactly the same; rather, every population displays enormous variability.

      Correct, but so are other answers. Darwin studied organisms from barnacles to birds, and he knew first hand of the variation among individuals. He also realized that much of that variation was heritable—offspring often looked like their parents. More importantly, inheritance explained many of the patterns he observed, from the existence of homologies to the fossil record. He used both of these facts to infer that natural selection would lead to evolution and the production of new species (see Section 2.3 and Figure 2.19).

   b. Much of the variation among individuals within a population is heritable.

      Correct, but so are other answers. For decades, Darwin meticulously studied organisms from barnacles to birds. He explored artificial selection, and he knew first hand of the variation among individuals and how traits could be passed from one generation to the next. He used both of these facts to infer that natural selection would lead to evolution and the production of new species (see Section 2.3 and Figure 2.19).

   c. Organisms can inherit characters that were acquired during their parents' lifetime.

      Incorrect. Jean-Baptiste Lamarck believed that individuals changed in response to their environment, much as a blacksmith became more muscular over time, and those acquired traits could be passed down to offspring. Lamarck postulated that the inheritance of acquired characteristics was a mechanism for the evolution of lineages (see Section 2.2). Darwin provided a different mechanism for the evolution of organisms that relied on variation and heritability (see Section 2.3). Today, scientists understand that the experiences that parents have over their lifetimes *can* sometimes influence how genes are expressed in future generations, a phenomenon known as epigenetics (see Box 2.1).

   d. Both a and b

      Correct. Darwin used two facts to infer that natural selection would lead to evolution and the production of new species: that no two individuals are exactly the same and much of the variation among individuals within a population was heritable (see Section 2.3 and Figure 2.19).

   e. Both b and c

      Incorrect. Although Darwin realized that heritable variation was necessary for evolution, he did not suggest that the characters were acquired within an individual's lifetime. The inheritance of acquired characteristics was postulated by Jean-Baptiste Lamarck as a mechanism for the evolution of lineages (see Section 2.2). Darwin provided a different mechanism for the evolution of organisms that relied on variation and heritability (see Section 2.3).

3. Can genes that respond to environmental stimuli be passed on to offspring?

   a. Yes. Individuals that learn how to respond to the environment can pass that information to their offspring.

      Incorrect. Lamarckism is the idea that traits acquired during an individual's lifetime could be passed down to offspring (see Section 2.2). However, some traits that are sensitive to environmental stimuli are heritable. Phenotypic plasticity is the changes in the phenotype produced by a single genotype in different environments (see Section 5.5). This chapter will explore the mechanism of phenotypic plasticity and its measurement (see Box 7.4 and Section 7.4).

b. Yes. Individuals can inherit the mechanisms that respond to the environment.

Correct. Some traits that are sensitive to environmental stimuli are heritable. Phenotypic plasticity is the changes in the phenotype produced by a single genotype in different environments (see Section 5.5). This chapter will explore the mechanism of phenotypic plasticity and its measurement (see Box 7.4 and Section 7.4).

c. No. The only "environmental" influences on gene expression that can be inherited come from other gene products, such as hormones, transcription factors, and cis- and trans-acting elements. Any effects of external environmental factors are not encoded by the genome, so they cannot be heritable.

Incorrect. Hormones, transcription factors, and cis- and trans-acting elements are part of the *internal* environment that can influence gene expression, and they are heritable. Other internal environmental factors, that are not genes, can also influence gene expression, including microRNA, ribosomal RNA, and transfer RNA, and these are also heritable (see Section 5.1). Many of these factors are sensitive to environmental factors *external* to an organism. These responses can affect gene expression, resulting in phenotypic plasticity, and, because the environment-sensitive factors are heritable, the capacity to respond to the external environment can be transmitted from parents to offspring. This chapter will explore the mechanism of phenotypic plasticity and its measurement (see Box 7.4 and Section 7.4).

d. No. Environmental factors, such as the amount of food available to an individual or the temperature an egg is exposed to, are not heritable.

Incorrect. Some traits that are sensitive to external environment stimuli are heritable. Phenotypic plasticity is the changes in the phenotype produced by a single genotype in different environments (see Section 5.5), and the environment can be external or internal. Internal environmental factors include microRNA, ribosomal RNA, and genetically coded factors such as cis- and trans-acting factors, and hormones. For example, morphogens are molecular signals that move through a field of cells, and their production can be affected by food or temperature stress. These factors are all heritable (see Chapter 5). This chapter will explore the mechanism of phenotypic plasticity and its measurement (see Box 7.4 and Section 7.4).

Identify Key Terms

1. a; 2. e; 3. b; 4. g; 5. f; 6. c; 7. h; 8. d

Link Concepts

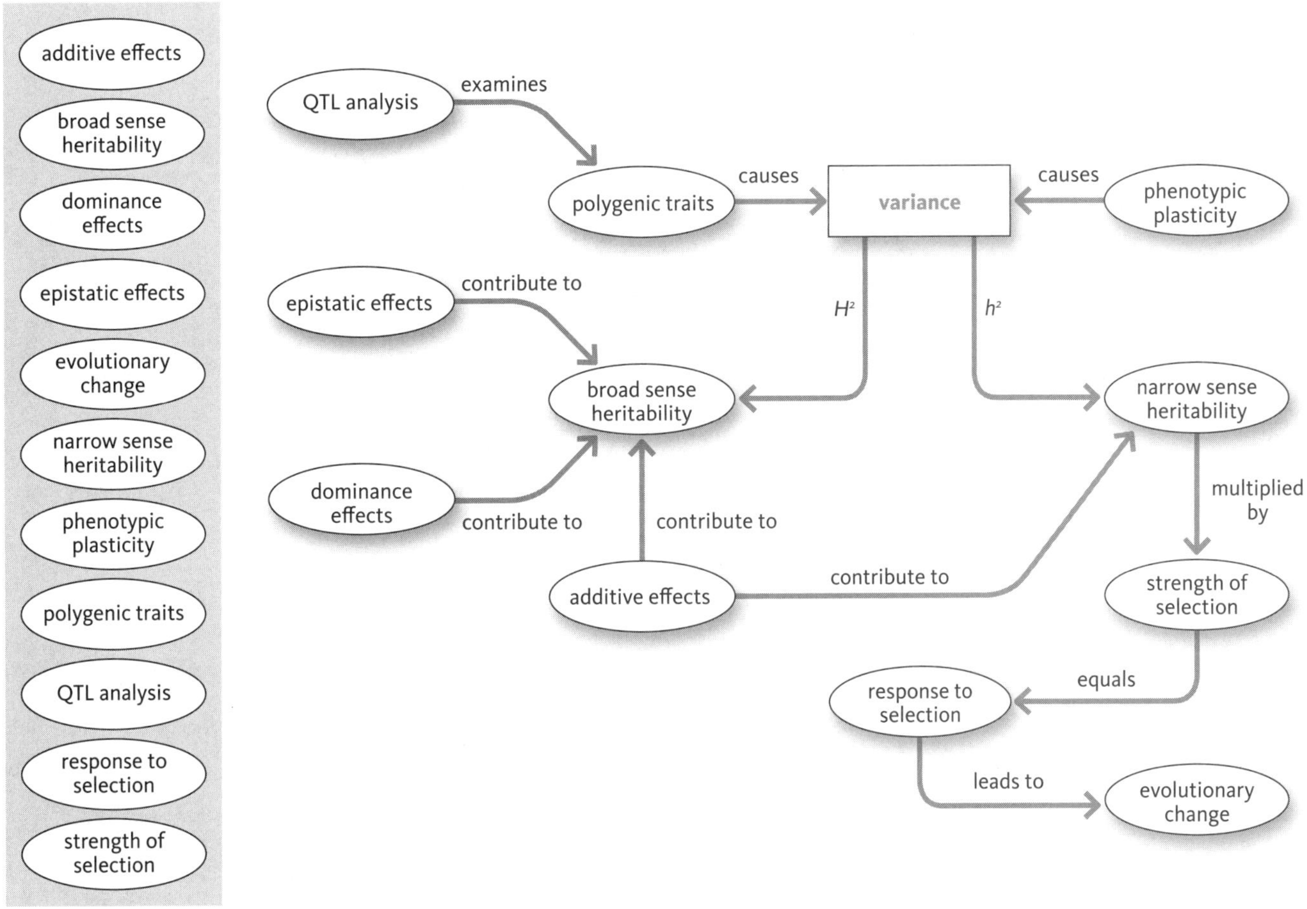

Interpret the Data

- Which region had the highest LOD for *Agouti*? For *Mc1r*?

  The tail and the rump, respectively

- Based on this evidence, why might the researchers looked more closely at Agouti?

  Because the LOD scores for *Agouti* were much higher (3 times or more) than for *Mc1r*

- How far along on the chromosome does the *Agouti* sequence lie?

  30.3 cM

- What does the diagram in B indicate about the phenotype associated with heterozygotes at this locus? Why is this important?

  That heterozygotes for this region have light coats, and homozygotes have dark coats. This pattern is consistent with a simple genetic basis for the trait, reflecting alternative alleles at a single locus with dominance effects.

Games and Exercises

Friend #	Height (cm)		
1	152.40	152.40 – 157.99 = -5.59	31.226
2	162.56	4.57	20.885
3	165.10	7.11	50.552
4	157.48	-0.51	0.260
5	152.40	-5.59	31.248
sum	789.94		134.171

Calculate s^2 and s for this example:
= 33.543
= 11.583

The Dating Game

- So using this tool, who is better for James, Sandi or Brenda?

  Sandi: = 9.4; = 7.30

  Brenda: = 7.3; = 6.98

Delve Deeper

1. Do populations that experience selection always evolve? Why or why not?

   No. Evolution of a population depends on how much phenotypic variation is attributable to additive genetic differences among individuals.

Test Yourself

1. e; 2. c; 3. a; 4. d; 5. d

8 Natural Selection
Empirical Studies in the Wild

Check Your Understanding

1. From which sources does variation among individuals ultimately arise?
   a. Genetic recombination

      Correct, but so are other answers. See Chapter 5.3 for a discussion of how recombination affects variation among even individual family members.

   b. Mutation

      Correct, but so are other answers. See Chapter 5.2 for a discussion of how both beneficial and deleterious mutations arise and their potential influence on individuals.

c. Independent assortment

Correct, but so are other answers. See Chapter 5.3 for a discussion of how independent assortment affects variation among gametes.

d. All of the above

Correct. Chapter 5 discusses how variation in the genotype arises and how that genetic variation can be translated into phenotypic variation.

e. None of the above

Incorrect. All of the above lead to variation among individuals. Chapter 5 discusses how variation in the genotype arises and how that genetic variation can be translated into phenotypic variation.

2. Within a population, allele frequencies can change as a result of genetic drift, natural selection, migration, and mutation. Which mechanism would you argue is the most sensitive to variation among individuals?

a. Genetic drift because it can only occur when individuals are very different from each other

Incorrect. Genetic drift is a random process; its effects depend on the number of individuals (drift is more likely to influence small populations than large ones) (see Section 6.4). It will occur regardless of how different those individuals are from each other, however.

b. Natural selection because variation among individuals is the foundation for relative success of individuals, and that relative success affects the frequencies of alleles within the population

Correct. Section 6.6 discusses relative fitness and how natural selection affects allele frequencies.

c. Migration because individuals with more allelic variation will be less likely to migrate than individuals with less variation

Incorrect. As a phenotypic trait, individuals may vary in their migratory tendencies, but allelic variation within individuals does not necessarily affect migration of the population as a whole. Individual variation in migratory tendencies can have major effects on allele frequencies within a population by affecting the proportion of individuals that leave (emigrate from) or enter (immigrate to) a population, however (also see Chapter 14 for the effect these processes can have on speciation).

d. Mutations because they damage individuals and remove them and their genes from the population

Incorrect. Mutations can be deleterious, but they can also be beneficial (see Section 5.2). However, mutations are random and rare, and the likelihood of a mutation within an individual is not affected by how much variation there is among individuals.

e. Allele frequencies within a population do not change.

Incorrect. Scientists have been able to measure changes in allele frequencies both in the lab and in the wild (see Section 6.6).

3. What is the selection differential (S)?

a. Negative selection

Incorrect. The selection differential can describe positive or negative selection. See Section 7.2 for a discussion of the selection differential (S) and its measurement.

b. The difference between the trait mean of reproducing individuals and the trait mean of the general population

Correct. The selection differential is a measure used by quantitative geneticists to examine the strength of selection on a phenotype (see Section 7.2).

c. A measure of evolution

Incorrect. See Section 7.2 for a discussion of the selection differential (S) and its measurement. Selection is a mechanism that can cause evolution, but it is not evolution. Whether evolution occurs in response to selection also depends on the narrow sense heritability of the trait (see Section 7.1).

d. The difference between variation in a phenotypic trait caused by the environment and variation caused by the underlying genetic architecture

Incorrect. See Section 7.2 for a discussion of the selection differential (S) and its measurement. See Section 7.1 for a discussion of the components of phenotypic variation.

e. The difference between narrow sense heritability and response to selection

Incorrect. See Section 7.2 for a discussion of the relationship between selection differential (S), narrow sense heritability, and the response to selection.

Identify Key Terms

1. f; 2. d; 3. g; 4. c; 5. a; 6. e; 7. b

Link Concepts

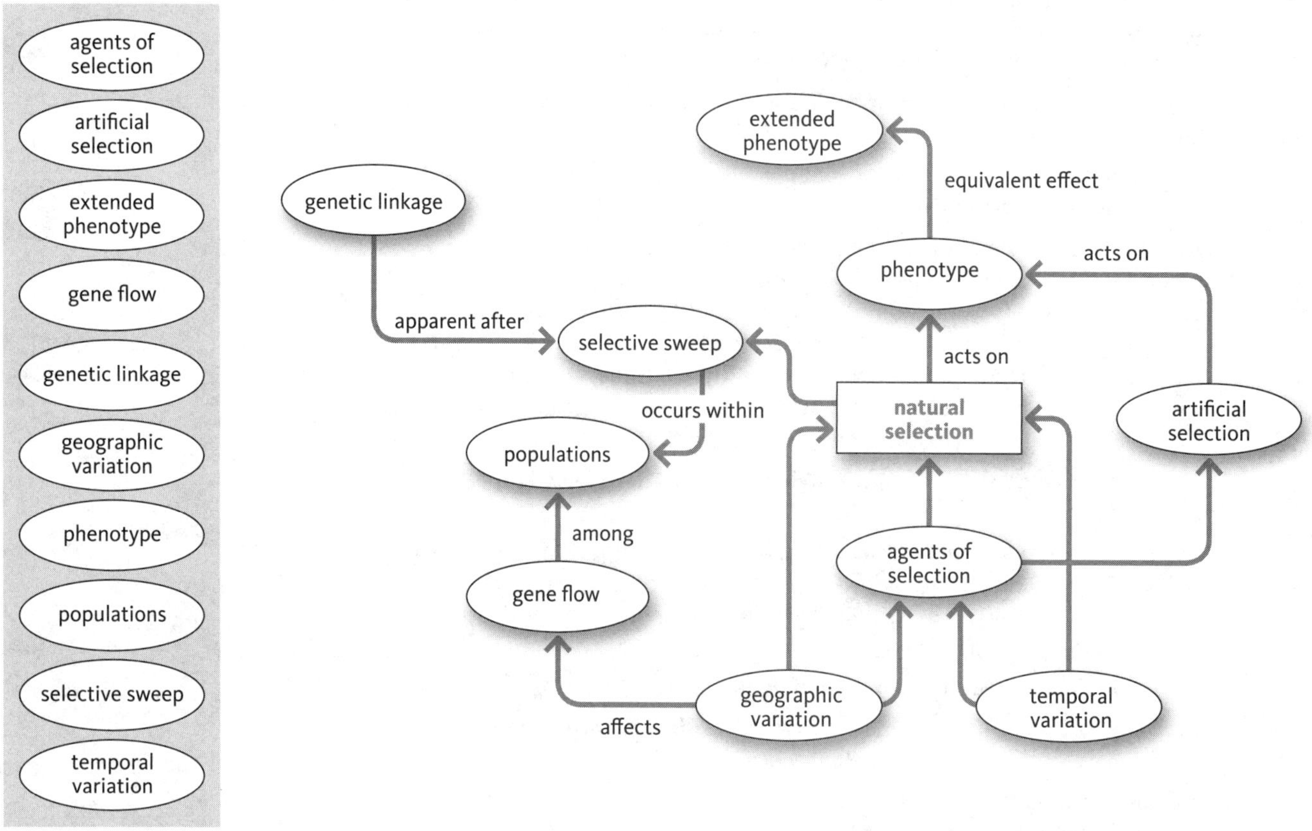

Interpret the Data

- At what frequency do scientists expect to find the lactase persistence phenotype in Australia?

  .3, .2, and .1
- Why might the frequency of the lactase persistence phenotype be high in Pakistan?

  Humans with the alleles of the lactase gene, *LCT*, that conferred lactase persistence migrated from Africa (or Europe) to Pakistan bringing cattle and a cultural shift to drinking their milk.
- What explains the pattern of frequencies around the world?

  A balance between strong selection in regions with a history of domestication, on the one hand, and gene flow, on the other.

Overcoming Misconceptions

Which of the following is a true statement?

c. Natural selection and genetic drift are both important mechanisms of evolution.

Delve Deeper

1. How many different lines of evidence allowed scientists to reconstruct the recent history of natural selection on sticklebacks in freshwater lakes?

   Scientists used at least three distinct lines of evidence to reconstruct stickleback history. Some of the evidence points to the necessary conditions for evolution by natural selection, and other evidence supports the evolution of the lineage over time. For example, scientists used population studies of living fish to determine how armor and spines affected survival of sticklebacks, and they looked for variation in the amount of armor among individuals. Other scientists examined the ecological circumstances under which armor was a favorable trait, and watched over time as natural selection rapidly favored individuals with little body armor in a freshwater lake. Other evidence came from QTL mapping, where scientists analyzed the source of the variation in the stickleback genome and found alternative alleles for the *ectodysplasin* (*Eda*) gene. They found a recessive "low" *Eda* allele that led to fewer lateral plates (low armor) during development. So, these lines of evidence pointed to heritable variation that could affect survival or reproductive success. More importantly, the appearance of the *Eda* allele could be traced to marine ancestors at least 2 million years ago. Evidence that the lineage evolved over time as a result of natural selection came from the fossil record. Scientists examined the degree of armor found in fossils and what time periods were associated with fossils with different degrees of armor.

   Taken together these lines of evidence provide enough information about the historical past of natural selection. However, another natural experiment in Lake Washington added additional evidence for the ecological circumstances leading to the evolution of armor.

Test Yourself

1. d; 2. b; 3. e; 4. b; 5. a; 6. d; 7. b; 8. e; 9. a; 10. c; 11. b

9 The History in Our Genes

Check Your Understanding

1. Which of the following is NOT a mutation that potentially alters DNA of future generations?
   a. Chromosome fusion

      Incorrect. Chromosome fusion occurs when two chromosomes are joined together as one. This anomaly can be passed down to future generations if it occurs within the germ line (see Section 5.2).

   b. Deletion

      Incorrect. Segments of DNA may be deleted accidentally, leading to the disappearance of a small portion of a gene or an entire set of genes. This anomaly can be passed down to future generations if it occurs within the germ line (see Section 5.2).

   c. Insertion

      Incorrect. A segment of DNA, as short as a single base or as long as thousands of bases (including entire genes), can be inserted into the middle of an existing sequence. This anomaly can be passed down to future generations if it occurs within the germ line (see Section 5.2).

   d. Point mutation

      Incorrect. Also known as a substitution, point mutations occur when a single base changes from one nucleotide to another. This anomaly can be passed down to future generations if it occurs within the germ line (see Section 5.2).

   e. Morphogen

      Correct. A morphogen is a signaling molecule that acts to alter expression of target genes—that is, it influences the expression of DNA, but it does not alter DNA.

2. Are homologous traits ancestral or derived traits?
   a. They are ancestral because homologous traits are inherited from a common ancestor.

      Correct, but so are other answers. Homologous traits are inherited from a common ancestor, but the homology depends on the clade of interest within the phylogeny. A homologous trait could be a shared derived trait (e.g., lobe-fins) in a clade that shares a common ancestor with other clades without that trait (e.g., teleosts). See Section 2.3 for an introduction to homology, and Chapter 4 to learn how homologies function in phylogenies.

   b. It depends on the clade of interest within the phylogeny.

      Correct. Within a phylogeny, homologous traits can be considered those inherited from a common ancestor, or those that are derived traits shared among members of a clade. Whether the trait in question is ancestral or derived depends on the area of interest. For example, the tetrapod limb can be considered an ancestral homology when looking at the evolution of mammals, but derived when looking at the classification of vertebrates (see Figure 4.22). Section 2.3 introduces homology, and Chapter 4 explains how homologies function in phylogenies.

   c. They are derived because homologous traits define a monophyletic clade.

      Correct, but so are other answers. Homologous traits can be used to define clades, but whether the homology is ancestral or derived depends on the area of interest within the phylogeny. For example, all members of the lobe-fins share a homologous limb structure, a derived trait, but that limb structure is ancestral if you are interested in the origin of weight-bearing elbows (Figure 4.22). See Section 2.3 for an introduction to homology, and Chapter 4 to learn how homologies function in phylogenies.

d. None of the above

Incorrect. Homologous traits can be considered those inherited from a common ancestor, or those that are derived traits shared among members of a clade. Whether the trait in question is ancestral or derived depends on the area of interest. For example, the tetrapod limb can be considered an ancestral homology when looking at the evolution of mammals, but derived when looking at the classification of vertebrates (see Figure 4.22). See Section 2.3 for an introduction to homology, and Chapter 4 to learn how homologies function in phylogenies.

3. Are synapomorphies important when developing a phylogeny?

a. Synapomorphies are not important when developing phylogenies because synapomorphies are traits shared by all the members of a species, population, or gene family, and they provide little useful information.

Incorrect. A synapomorphy is a trait that evolved in the immediate common ancestor of a clade and inherited by all the descendants. So, all the members of the clade do share the trait, whether that be a species, a population, or a gene family. But, because the ancestor of such a group does not possess the trait, a synapomorphy can be very useful in identifying the historical relationships within the phylogeny (see Section 4.3 for how to use synapomorphies for developing a phylogeny of carnivores).

b. Synapomorphies are not important when developing phylogenies because they are produced by convergent evolution or evolutionary reversal and as such, can confound the phylogeny.

Incorrect. Homoplasy is the similarity of traits that is *not* due to common ancestry. A synapomorphy is a trait that evolved in the immediate common ancestor of a clade and inherited by all the descendants. Because the ancestor of such a group does not possess the trait, a synapomorphy can be very useful in identifying the historical relationships within the phylogeny (see Section 4.3 for how to use synapomorphies for developing a phylogeny of carnivores).

c. Synapomorphies are important when developing phylogenies because they are a lineage of tetrapods that gave rise to mammals.

Incorrect. Synapsids are the lineage of tetrapods that gave rise to mammals. A synapomorphy is a trait that evolved in the immediate common ancestor of a clade and inherited by all the descendants. Because the ancestor of such a group does not possess the trait, a synapomorphy can be very useful in identifying the historical relationships within the phylogeny (see Section 4.3 for how to use synapomorphies for developing a phylogeny of carnivores).

d. Synapomorphies are important when developing phylogenies because they are the shared derived traits that can distinguish groups of species, populations, or genes from other groups sharing different characters.

Correct. A synapomorphy is a trait that evolved in the immediate common ancestor of a clade and inherited by all the descendants. Because the ancestor of such a group does not possess the trait, a synapomorphy can be very useful in identifying the historical relationships within the phylogeny (see Section 4.3 for how to use synapomorphies for developing a phylogeny of carnivores).

Identify Key Terms

1. i; 2. f; 3. c; 4. a; 5. e; 6. d; 7. h; 8. o; 9. k; 10. g; 11. b; 12. n; 13. j; 14. l; 15. p; 16. m

Link Concepts

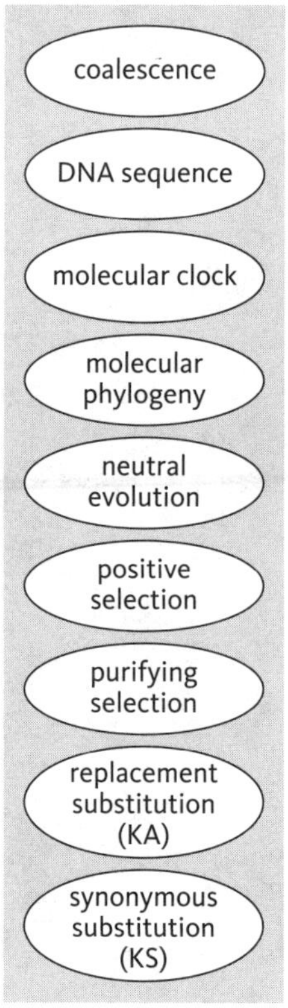

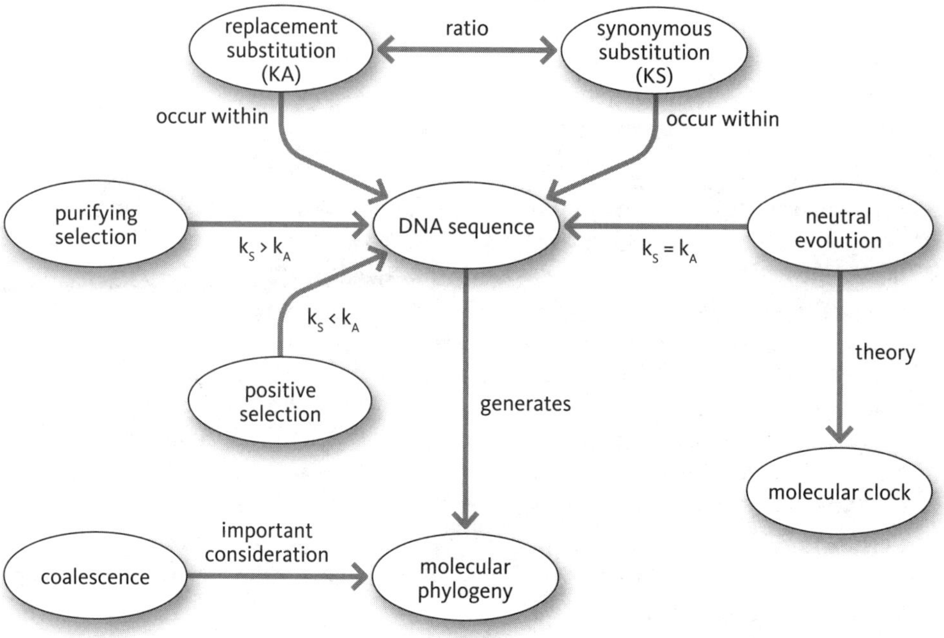

Interpret the Data

- What group shares the most recent common ancestor with the group of humans that colonized the Americas?

  Humans from East Asia

- According to Tishkoff et al. (2009), the longer branch lengths for groups of humans in the Americas, Oceania, and Pygmy, and some of the hunter-gatherers, indicate high levels of genetic drift. Why might those groups have experienced higher levels of genetic drift than groups that remained in Africa?

  Because those groups may have started with just a few individuals as small numbers migrated to new locations. The influence of genetic drift is greater in small populations than in large populations, so different initial population sizes should leave different genetic signatures.

Games and Exercises

	Horse	Donkey	Whale	Chicken	Penguin	Snake	Moth	Yeast	Wheat
Horse	0	1	5	11	13	21	29	46	46
Donkey		0	4	10	12	20	28	45	45
Whale			0	9	11	17	27	45	44
Chicken				0	3	17	28	47	46
Penguin					0	19	26	45	46
Snake						0	30	46	45
Moth							0	48	46
Yeast								0	48
Wheat									0

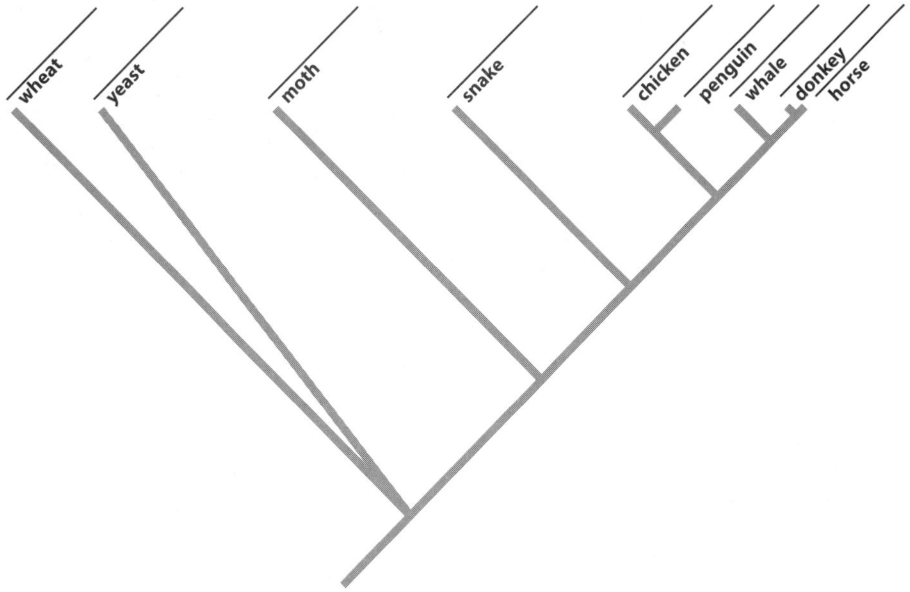

	Horse	Donkey	Whale	Chicken	Penguin	Snake	Moth	Yeast	Wheat
Human	12	11	10	13	13	14	29	44	44

Overcoming Misconceptions

Which of the following is a true statement?

a. The coalescent represents the lineage of an allele that arose and ultimately persisted within a population.

Delve Deeper

1. Do scientists use different lines of evidence to support and test phylogenetic hypotheses developed with molecular data?

   Yes. Scientists often check their results against different lines of evidence, such as the fossil record and morphological evidence. Molecular data can be used to develop phylogenetic hypotheses and tested with other lines of evidence, or molecular data can be used to support and/or test phylogenetic hypotheses developed with evidence from the fossil record or morphological data.

2. How can molecular data be used to understand natural selection in the past?

   Scientists can look for evidence of different kinds of selection in the past by comparing the ratios of synonymous to replacement substitutions within coding regions. The probability of a substitution at a synonymous site should be equal to the probability of a substitution at a replacement site. When they are not equal, scientists can reject the hypothesis of neutral evolution. Then they can look at the patterns of the ratios of synonymous to replacement substitutions. For example, strong positive selection on a gene can lead to unusually large numbers of replacement substitutions that change the structure of proteins, whereas purifying selection can eliminate replacement mutations to genes with essential functions that are easily disrupted by mutations, leading to a very low number of replacement substitutions compared to synonymous ones. For example, the FOXP2 gene has changed very little in our common ancestors, but in humans, two amino acids have changed in the protein in the past 6 million years and may contribute to our human-ness.

Test Yourself

1. d; 2. d; 3. c; 4. d; 5. e; 6. d; 7. a; 8. b; 9. d

10 Adaptation

From Genes to Traits

Check Your Understanding

1. Which of the following statements about genetically controlled traits is true?

   a. Interactions among alleles at different genetic loci (epistasis) can affect the expression of a trait, such as height, in different ways from individual to individual.

      Correct, but so are other answers. Many traits are polygenic—they are influenced by many genetic loci. The expression of human height, for example, may depend on interactions between alleles at different loci (epistasis), and/or it may result from interactions between alleles and the environment (phenotypic plasticity). So height can vary among individuals because individuals have different alleles for different genes, *and* these alleles can vary in their contributions to height because of the variation in the potential interactions among the genetic loci and the environment (Section 7.1).

b. A single genotype may produce different phenotypes depending on the environment.

Correct, but so are other answers. Phenotypic plasticity is the phenotypic variation among individuals that results when a single genotype responds to different environments (Section 5.5). The environment can be internal (e. g., regulatory genes and non-coding elements), or external (e. g., light or temperature). A single organism changing from brown to white fur, for example, is not necessarily evolution, however. The evolution of phenotypic plasticity comes when a *population*, such as snowshoe hares, becomes more or less sensitive to day length as a cue for the color change of their fur (Section 7.4).

c. A single mutation to a regulatory gene can affect many phenotypic traits.

Correct, but so are other answers. Traits are often highly interconnected and complex. Regulatory genes, for example, often influence the expression of more than just a single gene—they influence many genes, and therefore, potentially many phenotypic traits. A mutation to this kind of gene is likely to be pleiotropic (affecting many traits), and a single base change can have varied consequences. In fact, antagonistic pleiotropy is when a mutation causes beneficial effects for one trait and detrimental effects on other traits (Section 6.6).

d. All of the above

Correct. Many traits are polygenic—they are influenced by many genetic loci. Individuals can vary because they have different alleles for different genes, *and* because these alleles occur in different combinations in different individuals, resulting in diverse epistatic effects from individual to individual (Section 7.1). Phenotypic plasticity occurs when alleles interact with the environment in ways that affect the expression of a phenotype, and these genotype x environment interactions can contribute to variation as well (Section 5.5). Finally, a single mutation can affect many phenotypic traits if, for example, the mutation occurs in a regulatory gene (Section 6.6).

e. b and c only

Incorrect. Although both phenotypic plasticity and pleiotropy can affect phenotypic variation, many traits are polygenic—they are influenced by many genetic loci. The expression of human height, for example, may depend on interactions between alleles at different loci, and/or it may result from the interaction of the different alleles with the environment (phenotypic plasticity). So height can vary among individuals because individuals have different alleles for different genes, *and* these alleles can vary in their contributions to height because of the variation in the potential interactions among the genetic loci (Section 7.1).

f. None of the above

Incorrect. Many traits are polygenic—they are influenced by many genetic loci. Individuals can vary because they have different alleles for different genes, *and* these alleles can vary in their contributions because of variation in the potential interactions among the genetic loci (Section 7.1). Phenotypic plasticity occurs when alleles interact with the environment in ways that affect the expression of a phenotype, and these genotype x environment interactions can contribute to variation as well (Section 5.5). Finally, a single mutation can affect many phenotypic traits if, for example, the mutation occurs in a regulatory gene (Section 6.6).

2. Which effects of a mutation contribute most to the phenotypic resemblance among relatives: additive effects, dominant effects, or epistatic effects?

   a. Additive effects

      Correct. An additive allele will have twice the effect on a phenotype when two copies are present at a given locus than when only a single copy is present. Additive alleles are not influenced by the presence of other alleles, so their effects are not context (genotype) dependent the way dominant and epistatic effects are (see Box 7.1). Because of this, additive effects of alleles cause relatives to resemble each other. Scientists have characterized the influence of this effect on variation among individuals by narrow-sense heritability (see Section 7.1).

   b. Dominant effects

      Incorrect. A dominant allele overshadows the other allele at the same locus (see Box 5.2 for an introduction). The effects of dominant alleles on a phenotype are context (genotype) dependent, a context that changes every generation after meiosis. Because of this, dominance effects do not typically contribute to the phenotypic resemblance among relatives.

   c. Epistatic effects

      Incorrect. An epistatic allele interacts with alleles at other loci elsewhere in the genome, so the effect of an epistatic allele on the phenotype depends on the genotypic context (see Box 7.1). Moreover, that context changes every generation after meiosis. Because of this, epistatic effects do not typically contribute to the phenotypic resemblance among relatives.

   d. All are equally important to heritability.

      Incorrect. An additive allele will have twice the effect on a phenotype when two copies are present at a given locus than when only a single copy is present. Additive alleles are not influenced by the presence of other alleles, so their effects are not context (genotype) dependent the way dominant and epistatic effects are (see Box 7.1). A dominant allele overshadows the other allele at the same locus (see Box 5.2 for an introduction), and an epistatic allele interacts with alleles at other loci elsewhere in the genome, so the effect of an epistatic allele on the phenotype depends on the genotypic context (see Box 7.1). The effects of both dominant and epistatic alleles on a phenotype are context (genotype) dependent, a context that changes every generation after meiosis. Because of this, dominance and epistatic effects do not typically contribute to the phenotypic resemblance among relatives.

   e. None are important.

      Incorrect. Additive effects general cause phenotypic resemblance among relatives. An additive allele will have twice the effect on a phenotype when two copies are present at a given locus than when only a single copy is present. Additive alleles are not influenced by the presence of other alleles, so their effects are not context (genotype) dependent the way dominant and epistatic effects are (see Box 7.1). A dominant allele overshadows the other allele at the same locus (see Box 5.2 for an introduction), and an epistatic allele interacts with alleles at other loci elsewhere in the genome, so the effect of an epistatic allele on the phenotype depends on the genotypic context (see Box 7.1). The effects of both dominant and epistatic alleles on a phenotype are context (genotype) dependent, a context that changes every generation after meiosis. Because of this, dominance and epistatic effects do not typically contribute to the phenotypic resemblance among relatives.

3. What is a cis-acting regulatory element?

a. Stretches of DNA that are located far away from a focal gene (e.g., on another chromosome) that influence the expression of that gene

Incorrect. Trans-acting elements can be located away from a focal gene, even on another chromosome, and they tend to code for proteins, microRNAs, or other diffusible molecules. Cis-acting elements function to regulate nearby genes. They can be found up- or downstream from the focal gene, or even inside an intron. Cis elements often code for binding sites for one or more trans-acting factors. A mutation to either a cis-acting or a trans-acting element can influence gene expression (see Section 5.2).

b. Stretches of DNA located near a focal gene that influence the expression of that gene

Correct, but so are other answers. Cis-acting elements, also known as cis-regulatory regions, cis-regions, and cis-regulatory elements, function to regulate nearby genes. They can be found up- or downstream from the focal gene, or even inside an intron. Cis elements often code for binding sites for one or more transposable factors. Trans-acting elements are located away from a focal gene, even on another chromosome, and they tend to code for proteins, microRNAs, or other diffusible molecules. A mutation to either a cis-acting or a trans-acting element can influence gene expression (see Section 5.2).

c. A non-coding region of the genome that can be found either immediately upstream (adjacent to the promoter region), downstream, or inside an intron

Correct, but so are other answers. Cis-acting elements, also known as cis-regulatory regions, cis-regions, and cis-regulatory elements, are non-coding regions that function to regulate nearby genes. They can be found up- or downstream from the focal gene, or even inside an intron. Cis elements often code for binding sites for one or more transposable factors. Trans-acting elements are located away from a focal gene, even on another chromosome, and they tend to code for proteins, microRNAs, or other diffusible molecules. A mutation to either a cis-acting or a trans-acting element can influence gene expression (see Section 5.2).

d. All of the above

Incorrect. Trans-acting elements are located away from a focal gene, even on another chromosome, and they tend to code for proteins, microRNAs, or other diffusible molecules. Cis-acting elements function to regulate nearby genes. They can be found up- or downstream from the focal gene, or even inside an intron. Cis elements often code for binding sites for one or more transposable factors. A mutation to either a cis-acting or a trans-acting element can influence gene expression (see Section 5.2).

e. b and c only

Correct. Cis-acting elements, also known as cis-regulatory regions, cis-regions, and cis-regulatory elements, are non-coding regions that function to regulate nearby genes. They can be found up- or downstream from the focal gene, or even inside an intron. Cis elements often code for binding sites for one or more transposable factors. Trans-acting elements are located away from a focal gene, even on another chromosome, and they tend to code for proteins, microRNAs, or other diffusible molecules. A mutation to either a cis-acting or a trans-acting element can influence gene expression (see Section 5.2).

f. None of the above

Incorrect. Cis-acting elements, also known as cis-regulatory regions, cis-regions, and cis-regulatory elements, are non-coding regions that function to regulate nearby genes. They can be found up- or downstream from the focal gene, or even inside an intron. Cis elements often code for binding sites for one or more transposable factors. Trans-acting elements are located away from a focal gene, even on another chromosome, and they tend to code for proteins, microRNAs, or other diffusible molecules. A mutation to either a cis-acting or a trans-acting element can influence gene expression (see Section 5.2).

Identify Key Terms

1. f; 2. h; 3. j; 4. i; 5. g; 6. b; 7. c; 8. a; 9. d; 10. e

Link Concepts

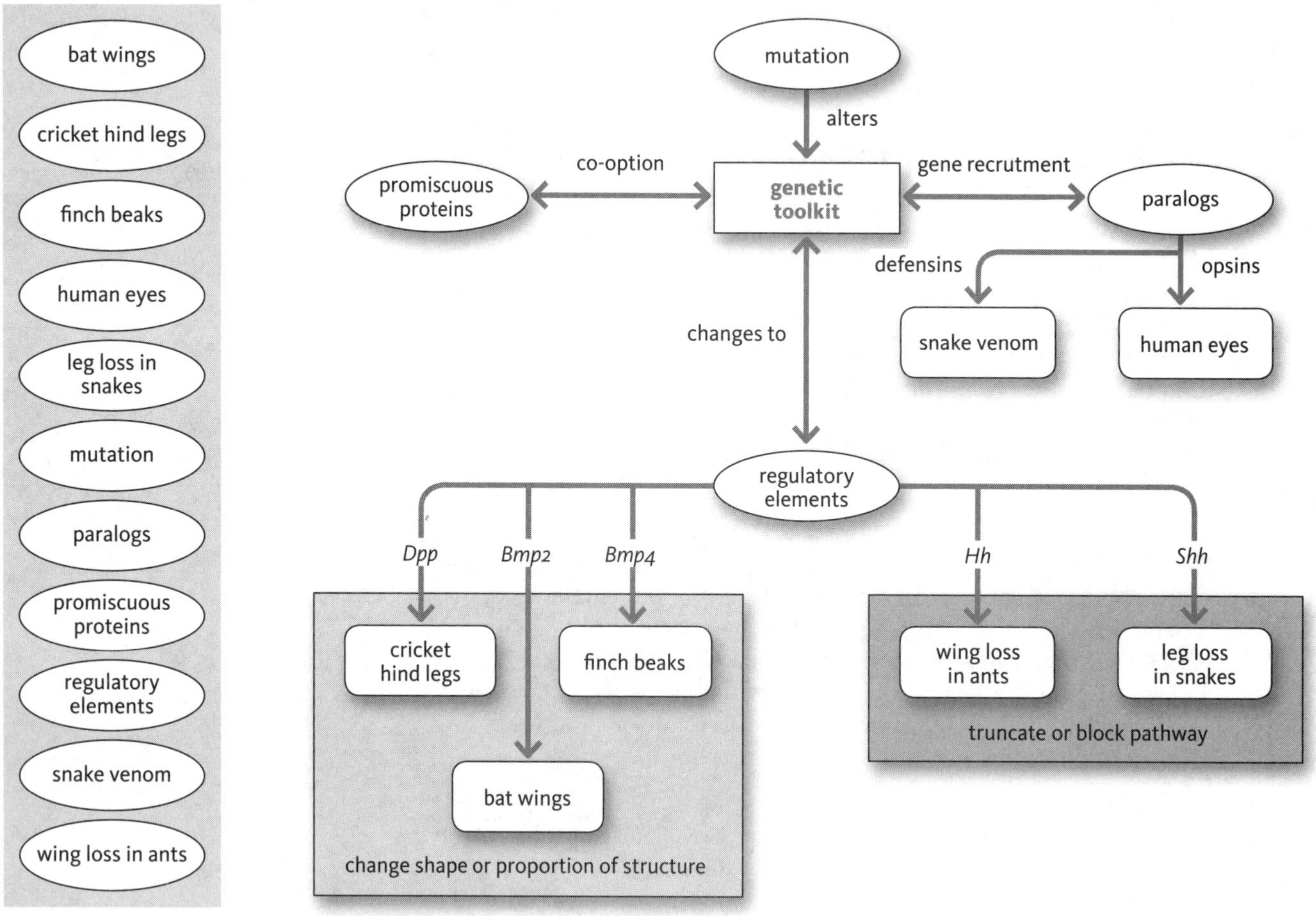

Interpret the Data

- Are all opsins used in vision?

  Not necessarily. Placopsins branched off before opsins could sense light.

- How many types of opsins can be found in the deuterostomes (a group of animals that includes humans)?

  Three - R-opsins, RGR/Go-opsins, and C-opsins

- Are Cnidarians sensitive to light?

  Yes. They have receptors for R-opsins, C-opsins, and RGR/Go-opsins.

- What mutation was necessary for the evolution of opsins? Why was that important?

  Duplication of the opsin gene. Duplication was important because the original gene could still function to produce some receptors while evolution acted on the duplicated gene to produce other receptors.

Overcoming Misconceptions

Which of the following is a true statement?

b. Through natural selection and time, evolution can produce seemingly complex adaptations through gene duplication and recruitment.

Delve Deeper

1.

Hox genes

Where are they located?
Within an organism, every cell contains exactly the same DNA. Gene regulatory networks are simply part of the DNA. In multicellular organisms, although the regulatory networks can be found in every cell, the same pathway may lead to different outcomes depending on the cell. For example, different cells in an organism's body may receive stronger or weaker signals in the pathway simply because they are closer to or farther from the origin of the signal (e.g., as a hormone diffuses across a body).

What do they do?
Gene regulatory networks can be thought of a series of interactions whose outcome depends on the interactions themselves, in essence like a flow chart. A simple network may start with an environmental signal that leads to the expression of a gene producing a transcription factor. The transcription factor may then bind to cis-acting element of other transcription factor genes. Those transcription factors may then bind to still other cis-acting regulatory elements, and so on and so on. Finally, protein-coding genes will either be activated or not, depending on the outcome of all that upstream regulation.

How might a mutation affect their function?
A mutation can affect any step along the way in the "flow chart." A mutation can alter the production of the protein-coding gene itself or any of the trans- and cis-acting elements in the hierarchy. So, a mutation that affects a transcription factor gene in the network can halt the entire remainder of the network, or it can shift it to another network, if, for example, it now binds to a new cis-acting element. The result could be production of an entirely new protein, a new amount of the protein, or production of in a new cell, and likewise for cis-acting elements.

2. If the DNA in every cell within an organism includes all the *Hox* genes for the development of that organism, why doesn't every cell develop into a cell in the eye, for example?

   Because every cell does not respond exactly the same to every regulatory network. For example, a cell may respond differently because of its location in the body of the organism. So, a diffusible signal, like a morphogen, may be more or less concentrated depending on how close the signal source is to the cell. Or the genes in a cell may respond differently depending on the amount of other gene products present that are present as the result of other regulatory networks. So, although every cell potentially has the capacity to develop into an eye cell, they will only do so if ALL of the other elements in the regulatory network are functioning to produce an eye cell.

Test Yourself

1. e; 2. e; 3. d; 4. a; 5. d; 6. c; 7. c; 8. d

11 Sex

Causes and Consequences

Check Your Understanding

1. What is the most important element necessary for natural selection to act?

   a. Individuals that can change to meet their needs

      Incorrect. Although individuals can change to meet their needs, these changes have no genetic basis—evolution cannot identify an individual's needs. Natural selection is a population-based phenomenon that requires heritable variation among individuals that confers some advantage or disadvantage. Within a population, those that do better, for example, are more likely to reproduce and have offspring that share some of the variable traits that led to that success. Over time, individuals with those successful traits become more common than those without—the essence of natural selection (see Sections 1.4 and 2.3 and Chapter 8, especially Section 8.2).

   b. Variation within a population

      Correct. Heritable variation among individuals serves as a vital raw material for natural selection. Natural selection is a population-based phenomenon that requires heritable variation among individuals that confers some advantage or disadvantage. Within a population, those that do better, for example, are more likely to reproduce and have offspring that share some of the variable traits that led to that success. Over time, individuals with those successful traits become more common than those without—the essence of natural selection (see Chapter 8).

   c. Thousands of years

      Incorrect. Although lineages can evolve in relatively short time periods, evolution often requires a significantly longer period than thousands of years. For example, fast generation times (such as in insects, viruses, and bacteria, Section 1.2) or when natural selection is particularly strong (such as selective sweeps, Section 8.6) can result in rapid evolution, but the evolution of hominids took millions of years.

d. All of the above

Incorrect. Individuals can change to meet their needs, but these changes have no genetic basis—evolution cannot identify an individual's needs (see Sections 1.4 and 2.3 and Chapter 8, especially Section 8.2). Also, evolution usually takes time. Although lineages can evolve in relatively short time periods, evolution often requires a significantly longer period than thousands of years. Variation within a population is essential for natural selection and can be considered the most important element for it to act (see Chapter 8).

e. None of the above

Incorrect. It is correct to say that individuals changing to meet their needs is not an important to natural selection, and that thousands of years is not *the most important* element. However, variation within a population is essential for natural selection and can be considered the most important element for it to act (see Chapter 8).

2. How does $R = h^2 \times S$ apply to artificial selection?

a. In artificial selection, breeders manipulate the entire reproductive process of all individuals, so the response to selection is simply the strength of that selection, $R = S$.

Incorrect. $R = h^2 \times S$ is called the breeders equation. It describes the evolutionary response (R) to selection that results from the interaction of the heritability of a trait (narrow sense) and the strength of selection (S) acting on the trait. In artificial selection, breeders serve as strong selective forces, picking and choosing individuals with the traits that they want to breed and produce the next generation. Heritability is still important in artificial selection. The greater the heritability of the trait, the greater response the breeder will see to his/her selection (Section 7.2).

b. In artificial selection, breeders create very strong selection on traits, which can result in a rapid evolutionary response if the heritability of the selected traits is high.

Correct. $R = h^2 \times S$ is called the breeders equation. It describes the evolutionary response (R) to selection that results from the interaction of the heritability of a trait (narrow sense) and the strength of selection (S) acting on the trait. In artificial selection, breeders serve as strong selective forces, picking and choosing individuals with the traits that they want to breed and produce the next generation. The greater the heritability of the trait, the greater response the breeder will see to his/her selection (Section 7.2).

c. In artificial selection, breeders manipulate narrow sense heritability, h^2, by selecting for specific traits.

Incorrect. $R = h^2 \times S$ is called the breeders equation. It describes the evolutionary response (R) to selection that results from the interaction of the heritability of a trait (narrow sense) and the strength of selection (S) acting on the trait. In artificial selection, breeders serve as strong selective forces, picking and choosing individuals with the traits that they want to breed and produce the next generation. They cannot manipulate the heritability of those traits, but the greater the heritability of the trait, the greater response the breeder will see to his/her selection (Section 7.2).

d. $R = h^2 \times S$ is the breeder's equation and therefore only applies to artificial selection.

Incorrect. $R = h^2 \times S$ is called the breeders equation, but this equation does not apply only to artificial selection. It describes the evolutionary response (R) to selection that results from the interaction of the heritability of a trait (narrow sense) and the strength of selection (S) acting on the trait in any population. In nature, both heritability and strength of selection can vary. By measuring any two of the variables in the equation (R, h^2, or S), scientists can solve for the third variable (Section 7.2).

3. Is phenotypic plasticity heritable?

   a. No. Phenotypic plasticity is the capacity of an individual to change in response to the environment, and characteristics acquired during an individual's lifetime are not heritable.

      Incorrect. Phenotypic plasticity is the capacity for a genotype to express more than one phenotype depending on the environment. So, like a hare turning white in winter, individuals can change. The difference between this kind of change and the Lamarckian concept of acquired characteristics is that phenotypic plasticity has a genetic basis—a basis that is heritable (see Section 7.4).

   b. Yes. Phenotypic plasticity is the capacity of an individual to change in response to the environment, and this ability to change in response to need is the very basis of evolution.

      Incorrect. Phenotypic plasticity is the capacity for a genotype to express more than one phenotype depending on the environment. The outcome is not a response to need, however. Overtime, individuals that responded in the right way at the right time fared better than those that did not. So hares that turned from white to brown as temperatures reached a certain threshold in spring produced more offspring than those that turned too late and became mismatched to their environment. Phenotypic plasticity is likely very important in the evolutionary responses (R) of lineages to selection, however (see Section 7.4).

   c. No. Phenotypes cannot be directly linked to any heritable component of genes, so phenotypic plasticity cannot be heritable.

      Incorrect. Although the link between phenotype and genotype is often complex, they can be linked. Indeed, the link can be observed with careful studies of phenotypic plasticity. Phenotypic plasticity is the capacity for a genotype to express more than one phenotype depending on the environment. Genetically identical plants can grow large leaves in high light and small leaves in low light. This reaction norm may be heritable, although it may vary among different genotypes (see Section 7.4).

   d. Yes. Individuals may vary in their plastic responses to the environment, and those responses can be heritable.

      Correct. Phenotypic plasticity is the capacity for a genotype to express more than one phenotype depending on the environment. Genetically identical plants can grow large leaves in high light and small leaves in low light. This reaction norm may be heritable, and it may vary among different genotypes (see Section 7.4).

Identify Key Terms

1. s; 2. e; 3. c; 4. o; 5. r; 6. a; 7. w; 8. p; 9. h; 10. u; 11. v; 12. d; 13. m; 14. j; 15. i; 16. k; 17. t; 18. g; 19. x; 20. q; 21. b; 22. n; 23. l; 24. f

Link Concepts

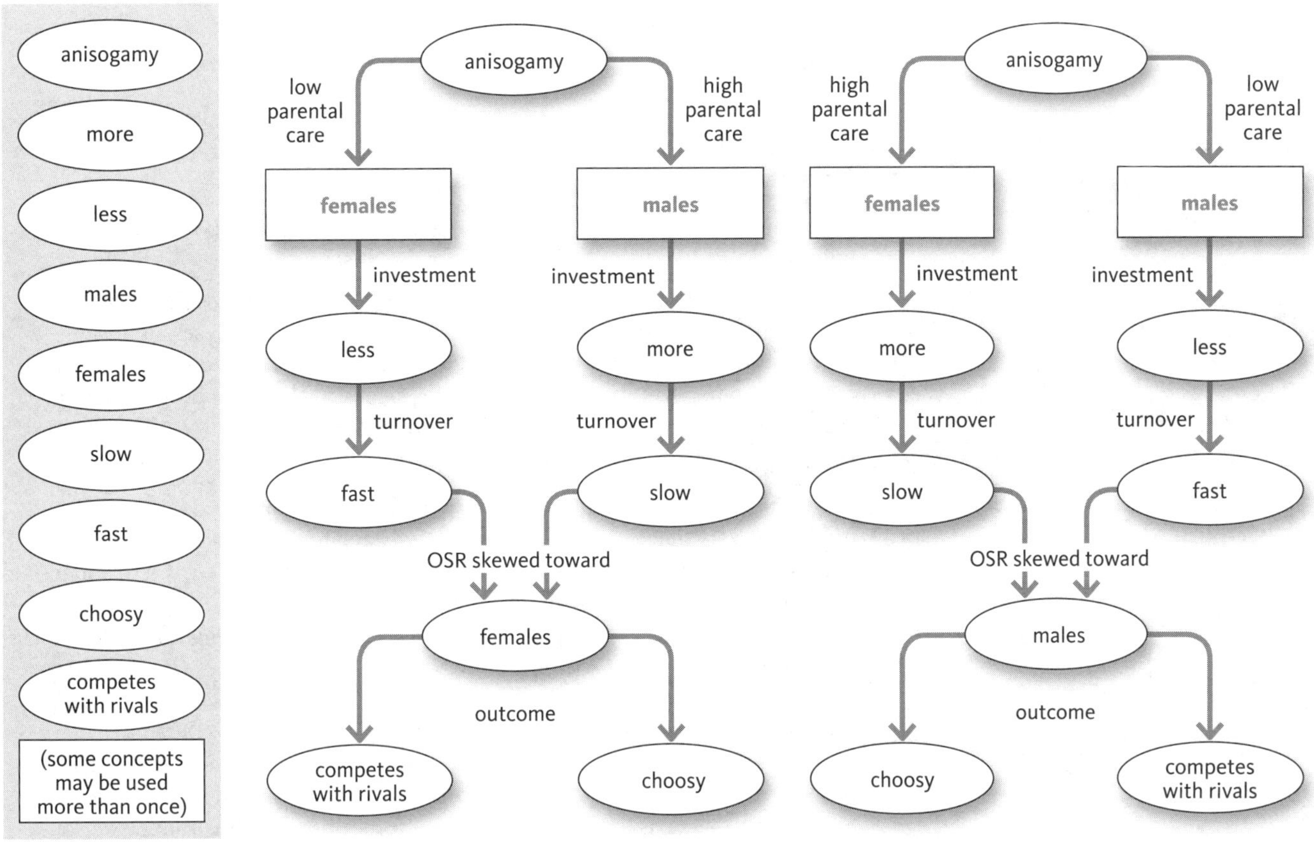

Interpret the Data

- What was the longest phallus found for a species?

  Approximately 19 cm
- What was the maximum number of pouches found in females for a species?

  3 pouches
- What was the maximum number of spirals found in females for a species?

  8 spirals
- From these graphs, could you predict the number of pouches or spirals you would expect to find in a species where the male phallus length was 9 cm?

  Yes, you could predict that in a species where the male had a mean phallus length of 9 cm, the female reproductive tract would likely have between 0 and 2 pouches, and 0 to 4 spirals.
- Do the relationships between phallus length and number of pouches/spirals shown in the above graphs indicate that male phallus length causes female reproductive tract characteristics?

  No. The relationships are correlations; they show a positive relationship between the variables (male phallus length and either number of pouches or number of spirals in the female reproductive tract) but not that male phallus length *causes* either one of the female characteristics. Selection is clearly acting on both males and females, but causative relationships would require a mechanism for how the male phallus length actually made the female reproductive tract change.

Overcoming Misconceptions

Which of the following is a true statement?

d. Ornaments will evolve through sexual selection when individuals vary in a sexually selected trait, that variation is heritable, and that variation affects reproductive success.

Delve Deeper

1. In Figure 11.27, the genus *Homo* is included. What does the position of the point indicate?

   Because we share a common ancestor with the other primates depicted in the graph, including our genus allows us to examine hypotheses about how evolution has shaped our traits. The position of our genus indicates that the relationship between body size and testes size is in the same range as primates that experience relatively little sperm competition.

2. What's the difference between sex, reproduction, and copulation?

   Sex is the male/female status of an individual, with females defined as the sex producing a small number of relatively large gametes, and males defined as the sex producing a large number of relatively small gametes. Often, the sex of an individual is determined by the compliment of chromosomes it inherits–for example, in humans a female develops when the sperm and egg that fuse to form a zygote each carry the X chromosome, whereas males develop when the egg carries an X and the sperm a Y chromosome. In other species, the heterogametic sex (the sex with un-matched chromosomes) is the female, not the male, such as in many birds which have a ZW system. In still other species, such as many turtles, sex is determined by external circumstances encountered by the animals as they develop, such as temperature. There are species where the same individual switches sex during its lifetime, and there are hermaphroditic species where individuals are both sexes simultaneously. And in some insects, females have paired chromosomes (they are diploid) and males have just a single chromosome (they are haploid); sex is determined by the act of reproduction itself (see below)!

   Reproduction is the act of producing offspring. Reproduction can occur with or without sex, depending on whether genetic material from multiple individuals is combined. In a typical species with sexual reproduction, chromosomal copies are separated and the ploidy (copy number) is halved prior to reproduction, in a special cell cycle called meiosis that produces eggs or sperm. These haploid gametes then fuse, restoring the full complement of chromosomal copies and bringing together genetic material from both a male and a female.

   Copulation is the physical act of mating, where males transfer sperm to females. For many species copulation is a necessary part of sexual reproduction as it brings male gametes into close proximity with eggs. However, it is not the same thing as reproduction. Many copulations do not lead to reproduction. This is especially true in humans because female ovulation is relatively concealed, making it difficult to time copulation with periods of fertility, but it is also true in species where females mate with multiple males and sperm compete for access to eggs. Reproduction can also occur without copulation. Asexual species reproduce without copulation, for example, as do many sexual species for which fertilization occurs outside of the bodies of females. Males of many springtails (an early lineage of flightless insects) deposit globs of sperm on stalks, so that females might bump into them as they walk by, and males of many fish and frogs spray their sperm onto clutches of eggs. In all of these cases, sexual reproduction is achieved without copulation.

Test Yourself

1. e; 2. e; 3. e; 4. e; 5. a; 6. d; 7. d; 8. c; 9. a

12 After Conception
The Evolution of Life History and Parental Care

Check Your Understanding

1. How does sexual selection differ from natural selection?

   a. Sexual selection only acts on males, whereas natural selection acts on populations.

   Incorrect. Natural selection and sexual selection both act on the heritable variation among individuals within a population (Section 2.3); sexual selection can be considered a subset of natural selection. It's the differences in fitness that arise from competition over access to reproduction. Although sexual selection typically involves males competing with rival males for access to females, occasionally it can work the other way around, with females competing for access to males (Section 11.2).

   b. Variation among the choosy sex is not necessary for sexual selection to operate; variation among all individuals is necessary for natural selection to operate.

   Incorrect. Natural selection and sexual selection both act on the heritable variation among individuals within a population (Section 2.3); the difference is that sexual selection is the differential reproductive success that results because individuals vary in traits that affect their ability to compete for fertilizations. Sexual selection affects the sex with the greater variance in reproductive success, but both the trait and the preference for that trait can vary (Section 11.2).

   c. Sexual selection does not really differ from natural selection; both require heritable variation in traits that confer some fitness differences to individuals possessing those traits.

   Correct, but so are other answers. Natural selection and sexual selection both act on the heritable variation among individuals within a population that leads to differential fitness (Section 2.3). Sexual selection refers specifically to the differential reproductive success that results because individuals vary in traits that affect their ability to compete for fertilizations. So sexual selection can be considered a specific type of natural selection (see Section 11.2).

   d. No. Sexual selection does not really differ from natural selection, but the optimum trait value under sexual selection can often be at odds with the optimum trait value under natural selection.

   Correct, but so are other answers. Natural selection and sexual selection both act on the heritable variation among individuals within a population that leads to differential fitness (Section 2.3). Sexual selection refers specifically to the differential reproductive success that results because individuals vary in traits that affect their ability to compete for fertilizations. Sexual selection can produce phenotypic traits that actually reduce fitness of many of the individuals that possess them, however (e.g., long, gangly tails that impair flight; bright colors or mating behaviors that attract the attention of predators), especially in populations where the operational sex ratio is highly skewed (Section 11.2).

e. Both c and d

Correct. Natural selection and sexual selection both act on the heritable variation among individuals within a population that leads to differential fitness (Section 2.3). Sexual selection refers specifically to the differential reproductive success that results because individuals vary in traits that affect their ability to compete for fertilizations. In that sense, sexual selection can be considered a specific type of natural selection. Sexual selection also can produce phenotypic traits that actually reduce the fitness of many of the individuals that possess them, however (e.g., long, gangly tails that impair flight; bright colors or mating behaviors that attract the attention of predators), especially in populations where the operational sex ratio is highly skewed (Section 11.2).

2. What is an/are important factor(s) influencing gene expression?

a. The gene control region

Correct, but so are other answers. The gene control region includes binding sites where different molecules can bind, affecting gene expression. For example, some proteins known as transcription factors, can bind to regulatory regions near hundreds of different genes, switching them all on or off at the same time. And transcription factors can affect the expression of other transcription factors, forming a cascading effect on gene expression (Section 5.1). Mutations to any part of the gene control region can alter gene expression by blocking transcription, influencing translation, or even altering the function of the protein (Section 5.2).

b. Cis- and trans-acting factors

Correct, but so are other answers. Cis- and trans-acting factors, such as hormones and microRNA, bind to regulatory regions either immediately adjacent to (cis-acting), or far away from (trans-acting) affected genes (Section 5.1). Mutations to cis-or trans-acting regulatory factors alter where, when, or to what extent genes are expressed (Section 5.2).

c. Internal environmental variables, such as hormone levels and microRNAs

Correct, but so are other answers. Hormone levels and microRNAs influence the expression of genes far removed from the regions that encode for the hormones and microRNAs themselves (Section 5.1). Mutations to these sequences can alter the timing, location, or level of expression of a gene or the developmental or environmental context in which the gene is expressed (Section 5.2).

d. The external environment in which the organism develops, such as sunlight or food availability

Correct, but so are other answers. The external environment of an organism can affect the expression of genes. For example, the amount of light a plant is exposed to can influence the expression of phenotypically plastic traits, such as leaf size—in high light, a genotype produces large leaves; in low light, that same genotype produces small leaves. Mutations can affect the timing or duration of this interaction between gene expression and the environment, or the sensitivity to the environmental variable (Section 7.4).

e. All of the above

Correct. The gene control region includes binding sites where different molecules can bind, affecting gene expression. Cis-regulatory factors bind to sites adjacent to the genes they affect. Trans-factors bind to sites far away from the genes they affect, coding for proteins, microRNAs, or other diffusible molecules that then influence expression of the focal genes. These gene products and microRNAs are components of the internal environmental variables that influence gene expression, but external environmental variables can also influence gene expression (see Section 5.1, Section 5.5, and Section 7.4).

f. None of the above

Incorrect. The gene control region includes binding sites where different molecules can bind, affecting gene expression. Cis-regulatory factors bind to sites adjacent to the genes they affect. Trans-factors bind to sites far away from the genes they affect, coding for proteins, microRNAs, or other diffusible molecules that then influence expression of the focal genes. These gene products and microRNAs are components of the internal environmental variables that influence gene expression, but external environmental variables can also influence gene expression (see Section 5.1, Section 5.5, and Section 7.4).

3. How may polyandry benefit females?

a. Females may be able to get the highest-quality genes possible for their offspring by mating with a number of males.

Correct, but so are other answers. Polyandry is a mating system where females mate (or attempt to mate) with multiple males. Polyandry is rare relative to polygyny because of constraints imposed by a female's fecundity. However, females that can obtain fertilizations for their eggs from a variety of males with good genes can hedge their bets, or upgrade sperm quality, getting the highest quality genes for at least some of their offspring (Section 11.4).

b. Females may benefit because they accumulate nutrients or other resources offered by each of the males.

Correct, but so are other answers. Polyandry is a mating system where females mate (or attempt to mate) with multiple males. Although polyandry is rare relative to polygyny because of constraints imposed by a female's fecundity, females can mate with a variety of males. By mating with multiple males, females can accumulate nutrients or other resources proffered by males in exchange for mating (Section 11.4).

c. Mating with multiple males may allow females to boost the health of their offspring by giving them a defense against a wider range of pathogens.

Correct, but so are other answers. Polyandry is a mating system where females mate (or attempt to mate) with multiple males. Females can benefit from this mating system because by mating with multiple males they can increase the likelihood that their offspring get a diversity of genes. Greater genetic diversity within offspring may provide a greater diversity of defenses against pathogens (Section 11.4).

d. Females may mate with multiple males simply because the costs of resisting courtship overtures are higher than the costs of mating multiple times.

Correct, but so are other answers. Polyandry is a mating system where females mate (or attempt to mate) with multiple males. Females may mate with a variety of males simply because it is energetically costly to resist their courtship overtures (Section 11.4).

e. All of the above

Correct. Polyandry is a mating system where females mate (or attempt to mate) with multiple males. Although polyandry is rare relative to polygyny, females benefit because their offspring may not only get good genes, but genes that are different from their mothers', or genes that provide a diversity of defenses against pathogens. Also, the cost of courtship may simply to energetically expensive that females simply mate rather than resist (Section 11.4).

Identify Key Terms

1. b; 2. d; 3. f; 4. g; 5. c; 6. a; 7. h; 8. e

Link Concepts

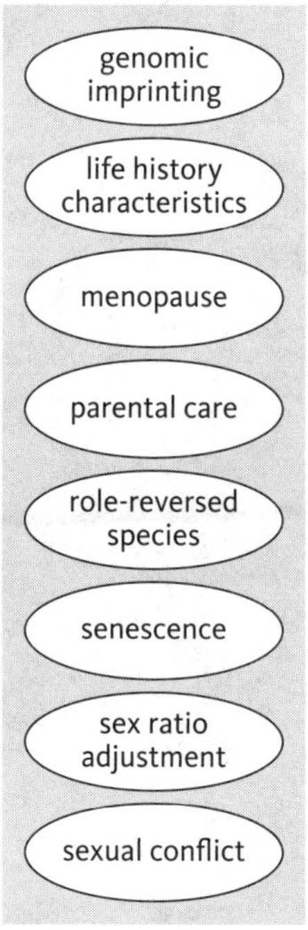

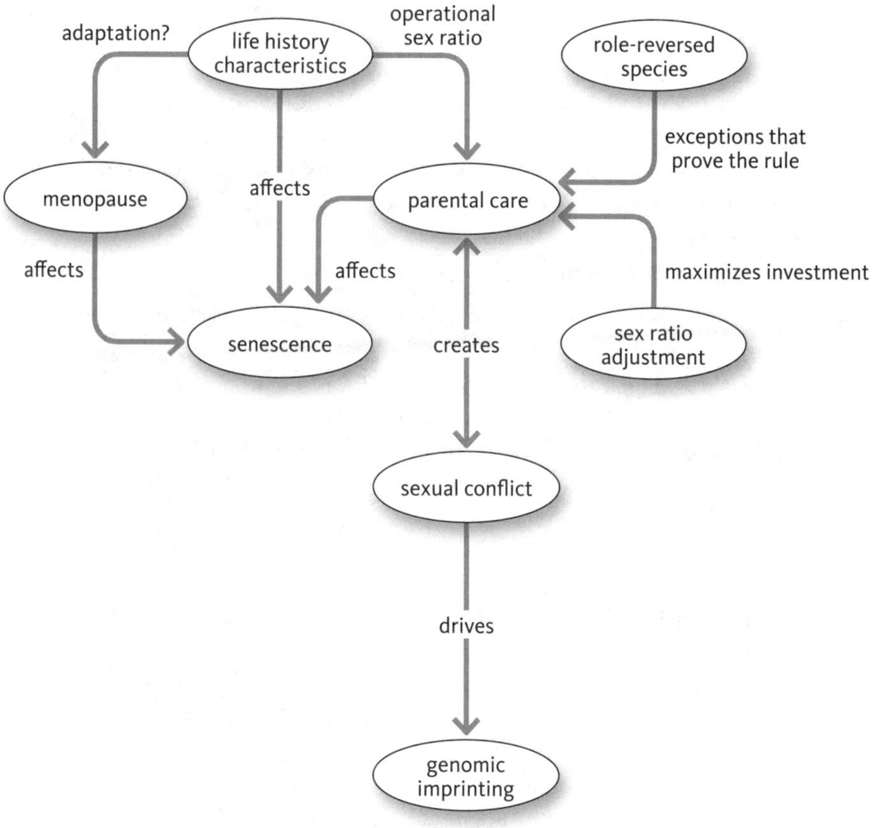

Interpret the Data

- Which trait(s) decreased in the predator-free environments?
  Brood size
- In which populations were the females younger and smaller at first parturition (when they give birth)? Why might female age and size be life-history traits?
  Predator-rich environments. Age can affect when a female begins reproducing and size can affect the number of offspring she can bear.
- Did males experience similar selection?
  Yes

Games and Exercises

- What does the sex ratio of 0.4 in the graph on the previous page mean?
  That there are four males born for every ten females
- What does the sex ratio of 0.1 in the graph on the previous page mean?
  That there is 1 male born for every 10 females
- How many males would you expect in a population of 75 wasps with a sex ratio of 0.2?
  15
  Cross multiplying gives 2 x 75/10 = 15

- What is the sex ratio for each population?

	Males	Population	Females	Sex ratio
Montana	47	181	134	0.350746
Idaho	27	135	108	0.25
Nevada	18	142	124	0.145161
Oregon	68	340	272	0.25

- Do any of the populations have equal sex ratios?

  Idaho and Oregon
- What does the ratio at birth indicate for these four countries?

  India has a higher ratio of males to females at birth than other countries. Tanzania has the lowest.
- How do sex ratios play out over time?

  They stay pretty consistent until the 55-64 years grouping when females become more prevalent in the populations in the United States, Germany, and Tanzania. In India, females start becoming more prevalent a little earlier (in the 25–54 years category).
- Roughly, what are the mutation rates for each category?

  90 point mutations/live birth

  10 small insertions and deletions (fewer than 50 base pairs)/live birth

  0.9 insertions of mobile elements/live birth

  0.05 large duplications and deletions (larger than 50 base pairs)/live birth
- So how often does a large duplication actually occur?

  Roughly once in every 50 live births (the exact rate is 0.42, so only once in every 42 live births)

Overcoming Misconceptions

Which of the following is a true statement?

d. Evolution is more than chance.

Delve Deeper

1. How might parent-offspring conflict factor into sex-ratio adjustment in Seychelles warblers?

   Parent-offspring conflict occurs when parents benefit from withholding care or resources from some offspring in order to invest in other offspring. Natural selection has favored helping behavior in Seychelles warblers—helping provided by female offspring. Adult females and their offspring would conflict, however, whenever the conditions changed so that one sex (e.g., males) may be preferable to another in terms of reproductive fitness. For example, females may withhold care to female offspring in deteriorating environmental conditions, when producing more male offspring would be more adaptive. Conflict would arise because the deprived female offspring would benefit more if they received the withheld care or resources.

2. What predictions might you make about the timing of senescence in the different populations of guppies that Dr. Reznick studies?

   Because the different populations are evolving with different life history strategies, they are likely to experience different tradeoffs as well. If senescence results because of antagonistic pleiotropic effects, populations with shorter lifespans are less likely to experience those effects than populations with longer lifespans. Guppy populations from low predation environments may start breeding later and experience longer lifespans than populations in high predation environments. They may also experience the harmful effects of antagonistic pleiotropic effects. In the lab, one might predict that populations from low predation environments would show signs of aging, such as immune system decline, more than populations from high predation environments.

Test Yourself

1. e; 2. d; 3. c; 4. d

13 The Origin of Species

Check Your Understanding

1. What are clades?

   a. Groups made up of organisms and all of their descendants

      Correct, but so are other answers. A clade is an organism and all its descendants (Sections 4.1).

   b. Hierarchies nested according to their synapomorphies

      Correct, but so are other answers. Clades are hierarchies nested according to their synapomorphies (Section 4.3).

   c. Groups of organisms that comprise a taxonomic unit

      Incorrect. Taxonomic units do not always align with clades because many well-known taxonomic units were based on different lines of evidence. Recently some taxonomic units, such as reptiles, have been revised because they are not monophyletic clades (see Figure 4.12).

   d. Groups of living organisms that share a phenotypic trait or character state

      Incorrect. Not all groups share traits because they are inherited them from a common ancestor. Sometimes, trait similarities arise by convergent evolution or evolutionary reversals. So, just because a group of organisms shares a phenotypic trait or character state doesn't make that group a clade. Additional evidence is necessary to know whether a phenotypic trait or character state was inherited from a common ancestor of the group (see Section 4.3).

   e. Both a and b

      Correct. A clade is an organism and all its descendants (Section 4.1), and clades are nested according to their synapomorphies (Section 4.3). Clades do not necessarily define taxonomic units, however, because many well-known taxonomic units were based on different lines of evidence. Recently some taxonomic units, such as reptiles, have been revised because they are not monophyletic clades (Section 4.2). Similarly, some organisms share characteristics because of convergent evolution, and these groups do not represent clades.

f. All of the above

Incorrect. A clade is an organism and all its descendants (Section 4.1), and clades are nested according to their synapomorphies (Section 4.3). However, clades do not necessarily define taxonomic units because many well-known taxonomic units were based on different lines of evidence. Recently some taxonomic units, such as reptiles, have been revised because they are not monophyletic clades (Section 4.2). Similarly, some organisms share characteristics because of convergent evolution, and these groups do not represent clades.

2. Which of the following may explain the origin of a particular female mate preference?

a. Male biased operational sex ratios

Incorrect. Male biased operational sex ratios often generate strong sexual selection because, as the abundant sex, males must compete over access to females. This strong selection can lead to the evolution of body size, weapons, and aggression in males, as well as traits that function as displays. Females may select mates based on characteristics that benefit her directly (such as nutrients, nest sites, protection, parental care) or indirectly (e.g., high genetic quality transmitted to her offspring) (Section 11.2). Although biased operational sex ratios set the stage for the evolution of female mate choice, they do not explain how specific preferences arise.

b. Anisogamy

Incorrect. Anisogamy describes the relative investment in gametes (eggs and sperm) by the different sexes, leading to distinct differences in how males and females maximize their reproductive success. Although anisogamy can set the stage for the evolution of female mate choice, it does not by itself explain how specific preferences arise (Section 11.2).

c. Pre-existing sensory biases in females

Correct. Some mate preferences may evolve from existing biases in the sensory system of females. Females may prefer certain colors, behaviors, or traits before the evolution of the male ornament. For example, orange may be associated with preferred foods, and males sporting bigger, brighter orange patches may do better mating than males without. These heritable preferences could drive the evolution of specific female mate preferences and male ornaments (see Figure 11.23 and Box 11.2).

d. Sexual conflict

Incorrect. Sexual conflict is the evolution of phenotypic characteristics that confer a fitness benefit to one sex but a fitness cost to the other. The conflict occurs when the two sexes have different optimal fitness strategies related to the production of offspring after mate choice (see Section 11.6).

3. How accurate is the current estimate for the age of the Earth (4.568 billion years)?

a. Not accurate at all because it is just an estimate

Incorrect. Scientists from a variety of disciplines have been measuring decay rates and incorporating these measurements into radioactive clocks that have been tested, reviewed, and retested. The probabilistic mathematical equations they use based on this information provide very narrow estimates of the ages of geological formations. Scientists have concluded with strong confidence that the Earth began to form from the solar system's primordial dust cloud 4.568 billion years ago (see Box 3.1).

b. Only slightly accurate because radiometric dating requires knowing decay rates, and no one has measured decay rates directly

   Incorrect. Scientists from a variety of disciplines have been measuring decay rates for decades. They've incorporated these measurements into radioactive clocks that have been tested, reviewed, and retested. The probabilistic mathematical equations they use based on this information provide very narrow estimates of the ages of geological formations. Scientists have concluded with strong confidence that the Earth began to form from the solar system's primordial dust cloud 4.568 billion years ago (see Box 3.1).

c. Somewhat accurate, but different dating techniques give different results

   Incorrect. Scientists often test their results using different dating techniques because the weight of evidence is important when making scientific claims. Thousands of research papers are published each year on radiometric dating, essentially all in agreement. Indeed, the fact that they can derive the same age with independent methods confirms that radiometric dating is a valid way to measure the age of rocks. Scientists have concluded with strong confidence that the Earth began to form from the solar system's primordial dust cloud 4.568 billion years ago (see Box 3.1).

d. Fairly accurate because scientists have been replicating the experiments and re-evaluating the evidence for decades

   Correct. Scientists often test their results using different dating techniques because the weight of evidence is important when making scientific claims. Thousands of research papers are published each year on radiometric dating, essentially all in agreement. Indeed, the fact that they can derive the same age with independent methods confirms that radiometric dating is a valid way to measure the age of rocks. Scientists have concluded with strong confidence that the Earth began to form from the solar system's primordial dust cloud 4.568 billion years ago (see Box 3.1).

Identify Key Terms

1. o; 2. j; 3. q; 4. r; 5. c; 6. n; 7. e; 8. i; 9. k; 10. f; 11. d; 12. b; 13. h; 14. g; 15. m; 16. l; 17. p; 18. s; 19. a

Link Concepts

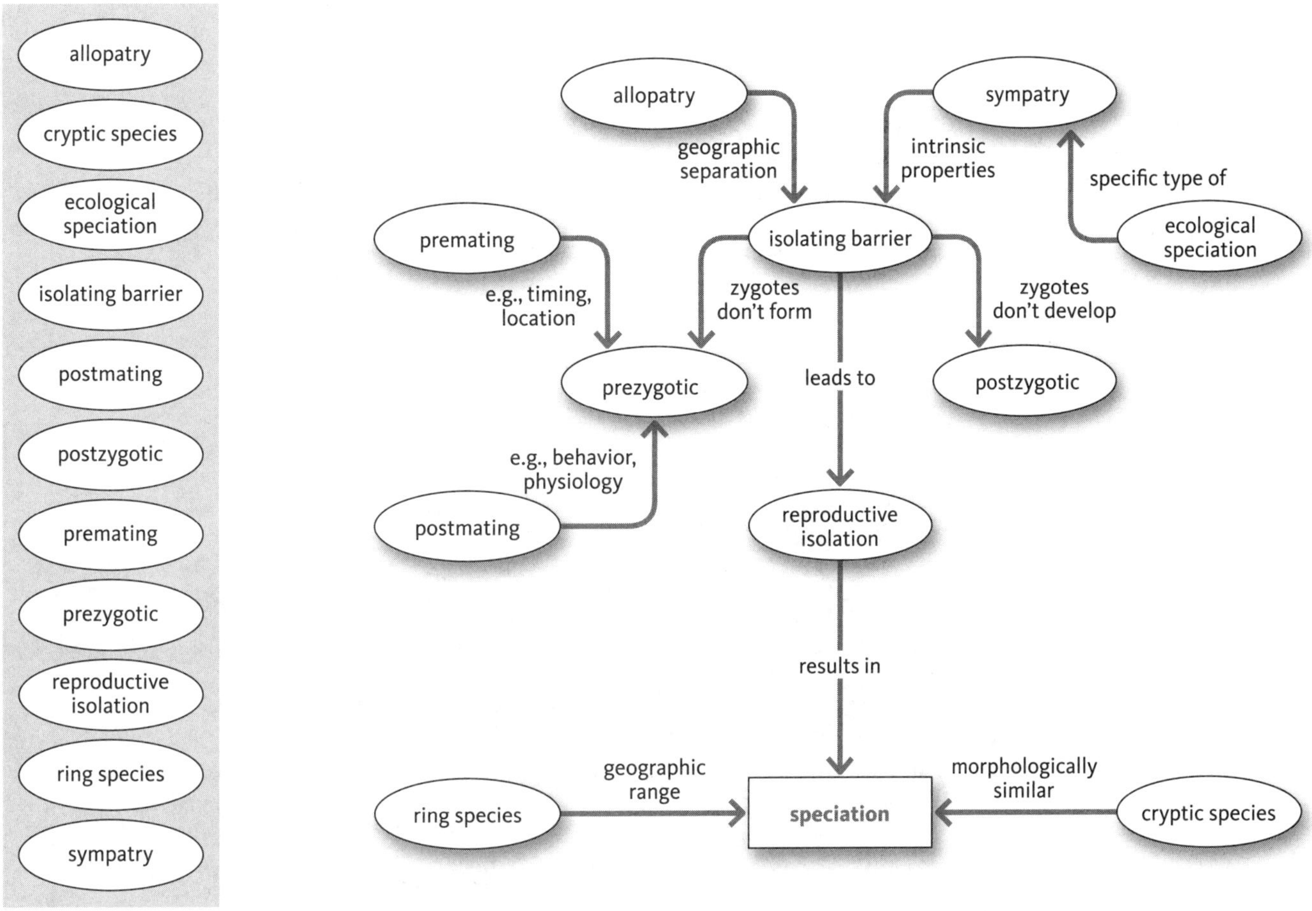

Interpret the Data

- Were *Drosophila* species pairs more likely to hybridize when they had a genetic distance < 0.5 than or >0.5?

  Drosophila species pairs were more likely to hybridize when they had a genetic distance < 0.5. For closely related species (those with a genetic distances less than 0.5), the amount of hybridization ranged from about 0.15 to 1. Some species pairs were hardly isolated from each other at all, and others were completely isolated, reproductively. For species pairs that were more distantly related (those with a genetic distances greater than 0.5), the amount of hybridization ranged from about 0.45 to 1. So, in general, species that were more closely related were more likely to hybridize than species that were less closely related - even though all of the species pairs did hybridize at some level.
- Did *Drosophila* species pairs hybridize even if they weren't closely related?

  Yes, even species that were very distantly related (values greater than 1) were not completely isolated reproductively.
- If it takes roughly a million years for D to reach a value of 1, when would you consider *Drosophila* species to be reproductively isolated?

  Drosophila species could be considered reproductively isolated in about ¾ of a million years (D between 0.5 and 1) because by then a typical pair of *Drosophila* species no longer interbred.

Overcoming Misconceptions

Which of the following is a true statement?

a. Random mutations are important to evolution because they provide the raw material for natural selection.

Delve Deeper

1. Explain the concept of ring species with this semantic map by describing what happens, why it happens, and when it happens.

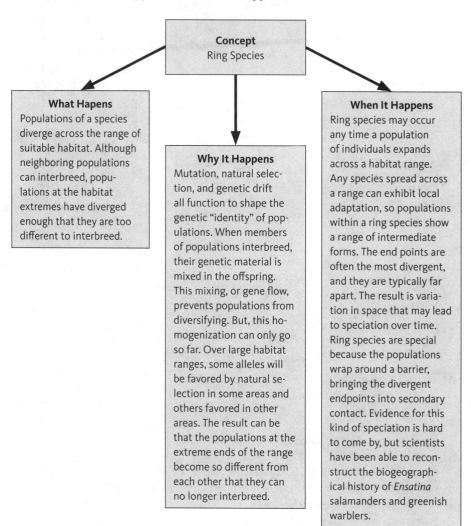

Concept
Ring Species

What Hapens
Populations of a species diverge across the range of suitable habitat. Although neighboring populations can interbreed, populations at the habitat extremes have diverged enough that they are too different to interbreed.

Why It Happens
Mutation, natural selection, and genetic drift all function to shape the genetic "identity" of populations. When members of populations interbreed, their genetic material is mixed in the offspring. This mixing, or gene flow, prevents populations from diversifying. But, this homogenization can only go so far. Over large habitat ranges, some alleles will be favored by natural selection in some areas and others favored in other areas. The result can be that the populations at the extreme ends of the range become so different from each other that they can no longer interbreed.

When It Happens
Ring species may occur any time a population of individuals expands across a habitat range. Any species spread across a range can exhibit local adaptation, so populations within a ring species show a range of intermediate forms. The end points are often the most divergent, and they are typically far apart. The result is variation in space that may lead to speciation over time. Ring species are special because the populations wrap around a barrier, bringing the divergent endpoints into secondary contact. Evidence for this kind of speciation is hard to come by, but scientists have been able to reconstruct the biogeographical history of *Ensatina* salamanders and greenish warblers.

Test Yourself

1. c; 2. c; 3. c; 4. c; 5. b; 6. d; 7. d

14 Macroevolution

The Long Run

Check Your Understanding

1. What is speciation?

   a. The process by which new species arise

   Correct, but so are other answers. Speciation is the evolutionary process by which new species arise, but understanding the concept requires thinking about the historical context, the fossil record, and relationships among lineages (Section 13.2).

   b. Cladogenesis

   Correct, but so are other answers. Cladogenesis is the splitting of one evolutionary lineage into two or more lineages; it is the process by which new species arise. Understanding speciation, however, requires thinking about the historical context, the fossil record, and relationships among lineages (Section 13.2).

   c. When one lineage splits into two or more lineages

   Correct, but so are other answers. Speciation occurs when one lineage splits into two or more lineages, also known as cladogenesis. So speciation is the evolutionary process by which new species arise, but understanding the concept requires thinking about the historical context, the fossil record, and relationships among lineages (Section 13.2).

   d. All of the above

   Correct. Speciation is the evolutionary process by which new species arise. It occurs when one lineage splits into two or more lineages, also known as cladogenesis. So understanding the concept of speciation requires thinking about the historical context, the fossil record, and relationships among lineages (Section 13.2).

   e. a and c only

   Incorrect. Speciation is the evolutionary process by which new species arise. It occurs when one lineage splits into two or more lineages, also known as cladogenesis. So understanding the concept of speciation requires thinking about the historical context, the fossil record, and relationships among lineages (Section 13.2).

2. Why don't scientists agree on a single definition of species?

   a. Because research methods can dictate which definition is most useful

   Correct, but so are other answers. Defining a species as a unit is an artifact of our human need to classify, but scientists generally agree that there is something special about species. Alleles flow among populations within a species differently then they flow between species—different species behave like independent evolutionary units, following separate trajectories. The precise definition of a species, however, is not clear, and research methods often dictate which definition is most useful. For example, paleontologists deal with morphological differences, and molecular phylogeneticists deal with genetic differences (Section 13.1).

   b. Because they have not discovered the true definition of a species

   Incorrect. Scientists don't expect to find a "true" definition of species. Indeed, a single definition for species that covers all taxa may not be possible. Scientists generally agree that there is something special about species, however. Alleles flow among populations within a species differently then they flow between species—different species behave like independent evolutionary units, following separate trajectories differences (Section 13.1).

c. Because different scientists have different philosophies about defining species

Correct, but so are other answers. Defining a species as a unit is an artifact of our human need to classify, but scientists generally agree that there is something special about species. Alleles flow among populations within a species differently then they flow between species—different species behave like independent evolutionary units, following separate trajectories. The precise definition of a species, however, is not clear, and different scientists have different philosophies about defining species. For example, some scientists may not believe that species is the most important taxonomic unit of concern (Section 13.1).

d. All of the above

Incorrect. Defining a species as a unit is an artifact of our human need to classify, but scientists generally agree that there is something special about species. Alleles flow among populations within a species differently then they flow between species—different species behave like independent evolutionary units, following separate trajectories differences. There may not be a "true" definition. Research methods often dictate which definition is most useful. Nor do scientists necessarily share the same philosophies about defining species (Section 13.1).

e. a and b only

Incorrect. Defining a species as a unit is an artifact of our human need to classify, but scientists generally agree that there is something special about species. Alleles flow among populations within a species differently then they flow between species—different species behave like independent evolutionary units, following separate trajectories differences. Research methods often dictate which definition is most useful, but there may not be a "true" definition (Section 13.1).

f. a and c only

Correct. Defining a species as a unit is an artifact of our human need to classify, but scientists generally agree that there is something special about species. Alleles flow among populations within a species differently then they flow between species—different species behave like independent evolutionary units, following separate trajectories differences. Research methods often dictate which definition is most useful. Nor do scientists necessarily share the same philosophies about defining species. However, there may not be a "true" definition (Section 13.1).

3. Why isn't the fossil record a complete record of life on Earth?

a. Because organisms eat other organisms

Correct, but so are other answers. A very big part of the reason the fossil record is incomplete stems from the fact that organisms eat other organisms, scattering remains and destroying evidence of their existence, let alone leaving anything to fossilize (Section 3.2).

b. Because conditions have to be just right in order to preserve fossils

Correct, but so are other answers. The conditions have to be just right for an organism to fossilize, many physical processes (wind, waves, running water) and biological processes (fungi, algae) can destroy remains of organisms before they can mineralize. It takes an even rarer set of circumstances for soft tissues, such as skin, to fossilize (Section 3.2).

c. Because wind and rain can erode fossils from the substrate

Correct, but so are other answers. The same processes that prevent organisms from fossilizing can also destroy fossils that have formed. Substrates, such as sandstones and other sedimentary rocks containing fossils, can become exposed through uplift or erosion. Physical processes, such as wind and rain, can quickly destroy millions of years of mineralization (Section 3.2).

d. Because rock-bearing fossils can be difficult to access

Correct, but so are other answers. Many fossils may simply be buried far below the surface currently, or they may be buried under snow and ice, or ocean sediments. Scientists have a limited understanding of the diversity and extent of rock strata, but they have been very successful identifying rocks that should bear fossils (see Section 4.4).

e. All of the above

Correct. A very big part of the reason the fossil record is incomplete stems from the fact that organisms eat other organisms, scattering remains and destroying evidence of their existence, let alone leaving anything to fossilize. On the off chance nothing scavenges the dead organism, the conditions have to be just right for fossilization. Many physical processes (wind, waves, running water) and biological processes (fungi, algae) can destroy remains of organisms before they can mineralize. It takes an even rarer set of circumstances for soft tissues, such as skin, to fossilize. The same processes that prevent organisms from fossilizing can also destroy fossils that have formed. Substrates, such as sandstones and other sedimentary rocks containing fossils, can become exposed through uplift or erosion. Physical processes, such as wind and rain, can quickly destroy millions of years of mineralization. But ultimately, many fossils may simply be buried far below the surface, or they may be buried under snow and ice, or ocean sediments (Section 3.2).

f. a and b only

Incorrect. A very big part of the reason the fossil record is incomplete stems from the fact that organisms eat other organisms, scattering remains and destroying evidence of their existence, let alone leaving anything to fossilize. And, on the off chance nothing scavenges the dead organism, the conditions have to be just right for fossilization. However, the same processes that prevent organisms from fossilizing can also destroy fossils that have formed, and other fossils may simply be buried far below the surface, under snow and ice, or beneath ocean sediments (Section 3.2).

g. c and d only

Incorrect. The same processes that prevent organisms from fossilizing can also destroy fossils that have formed, and other fossils may simply be buried far below the surface, under snow and ice, or beneath ocean sediments. However, a very big part of the reason the fossil record is incomplete stems from the fact that organisms eat other organisms, scattering remains and destroying evidence of their existence, let alone leaving anything to fossilize. And, on the off chance nothing scavenges the dead organism, the conditions have to be just right for fossilization (Section 3.2).

Identify Key Terms

1. f; 2. e; 3. j; 4. d; 5. i; 6. a; 7. b; 8. l; 9. k; 10. c; 11. h; 12. g

Link Concepts

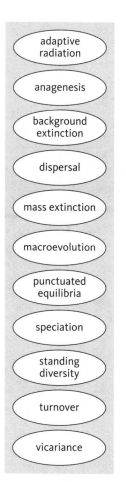

- adaptive radiation
- anagenesis
- background extinction
- dispersal
- mass extinction
- macroevolution
- punctuated equilibria
- speciation
- standing diversity
- turnover
- vicariance

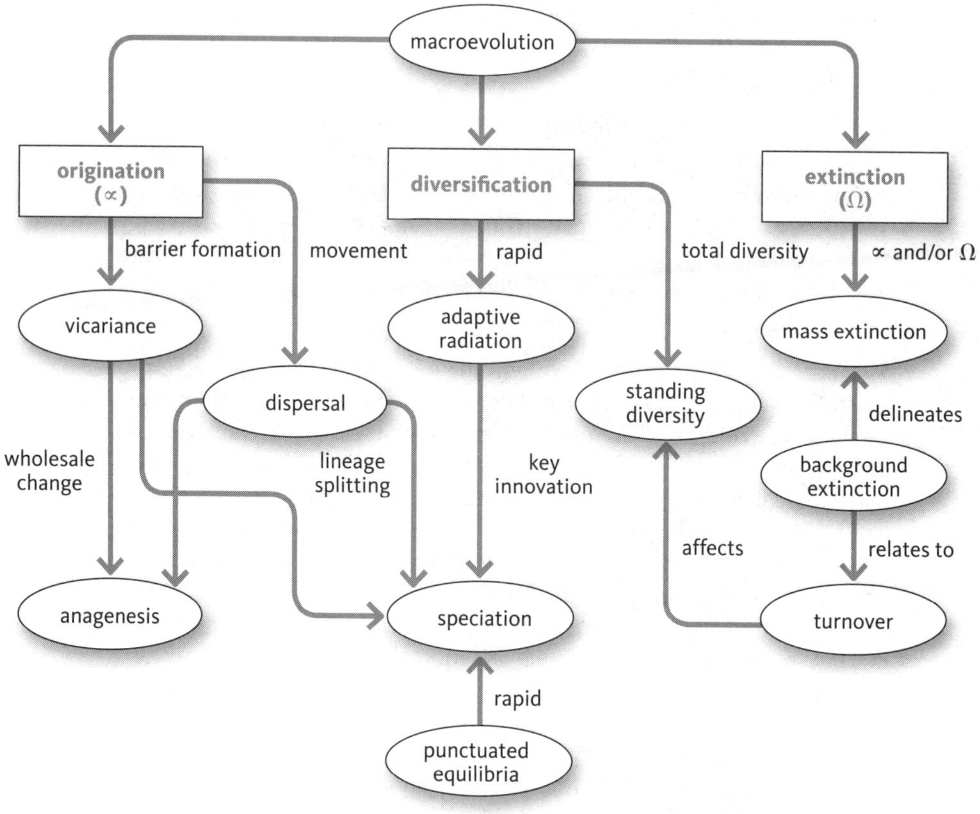

Interpret the Data

- According to Graph A, how many times has the Earth experienced warming in the last 600,000 years?

  Seven

- Prior to 2012, what was the highest carbon concentration the Earth experienced?

  Approximately 300 parts per million

- According to Graph B, what is the baseline against which other temperature deviations are being measured? Why might a scientist choose this method to make comparisons?

  The baseline is the average temperature measured over 20 years, from 1960 to 1980. Temperature anomalies are shown as the difference between that baseline and the average temperature for a given year.

  Establishing a baseline allows scientists to understand how observational data may change over time. The baseline establishes a point of reference, and calculating differences from the baseline can help determine *relative* changes—to understand trends.

  Average annual temperatures are one line of evidence for global climate change, and calculating the statistic—average annual temperature—requires actual measurement of temperatures. (Scientists can also determine estimates of temperature through indirect measures.) Data on global temperatures can be limiting because of our history and ability to measure global temperature from a diversity of locations. Turns out, the period from 1960 to 1980 has generally better coverage and data availability than earlier years (even though data from earlier years might have been less affected by anthropogenic changes than later years). Calculating differences from this baseline shows that the years before 1960 were relatively cooler and the years after 1980 were relatively warmer. More importantly, the extent of the differences is readily apparent.

- Does the variation in projected temperatures for the year 2100 mean that scientists don't know what they are talking about?

  No. Models take into account a diversity of variables, conditions, and scenarios. Different scientists weigh the evidence that supports those variables differently, so variation is expected. Indeed, having different models with different variables and different conditions is an important part of the scientific process, allowing for rigorous debate and consensus building. The computer models used to make predictions about temperature in the year 2100 consistently show that the planet will warm—even if carbon levels remain at levels measured in the year 2000.

Overcoming Misconceptions

Which of the following is a true statement?

d. Scientists have discovered many fossils that show the transition between dinosaurs and birds and between our primate ancestors and modern humans.

Delve Deeper

1. Are punctuated equilibria and Darwin's theory of natural selection at odds in evolutionary theory?

   No. Punctuated equilibria is a model of change over time. It proposes that although most lineages do not change much over the course of their geologic history, every once in a while there are periods of rapid change, often leading to speciation events. Darwin proposed that gradual changes accumulate over extended periods of time, ultimately leading to speciation events.

   Many of those that try to fault evolution in general, and macroevolution in particular, set up a false argument that punctuated equilibria and Darwin's idea of gradual change are alternative theories. They imply that both cannot be true. But just like macroevolution and microevolution, this argument focuses on the patterns—not the underlying mechanisms. Darwin proposed that organisms struggle for existence—those with traits that help them do better (that is, survive and/or reproduce) have more offspring, and the traits become more common in the population in future generations. Although he didn't have an understanding of mutations, genes, or genetic toolkits, he understood that traits were valuable commodities in the struggle. If a mutation, say to an important gene in the genetic toolkit, led to an altered development of fins, that trait may rapidly spread through a portion of the population if those that possess the new limb structure do well. (Maybe they can access a new food source that they couldn't get to without limbs they could put weight on and use to muck around at the water's edge.) Changes may continue to accumulate in the genetic structure of the limb relatively rapidly (remember, rapid change in the geologic record can still be hundreds of thousands of years), again, because individuals with incrementally better and better limbs do relatively better than the others in the population. That set of traits may work for a while—maybe with small gradual changes accumulating that are not as detectable (especially in the fossil record)—a long period of stasis. Then another structure-changing mutation may occur that allows some individuals to take in oxygen differently. And rapid change occurs again.

Also, the tree of life is not linear. It is not a simple path from fish to frogs to lizards to birds to mammals, and finally, to humans (see Figure 4.6). A trait that allows exploitation of a new habitat may lead to rapid adaptive radiations—radiations that can occur on small or large scales. But because of the fossil record, scientists may never find more than one or two representatives of any radiation. They have to piece together the tree of life from an entirely incomplete fossil record, but the more they look, the more they find. They are gradually filling in gaps in our understanding, and both models offer insight to deciphering the patterns of the history of life.

2. How is the term *Cambrian Explosion* misleading?

   The term *Cambrian Explosion* is misleading because it implies that a diversity of animals sprang into existence all at once. First, the Cambrian Explosion did not occur on a single date 542 million years ago, nor did it occur within a single year, decade, century, or even millennia. The Cambrian Explosion can be thought of as a 23-million-year (plus or minus) event where animal taxa rapidly diversified (again, "rapid" is a relative term in geology). Second, animals did not suddenly pop into existence. Animals began appearing in the fossil record more than 635 million years ago—that's nearly 200 million years before the date humans attach to the Cambrian Explosion. In fact, estimates based on molecular clocks indicate that animals likely first appeared on Earth around 800 million years ago, with major groups splitting off 600 and 700 million years ago. Sometime around 542 million years ago, however, evolutionary processes led to the huge diversification of groups found in the fossil record. The molecular clock estimates predict that scientists should be able to find fossil animals in rocks up to or older than 800 million years old; macroevolution predicts that the Cambrian Explosion may be related to changing physical conditions and a new kind of ecosystem; and microevolution predicts that the Cambrian Explosion may be related to the evolution of a versatile genetic toolkit.

Test Yourself

1. a; 2. c; 3. b; 4. b; 5. c; 6. d; 7. a; 8. e

15 Intimate Partnerships

How Species Adapt to Each Other

Check Your Understanding

1. Which of the following statements are depicted by this phylogeny?

   a. Otters (*Lutrinae*) evolved from martins (Martes group).

      Incorrect. Phylogenies are often incorrectly interpreted as a ladder of evolution—one species evolving into the next. However, think of the phylogeny as fluid, with the parts at every intersection swinging independently. The *Lutrinae/Procyonidae* (raccoons, ringtails, etc.) branch could easily flip, so that it might appear that otters evolved "from" skunks (*Mephitidae*). So, the correct way to interpret the relationships is to follow the lineage back to the node—those groups share a common ancestor (Section 4.1).

b. The ancestors of otters (*Lutrinae*) became gradually more "otter-like" over time.

Incorrect. Phylogenies represent the relationships among populations, genes, or species. The clades are organized so that the taxa within them share derived characteristics. So, otters share more derived characteristics with the *Mephitidae* (skunks) than they do with *Canidae* (wolves). That doesn't mean that organisms were becoming more "otter-like" and less "wolf-like." The visualization of the phylogeny is fluid, with the parts at every intersection swinging independently. Any of the branches from *Ursidae* (bears) to *Lutrinae* can easily be flipped within the phylogeny, which eliminates that interpretation (Section 4.1).

c. Living otters (*Lutrinae*) represent the end of a lineage of animals whose common ancestor was wolf-like.

Incorrect. Phylogenies represent the relationships among populations, genes, or species. Each branch represents a lineage. Lineages are organized in clades, so that the taxa within them share derived characteristics. Otters (*Lutrinae*) share more derived characteristics with walruses (*Odobenidae*) than they do with wolves (*Canidae*). But, otters do not represent the end of the Caniformia lineage, and the common ancestor was not necessarily wolf-like. Many lineages of Caniformia have gone extinct, and many still exist. The trick is figuring out how all these lineages are related (Section 4.1).

d. Otters (*Lutrinae*) share a common ancestor with *Odobenidae* (walruses).

Correct. Phylogenies represent the relationships among populations, genes, or species. Each branch represents a lineage. Lineages are organized in clades, so that the taxa within them share derived characteristics. The visualization of the phylogeny is fluid, with the parts at every intersection swinging independently. Any of the branches from *Ursidae* (bears) to *Lutrinae* can easily be flipped within the phylogeny. The correct way to interpret the relationships is to follow the lineage back to the node—those groups share a common ancestor (Section 4.1).

e. All of the above are depicted by this phylogeny.

Incorrect. Phylogenies are often incorrectly interpreted as a ladder of evolution or then end of a lineage because they can appear very ladder-like. However, think of the phylogeny as fluid, with the parts at every intersection swinging independently. Any of the branches from *Ursidae* (bears) to *Lutrinae* can easily be flipped within the phylogeny. Phylogenies represent the relationships among populations, genes, or species. Each branch represents a lineage. Lineages are organized in clades, so that the taxa within them share derived characteristics. The correct way to interpret the relationships is to follow the lineage back to the node—those groups share a common ancestor. Lineages below that node share derived characteristics (Section 4.1).

2. What is a species?

a. Groups of actually (or potentially) interbreeding natural populations that are reproductively isolated from other such groups.

Correct, but so are other answers. The biological species concept defines a species based on the ability of individuals to interbreed (Section 13.1). It is commonly used for organisms that reproduce sexually, although this concept has its limitations. Species that are assumed to potentially interbreed, such as giraffes, may not comprise a single species according to other species concepts (Section 13.5).

b. The smallest possible groups whose members are descended from a common ancestor and who all possess defining or derived characteristics that distinguish them from other such groups.

   Correct, but so are other answers. The phylogenetic species concept defines species as recognizable geographic forms that have a unique evolutionary history—species are the "tips" of a phylogenetic tree. This concept is useful when reconstructing the relationships among organisms using DNA, where interbreeding may or may not be a significant factor (Section 13.1).

c. Metapopulations of organisms that exchange alleles frequently enough that they comprise the same gene pool, and therefore, the same evolutionary lineage.

   Correct, but so are other answers. The general lineage species concept defines species based on the boundaries of the gene pool. The general lineage species concepts is more a conceptual definition, rather than methodological (Section 13.1).

d. All are valid definitions of a species.

   Correct. The biological species concept defines a species based on the ability of individuals to interbreed. It is commonly used for organisms that reproduce sexually. The phylogenetic species concept defines species as recognizable geographic forms that have a unique evolutionary history. And the general lineage species concept defines species based on the boundaries of the gene pool (Section 13.1).

e. Scientists don't know what a species actually is.

   Incorrect. All are valid definitions, reflecting scientists' understanding that a single definition may not be applicable to all organisms. The biological species concept defines a species based on the ability of individuals to interbreed. It is commonly used for organisms that reproduce sexually. The phylogenetic species concept defines species as recognizable geographic forms that have a unique evolutionary history. And the general lineage species concept defines species based on the boundaries of the gene pool (Section 13.1).

3. What is a mobile genetic element?

   a. A bacterium or viral parasite

      Incorrect. Both bacteria and viruses can be considered parasites, and they can and do infect other organisms. However, only some viruses were able to integrate their genes into the genomes of their hosts and become mobile genetic elements. The origins of many mobile genetic elements are unknown. Mobile genetic elements can be considered genomic parasites—they make new copies of themselves that can then be reinserted into the genome. They can literally dominate a genome. Like bacteria and viruses, however, mobile genetic elements can have devastating effects on the organism. For example, if a mobile genetic element disrupts a cell's normal rhythms of growth and division, it may begin to multiply out of control, giving rise to cancer (Section 5.1).

   b. A gene with no known function

      Incorrect. Mobile genetic elements are not genes. Mobile genetic elements can be considered genomic parasites—they make new copies of themselves that can then be reinserted into the genome. They can literally dominate a genome. Mobile genetic elements can disrupt gene expression within a cell, however. For example, if a mobile genetic element disrupts a cell's normal rhythms of growth and division, it may begin to multiply out of control, giving rise to cancer (Section 5.1).

c. Types of DNA that can move around in the genome

Correct. Mobile genetic elements include transposons ("jumping genes") and plasmids. They are considered genomic parasites—they make new copies of themselves that can then be reinserted into the genome. They can literally dominate a genome. Mobile genetic elements can have devastating effects on the organism. For example, if a mobile genetic element disrupts a cell's normal rhythms of growth and division, it may begin to multiply out of control, giving rise to cancer (Section 5.1).

d. Noncoding sections of DNA that function in alternative splicing

Incorrect. The noncoding sections of DNA that function in alternative splicing are known as introns, and although alternative splicing provides amazing flexibility to generate proteins, introns are not mobile genetic elements. Mobile genetic elements can be considered genomic parasites—they make new copies of themselves that can then be reinserted into the genome. They can literally dominate a genome. Mobile genetic elements can disrupt gene expression within a cell, however. For example, if a mobile genetic element disrupts a cell's normal rhythms of growth and division, it may begin to multiply out of control, giving rise to cancer (Section 5.1).

Identify Key Terms

1. c; 2. d; 3. h; 4. f; 5. a; 6. g; 7. i; 8. b; 9. e

Link Concepts

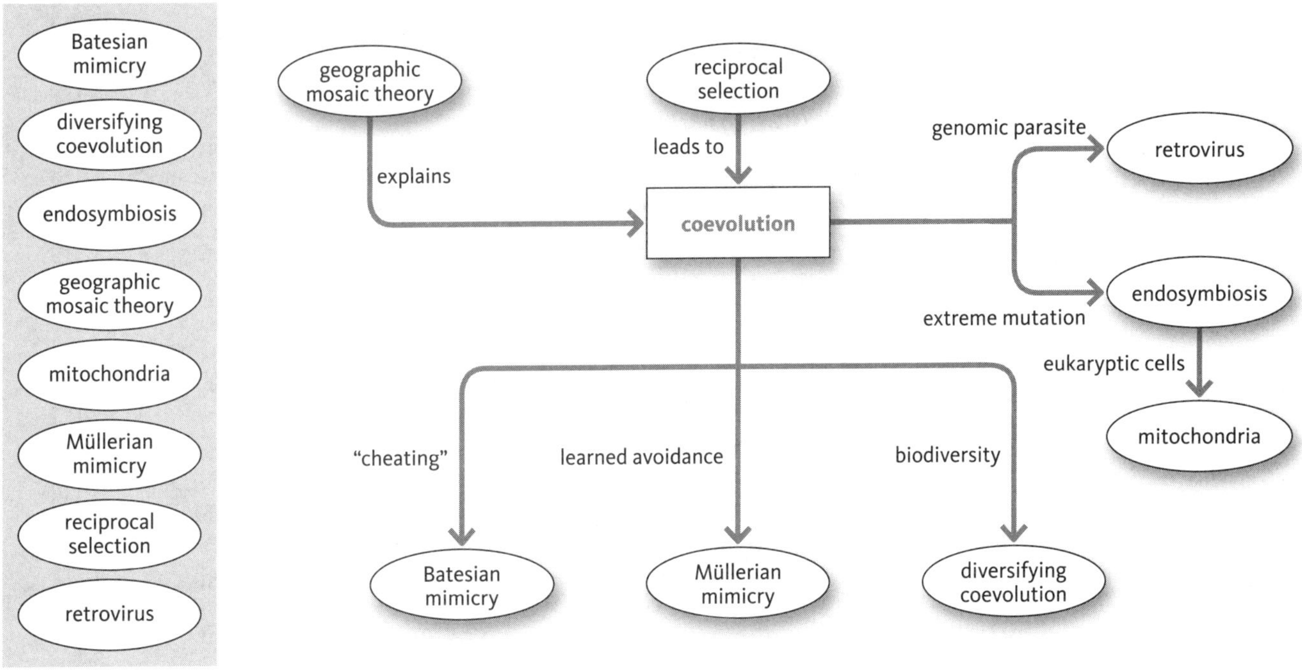

Interpret the Data

- What was the most common virulence grade in 1951? In 1952?

  Grade I and Grade III

- Explain how the virulence grades could change so quickly between 1951 and 1952?

  Myxoma virus is not a single individual, nor a single strain. Viruses can evolve quickly because of high reproductive rates—mutations can spread through a virus population producing new strains rather quickly. Any particular rabbit is likely to be infected by more than one strain of myxoma virus, and competition among strains within that rabbit would lead to increased virulence. However, individual rabbits infected with strains that become too virulent may die before the virus can be transmitted—any virus population within a dead individual is a dead strain. So strains that aren't too virulent will be more successful because their populations can grow and still be transmitted to new hosts.

  When the virus was first introduced, rabbits were abundant, and virulence was high because selection favored competition among strains. As rabbit populations plummeted, hosts became scarce and selection for increased transmission became dominant. Those strains that did not kill their hosts immediately (grades II–V) did better and became more common in the rabbit population.

- How stable is the intermediate grade of virulence in the myxoma virus?

  Fairly stable, but not absolute. Selection clearly favors the intermediate grades, but that doesn't mean something couldn't alter the balance. Virus populations are still experiencing mutations, and one of those could lead to entirely unknown trajectories. In fact, the data show that the balance does indeed shift—a more virulent grade appeared for a short time from 1970 to 1974, although it was only a small proportion of the total virus population.

Overcoming Misconceptions

Which of the following is a true statement?

d. Coevolution can lead to the loss of biodiversity.

Delve Deeper

1. How do variation among individuals, differential survival or reproduction, and heredity act to generate the patterns of newt toxicity and snake resistance observed by the Brodies?

   Variation among individuals, differential survival or reproduction, and heredity are all necessary for the coevolution of newts and snakes. Within populations of newts, some individuals may be more toxic than other individuals, and within snake populations, some individuals may be more resistant to that toxin than others. In some areas where the various individuals of each species interact, some newts may be less likely to be eaten by snakes (and therefore produce more offspring) than other newts, and some snakes may be able to eat more (and therefore produce more offspring) than other snakes. But because resistance affects a snake's ability to move, snakes that are highly resistant to the toxin may be more likely to be eaten by their predators (and therefore produce fewer offspring) than other snakes. So differential survival and reproduction affect the relative number of offspring individuals produce in the next generation. Heredity is the means by which the traits that influence survival and reproduction are transmitted to the next generation. Offspring of toxic newts are likely to be toxic, and offspring of resistant snakes are likely to be resistant.

More importantly, toxic newts influence the survival and reproduction of snakes, and resistant snakes influence the survival and reproduction of newts. The interaction of the two species affects the frequencies of the alleles for each of these traits in the next generation. Hotspots occur where toxicity and resistance are evolving rapidly because this kind of natural selection is strong. Coldspots occur where the two species are not coevolving (perhaps because predation on slow, resistant snakes is high enough to overcome the survival and reproductive benefits of being able to eat newts).

Test Yourself

1. b; 2. a; 3. e; 4. c; 5. b; 6. d; 7. e; 8. b; 9. d; 10. d; 11. c

16 Brains and Behavior

Check Your Understanding

1. What kind of variation is necessary for natural selection to occur?
   a. Variation in the expression of a trait

      Correct, but so are other answers. Variation among individuals in the expression of a trait is one of three conditions necessary for evolution by natural selection. The others include variation that must be at least partially heritable; and differential survival and reproduction as a result of these differences (see Sections 2.3 and 8.2).

   b. Phenotypic variation that has a genetic component

      Correct, but so are other answers. Phenotypic variation that has a genetic component is one of three conditions necessary for evolution by natural selection because the genetic component is the basis of heritability. But some individuals must also survive and reproduce more effectively than others because of these differences (see Sections 2.3 and 8.2).

   c. Variation in additive alleles that contribute to a phenotypic trait

      Correct, but so are other answers. Additive alleles are a significant component of the heritability of a trait; the additive effects of alleles cause relatives to resemble each other in their phenotypes. Heritability is one of three conditions necessary for evolution by natural selection. But, individuals also must differ in their expression of a trait, and some individuals must survive and reproduce more effectively than others because of these differences (see Sections 2.3 and 8.2).

   d. a and b only

      Incorrect. Both variation in the expression of a trait and phenotypic variation that has a genetic (i.e., heritable) component are necessary for natural selection, but that heritability is affected by whether alleles are additive or not. The additive effects of alleles cause relatives to resemble each other in their phenotypes. Some individuals also must survive and reproduce more effectively than others because of these differences (see Sections 2.3 and 8.2).

e. All of the above

Correct. Variation in the expression of a trait, phenotypic variation that has a genetic (i.e., heritable) component, and variation in additive alleles that contribute to a phenotypic trait are all necessary for natural selection. Some individuals also must survive and reproduce more effectively than others because of these differences (see Sections 2.3 and 8.2).

f. None of the above

Incorrect. Variation in the expression of a trait, phenotypic variation that has a genetic (i.e., heritable) component, and variation in additive alleles that contribute to a phenotypic trait are all necessary for natural selection. Some individuals also must survive and reproduce more effectively than others because of these differences (see Sections 2.3 and 8.2).

2. What factors influence a population's evolutionary response to selection?

a. Differential reproductive success of individuals in the population and the strength of selection

Incorrect. Differential reproductive success of individuals certainly influences the strength of selection, but the evolutionary response of a population to that selection also depends on the heritability of the traits that influence that differential reproductive success. The evolutionary response to selection is determined by the phenotypic variation that influences fitness and the ability to transmit those phenotypic characteristics to offspring: $R = S \times h^2$. So, if a trait's heritability is high, even weak selection can lead to significant evolutionary change. But if selection is strong, a population can respond even if a trait is only weakly heritable. Obviously, the most rapid evolutionary responses occur when both selection is strong and heritability is high (Section 7.2).

b. The strength of selection and the amount of phenotypic variation among individuals

Incorrect. The evolutionary response to selection is determined in part by the strength of selection, but the amount of phenotypic variation among individuals is only important if that variation is heritable. So, evolutionary response to selection is influenced by the phenotypic variation that influences fitness (strength of selection) and the ability to transmit those phenotypic characteristics to offspring (heritability): $R = S \times h^2$. If a trait's heritability is high, even weak selection can lead to significant evolutionary change. And if selection is strong, a population can respond even if a trait is only weakly heritable. Obviously, the most rapid evolutionary responses occur when both selection is strong and heritability is high (Section 7.2).

c. The strength of selection and how much of the variation in a phenotypic trait is heritable

Correct. The evolutionary response to selection is determined by the phenotypic variation that influences fitness (strength of selection) and the ability to transmit those phenotypic characteristics to offspring (heritability): $R = S \times h^2$. So, if selection is strong, a population can respond even if a trait is only weakly heritable. But if a trait's heritability is high, even weak selection can lead to significant evolutionary change. Obviously, the most rapid evolutionary responses occur when both selection is strong and heritability is high (Section 7.2).

d. The amount of genotypic variation among individuals and how much of that variation is heritable

Incorrect. Genotypic variation is important, because the genotype is the basis for heritable variation, but selection can only act if variation in the genotype affects the phenotype. The amount of that variation that is heritable, along with the strength of selection, determines the evolutionary response of a population. The evolutionary response to selection results from the phenotypic variation that influences fitness and the ability to transmit those phenotypic characteristics to offspring: $R = S \text{ x } h^2$. So, if a trait's heritability is high, even weak selection can lead to significant evolutionary change. But if selection is strong, a population can respond even if a trait is only weakly heritable. Obviously, the most rapid evolutionary responses occur when both selection is strong and heritability is high (Section 7.2).

e. The reproductive success of some individuals in the population versus the reproductive success of other individuals in the population

Incorrect. Differential reproductive success an important component of natural selection, but the evolutionary response of a population to that selection also depends on the heritability of the traits that influence that differential reproductive success. The evolutionary response to selection is determined by the phenotypic variation that influences fitness and the ability to transmit those phenotypic characteristics to offspring: $R = S \text{ x } h^2$. So, if a trait's heritability is high, even weak selection can lead to significant evolutionary change. But if selection is strong, a population can respond even if a trait is only weakly heritable. Obviously, the most rapid evolutionary responses occur when both selection is strong and heritability is high (Section 7.2).

3. Why is an understanding of the genetic toolkit important to understanding the evolution of traits?

a. Because the same underlying networks of genes govern the development of all animals, and new traits evolve as a result of mutations to genes within that network

Correct, but so are other answers. Patterning genes, such as *Hox* genes and limb-patterning networks, demarcate the geography of developing animals, determining the relative locations and sizes of body parts. For 570 million years, mutations, even those that lead to relatively subtle changes to the toolkit, have been able to generate tremendous diversity in the animal kingdom (Section 10.3).

b. Because the genetic toolkit consists of all of the genes scientists have identified so far

Incorrect. Scientists are just beginning to understand how genes in gene networks function to produce complex adaptations in the phenotype. The genetic toolkit refers to the complex of networks shared by an ancient common ancestor of all animals 570 million years ago (plants also appear to share a genetic toolkit). Patterning genes, such as *Hox* genes, function within networks that can be deployed in new developmental contexts as a result of mutation. For example, limb-patterning networks can be turned on or off leading to the development of long legs or no legs at all (Section 10.3).

c. Because gene networks within the toolkit act like "modules" that can be deployed in new developmental contexts, yielding novel traits

Correct, but so are other answers. Patterning genes, such as *Hox* genes, function within networks that can be deployed in new developmental contexts as a result of mutation, yielding novel traits. For example, limb-patterning networks can be turned on or off leading to the development of long legs or no legs at all (Section 10.3).

d. a and c only

Correct. Patterning genes, such as *Hox* genes and limb-patterning networks, demarcate the geography of developing animals, determining the relative locations and sizes of body parts. Within the toolkit, gene networks act as modules that can be deployed in new developmental contexts, yielding novel traits. For 570 million years, mutations, even those that lead to relatively subtle changes to the toolkit, have been able to generate tremendous diversity in the animal kingdom (Section 10.3).

e. All of the above

Incorrect. The genetic toolkit is a subset of described genes. Specifically, it refers to the complex of networks shared by an ancient common ancestor of all animals 570 million years ago (plants also appear to share a genetic toolkit). Patterning genes, such as *Hox* genes and limb-patterning networks, demarcate the geography of developing animals, determining the relative locations and sizes of body parts. Within the toolkit, gene networks act as modules that can be deployed in new developmental contexts, yielding novel traits (Section 10.3).

f. None of the above

Incorrect. Patterning genes, such as *Hox* genes and limb-patterning networks, demarcate the geography of developing animals, determining the relative locations and sizes of body parts. Within the toolkit, gene networks act as modules that can be deployed in new developmental contexts, yielding novel traits. For 570 million years, mutations, even those that lead to relatively subtle changes to the toolkit, have been able to generate tremendous diversity in the animal kingdom (Section 10.3).

Identify Key Terms

1. g; 2. d; 3. a; 4. e; 5. l; 6. b; 7. c; 8. k; 9. j; 10. f; 11. i; 12. h

Link Concepts

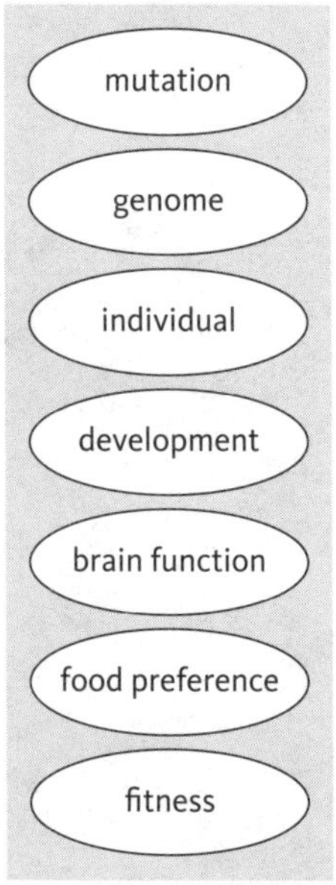

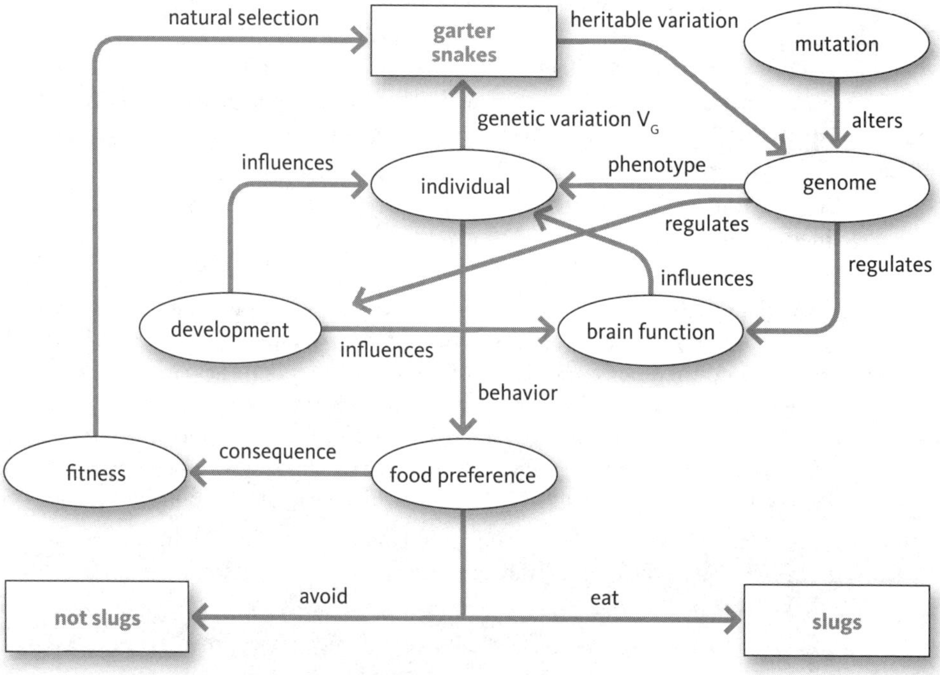

Interpret the Data

- What proportion of high-learning flies were alive at 50 days? What proportion of control flies were alive at 50 days?

  At 50 days, about 40 percent of high-learning flies, and about 60 percent of control flies were still alive. That's 20 percent fewer high-learning flies alive than control flies.
- What was the learning index for long-lived flies at 5 days? For normal flies?

  Approximately 0.21 and approximately 0.34
- Was the learning advantage of normal flies consistent over the lifetime of the flies?

  No. Normal flies were better learners early in their lives, but that advantage disappeared as they aged. In fact, the learning index was not that different between normal flies and long-lived flies after flies reached 19 days old (notice that the error bars overlap, indicating that there was a lot of variation in the populations at that age).
- Why did the scientists test the effects of learning with both a higher-learning experiment and a longer-lived experiment?

  Because the results are correlations, the scientists could not determine that higher learning causes shorter life spans. By approaching the experiment from two directions—one that examines the effect of higher learning on life span and one that examines the effect of life span on learning—the scientists could compare the results. Each experiment supports the hypothesis that learning negatively affects life span, so the evidence is that much stronger.

Overcoming Misconceptions

Which of the following is a true statement?

c. Some mutations can result in alleles that enhance the fitness of a group at the expense of the fitness of the individual with those alleles.

Delve Deeper

1. What's the difference between learning and the evolution of learning behavior?

   Learning can be cultural—individuals can change their behavior based on observing other individuals. So information can be passed on from generation to generation, individuals can vary in what they've learned, and having learned can alter their behavior in ways that influence their fitness. But the learned behavior itself is, by definition, not directly inherited. But learning itself is not genetically controlled. Once you learn math, your offspring don't know math. You may have a better capacity to learn math, however, and that capacity can be transferred to some or all of your offspring—if it has a genetic component. The evolution of learning behavior is a population-level process. When learning behavior has a genetic component, that component can be passed on to future generations. If individuals vary in their capacity for complex cognition (say, to count the number of lions in a pack), and that ability confers some fitness advantage, evolution can shape learning behavior through natural selection.

2. If the pack size of 20 animals was thrown out of the analysis presented in Figure 16.25C, do you think the relationship would hold? Why or why not?

   The relationship would hold because each individual would still get more energy per day as pack size increases. The pack size of 20 (x-axis) could be considered an outlier because the amount of food that particular pack was able to take down was so high. Each dog was able to procure over 80,000 kJ/day (y-axis) and is so much higher than individual members of other packs, no matter what the size. But the relationship still holds (and the line actually reflects more of an influence of all the other points anyway). The amount of food each individual dog gets increases with increasing pack size.

Test Yourself

1. b; 2. d; 3. c; 4. d; 5. a; 6. b; 7. c; 8. a; 9. b; 10. a; 11. a

17 Human Evolution

A New Kind of Ape

Check Your Understanding

1. Why are phylogenies such important tools in evolutionary biology?
   a. Because the relationships described by phylogenies are based on the best available evidence

      Correct, but so are other answers. Phylogenies are developed using nested sets of shared derived characteristics based on the best available evidence at the time. However, scientists continue to discover new fossils, new species, and new ways to think about morphological and molecular evidence. They understand that phylogenies may shift and change, and they can use phylogenies to make predictions about future relationships (see Section 4.4 for an example).

   b. Because the relationships described by phylogenies are based on different lines of evidence, including morphology, DNA, and fossils

      Correct, but so are other answers. Phylogenies can be developed using morphological, molecular, and even cultural evidence. Fossils (and other historical artifacts) can provide additional evidence for morphological relationships, DNA relationships, and the timing of branching events. Phylogenies are developed using nested sets of shared derived characteristics; scientists can compare phylogenies and map other historical data onto phylogenies. The more independent lines of evidence used to build the relationships, the greater the likelihood that the relationships are accurate (see Section 14.2 for an example). Scientists understand that phylogenies may shift and change, however, and they can use phylogenies to make predictions about future relationships.

   c. Because the relationships described by phylogenies are hypothetical relationships that can be tested with additional evidence

      Correct, but so are other answers. Phylogenies represent hypothetical historical relationships based on currently available evidence. They are explanations for patterns in nature that scientists can test with further evidence. Scientists understand that phylogenies may shift and change, however, and they can use phylogenies to make predictions about future relationships (see Box 4.1 for an example).

d. Because the relationships described by phylogenies are developed with advanced statistical tools that can clarify complex relationships and generate additional hypotheses

   Correct, but so are other answers. The information used to generate phylogenies can be voluminous, investigating a diversity of genes, populations, or species and involving numerous character states. Statistical methods, such as maximum parsimony, bootstrapping, distance matrix methods, maximum likelihood methods, and Bayesian methods, help scientists resolve their data and develop the best possible hypotheses. These hypotheses can then be tested with additional evidence (see Section 9.3).

e. All of the above

   Correct. Phylogenies are developed using nested sets of shared derived characteristics based on the best available evidence at the time (see Section 4.4 for an example). They use different lines of evidence wherever possible to strengthen inferences (see Section 14.2 for an example). But phylogenies represent hypothetical historical relationships; they are explanations for patterns in nature that scientists can test with further evidence (see Box 4.1 for an example). And the information used to generate phylogenies can be voluminous, investigating a diversity of genes, populations, or species and involving numerous character states. Statistical methods, such as maximum parsimony, bootstrapping, distance matrix methods, maximum likelihood methods, and Bayesian methods, help scientists resolve their data and the best possible hypotheses (see Section 9.3).

f. a, b, and c only

   Incorrect. Phylogenies are developed using nested sets of shared derived characteristics based on the best available evidence at the time (see Section 4.4 for an example). They use different lines of evidence wherever possible to strengthen inferences (see Section 14.2 for an example). Phylogenies represent hypothetical historical relationships; they are explanations for patterns in nature that scientists can test with further evidence (see Box 4.1 for an example). But the information used to generate phylogenies can be voluminous, investigating a diversity of genes, populations, or species and involving numerous character states. Statistical methods, such as maximum parsimony, bootstrapping, distance matrix methods, maximum likelihood methods, and Bayesian methods, help scientists resolve their data and develop the best possible hypotheses (see Section 9.3).

2. Which of the following statements about molecular clocks is FALSE?

   a. Molecular clocks cannot be used to measure divergence in species separated by more than a few hundred million years.

      Correct. Molecular clocks can be used to measure divergence in species separated by hundreds of millions of years, but clocks must be calibrated because different types of DNA segments evolve at different rates. For species separated by hundreds of millions of years slow-evolving segments of DNA will provide greater accuracy than fast-evolving segments because over time, patterns in fast-evolving segments will be more and more difficult to discern (Section 9.6).

   b. Molecular clocks must be calibrated because different types of DNA segments evolve at different rates.

      Incorrect. Molecular clocks result because within DNA, base-pair substitutions accumulate at a roughly clock-like rate, but different types of DNA segments evolve at different rates. This type of variation is predicted by neutral theory (Section 9.6).

c. Molecular clocks result because within DNA, base-pair substitutions accumulate at a roughly clock-like rate, although substitution rates may differ between lineages.

   Incorrect. Base-pair substitutions do accumulate at a roughly clock-like rate, and these substitution rates can differ between lineages. This type of variation is predicted by neutral theory (Section 9.6).

d. Molecular clocks can be used to predict and test the ages of unknown samples of DNA.

   Incorrect. Molecular clocks can and have been used to predict and test ages of unknown samples of DNA. In fact, scientists were able to accurately predict the age of an HIV-1 sample dating back more than 40 years using a molecular clock (Section 9.6).

e. None of the above is a false statement about molecular clocks.

   Incorrect. Molecular clocks can be used to measure divergence in species separated by hundreds of millions of years. They result because base-pair substitutions accumulate at a roughly clock-like rate, but different types of DNA segments evolve at different rates. Also, substitution rates can differ between lineages. More importantly, molecular clocks can and have been used to accurately predict and test ages of unknown samples of DNA (Section 9.6).

3. Which is NOT a benefit of living in a social group?

   a. Better defense capabilities

      Incorrect. One of the benefits of living in a social group is that a group may be able to defend critical resources, such as a hunting territory, better than an individual. This benefit may come with costs, however. Group living can increase competition for food or mates, can increase conspicuousness to predators, or decrease confidence in paternity (Section 16.4).

   b. Large brain size

      Correct. Brain size may have evolved in response to selective pressures associated with living in a social group. Indeed, scientists have found that primate species that live in large groups tend to have proportionately larger neocortexes than those species that live in small groups (Figure 16.32), but large brain size is not necessarily a benefit of living in a social group.

   c. Cooperative hunting

      Incorrect. The ability to hunt and forage cooperatively is a benefit of living in a social group (Section 16.4). Despite having to share, the amount of food an individual eats per day may actually increase as group size increases (Figure 16.20).

   d. Smaller risk to individuals of being killed by a predator

      Incorrect. Living in a group can reduce the risk that any particular individual within the group is killed by a predator. That reduced risk is called the dilution effect, and it can be a great benefit of living in a social group. However, group living incurs costs such as increased conspicuousness to predators, increased competition for food or mates, or decreased confidence in paternity (Section 16.4).

   e. Enhanced opportunities for learning

      Incorrect. Learning may be a benefit of living in social groups because opportunities to observe may be enhanced. For example, juvenile New Caledonian crows stay with their parents for over a year, and that extended stay may allow them time to learn how to use tools instead of having to invent tools themselves (Section 16.7). Group living can incur costs, however, such as increased competition for food or mates and can increased conspicuousness to predators (Section 16.4).

Identify Key Terms

1. b; 2. d; 3. a; 4. c

Interpret the Data

- If ultra-social relationships acted as selective agents on human brain development, would you predict that children, chimpanzees, and orangutans should differ in their understanding of basic math? Why or why not?

  No. Children should not do better than chimpanzees and orangutans with tests of skills like math because the ultra-social hypothesis makes no prediction about humans, math skills, and development. Humans and other apes share a common ancestor, a common ancestor that experienced selection for complex cognition, including tool use and sophisticated forms of communication and perhaps even recognition. But the ultra-social hypothesis predicts that humans are *fundamentally different* from other apes in social cognition. Children should do better than chimpanzees and orangutans on tests related to observation and learning because selection favored rapid development of social cognition in humans.

- What do the width of the gray bars and the "whiskers" for children, chimpanzees, and orangutans indicate?

  Variation among individuals in test scores

- If the whiskers indicate individuals with test results at the ends of the distributions of each species, how would you interpret the differences among children and chimpanzees in social skills test?

  A few individual children had test results that overlapped with a few chimpanzee test results, but the majority of children scored higher than the majority of chimpanzees.

Overcoming Misconceptions

Which of the following is a true statement?

d. As scientists discover more and more hominin fossils, they may be able to define the genes that distinguish humans from other species.

Delve Deeper

1. Which molecule(s) related to our emotions likely evolved early in our mammalian history?

   Dopamine is a neurotransmitter that arouses an animal's attention. Rats and other mammals share this neurotransmitter, although they may respond differently than humans. Oxytocin is a hormone that elicits bonding behavior in mammals, including humans. Smell usually causes the release of this hormone in other mammals, but in humans, release is related to sight, which is consistent to other shifts toward sight in the apes.

2. Why is the discovery of stone tools in the fossil record important in understanding the evolution of our genus, *Homo*?

   Stone tools may have been used as early as 3.4 million years ago, although the strongest evidence indicates that stone tools were in use commonly around 2.6 million years ago. This new technology indicated a major transition in how the brains of our ancestors worked. The capacity to recognize and manipulate our surroundings is a function of complex cognition. The discovery of tools potentially as old as 3.4 million years ago indicates that species of Homo that branched off much earlier than our own, such as *Homo erectus* and possibly *H. habilis*, were using tools.

Stone tool technology has also evolved, and the tools discovered in the fossil record provide additional evidence of the course of our evolution, including how our ancestors interacted with the environment as well as other species, and the evolution of components of culture such as trade and migration.

3. What are some of the important factors influencing gene expression in humans?

   The external environment a gene is exposed to during development: The external environment a developing individual is exposed to can play an important role in gene expression. The temperature an individual develops in, the food available to it, and the infections it suffers can all influence gene expression and development. Even the social environment, such as presence or absence of affection from parents, can alter gene expression and development. In fact, different environments can lead to very different phenotypes from a single genotype. The "environment" is not restricted to the external conditions around us, however. The environment can be anything that interacts with the promoter region of a gene in a way that influences whether that gene is expressed.

   Pleiotropy: Pleiotropy occurs when a single gene affects the expression of many different phenotypic traits; antagonistic pleiotropy is when the beneficial effects for one trait cause detrimental effects on other traits. So pleiotropic effects can be very important factors influencing gene expression in humans. For example, animals as different from us as insects use the same "genetic toolkit" to regulate development. Genes at the top of these regulatory hierarchies influence many others, and mutations in a few of these regulatory genes can produce far-reaching changes, including limb loss and the development of novel traits.

Test Yourself

1. e; 2. b; 3. c; 4. e; 5. d; 6. e

18 Evolutionary Medicine

Check Your Understanding

1. Which is NOT one of the three conditions necessary for evolution by natural selection to occur?

   a. Individuals must differ in the characteristics of a trait.

      Incorrect. Individual variation in the expression or characteristics of a trait is the raw material for evolution by natural selection. The ultimate source of that variation is mutations to the genetic architecture within an individual. Some mutations are detrimental, some are beneficial, and some are neutral (although additional mutations can change that). How the genotype (the genetic makeup of an individual) affects the phenotype (the observable, measurable characteristic) can be complex, but the phenotype is the manifestation of individual differences on which natural selection can act (see Chapter 5). Heritable variation is integral to natural selection for most organisms because it is an effective mechanism by which beneficial mutations may be transmitted. Then, natural selection acts to alter the abundances of those characteristics in the population based on the relative reproductive success of individuals possessing them (see Section 2.3). Also see Chapter 8 for evidence of natural selection in the wild.

b. The differences among individuals in a trait must be at least partially heritable.

Incorrect. Heritable variation is integral to natural selection for most organisms because it is an effective mechanism by which genotypic variation may be transmitted. Genotypic variation arises from mutations. Whether they be detrimental, neutral, or beneficial, mutations that affect the phenotype can affect survival and reproduction and therefore be passed to offspring or not (see Chapter 7). Natural selection acts to alter the abundances of those mutations in the population based on the relative reproductive success of individuals (see Section 2.3).

c. Some individuals survive and reproduce more successfully than others because of differences in a trait.

Incorrect. Natural selection acts on the relative success of individuals within a population. Individuals with heritable phenotypic traits that confer some advantage or disadvantage within the population will, on average, have higher or lower survival and/or reproductive rates than other individuals who possess different versions of those traits. Natural selection acts to alter the abundances of those traits, leading to evolution of the population (see Section 2.3). Also see Chapter 8 for evidence of natural selection in the wild.

d. The more an individual needs a trait, the more quickly it will adapt to its environment.

Correct. Evolution by natural selection is not driven by need. An individual cannot compel itself or its offspring to do better. A whale may need fins to swim, but the ancestors of whales did not necessarily need fins to swim. Fins in whales evolved over hundreds of thousands of years as mutations that affected the ancestral whale phenotype influenced the relative survival and reproductive success of individuals as they hunted in a new, watery habitat. Certainly, the greater the influence those phenotypic changes had on the relative success of individuals, the more likely individuals in the next generation carried those mutations, and the more likely they were to be more successful in that new habitat than individuals without those mutations. However, evolution cannot identify needs, and need does not influence the evolutionary response to selection (see Section 1.4 for misconceptions about evolution and Chapter 7 for information about the evolution of phenotypes).

e. All are necessary for natural selection to occur.

Incorrect. For evolution by natural selection to occur, individuals must differ in the characteristics of a trait, and those differences must be heritable. Ultimately, some individuals survive and reproduce more successfully than others because of those differences (see Section 2.3). However, evolution by natural selection is not driven by need. An individual cannot compel itself or its offspring to do better (see Section 1.4). And need does not influence the evolutionary response to selection (see Chapter 7 for information about the evolution of phenotypes).

2. Which of the following is NOT an important factor influencing gene expression in humans?

   a. The external environment a gene is exposed to during development.

      Incorrect. The external environment a developing individual is exposed to can indeed play an important role in gene expression. The temperature an individual develops in, the food available to it, the infections it suffers can all influence gene expression and development. In fact, different environments can lead to very different phenotypes from a single genotype, a phenomenon known as phenotypic plasticity (see Section 5.5). In humans, conducting the critical experiments to identify phenotypically plastic traits is difficult, but evidence from other organisms indicates that we should expect such developmental plasticity (see Section 7.4). The "environment" is not restricted to the external conditions around us, however. The environment can be anything that interacts with the promoter region of a gene in a way that influences whether that gene is expressed (see Section 5.5).

   b. Pleiotropy

      Incorrect. Pleiotropy occurs when a single gene affects the expression of many different phenotypic traits; antagonistic pleiotropy is when the beneficial effects for one trait cause detrimental effects on other traits (see Section 6.6). So, pleiotropic effects can be very important factors influencing gene expression in humans. For example, animals as different from us as insects use the same "genetic toolkit" to regulate development. Genes at the top of these regulatory hierarchies influence many others, and mutations in a few of these regulatory genes can produce far-reaching changes, including limb loss and the development of novel traits (see Chapter 10 for an overview of evo-devo).

   c. Epigenetic effects

      Incorrect. Epigenetic factors, such as DNA methylation and coiling of DNA onto histones, can affect which genes can be expressed. Both methylation and coiling effectively silence genes, preventing their expression for extended periods. Early life experiences such as severe trauma can silence genes for the lifetime of an individual. In rare instances, these effects can even be transmitted to offspring, though epigenetic effects rarely persist for more than one or two generations (See Chapter 5).

   d. Genomic imprinting

      Incorrect. Genomic imprinting can be an extremely important factor influencing gene expression. It occurs when one parent contributes copies of genes that are silenced through an epigenetic process known as methylation. Imprinting then can result in offspring who express either the maternal or paternal copy of the gene, but not both (see Figure 12.13). This kind of parental conflict can lead to evolutionary arms races over the control of gene expression in offspring through genetic imprinting (see Section 12.2).

   e. All of the above are important factors influencing gene expression.

      Correct. The temperature an individual develops in, the food available to it, the infections it suffers can all influence gene expression and development. In fact, different environments can lead to very different phenotypes from a single genotype, a phenomenon known as phenotypic plasticity (see Section 5.5). Pleiotropy occurs when a single gene affects the expression of many different phenotypic traits (see Section 6.6). Genes at the top of regulatory hierarchies influence many others, and mutations in a few of these regulatory genes can produce far-reaching changes (see Chapter 10 for an overview of evo-devo). Epigenetic factors, such as DNA methylation and coiling of DNA onto histones, can affect which genes can be expressed. Both methylation and coiling effectively silence genes, preventing their expression for extended periods. And genomic imprinting occurs when one parent contributes copies of genes that are silenced through methylation. It also can be an extremely important factor influencing gene expression (see Section 12.2).

f. Only b and c are important factors influencing gene expression.

Incorrect. Pleiotropy and epigenetics are clearly important factors influencing gene expression in humans. However, external environmental factors, such as temperature an individual develops in, the food available to it, the infections it suffers, can all influence gene expression and development. In fact, different environments can lead to very different phenotypes from a single genotype, a phenomenon known as phenotypic plasticity (see Section 5.5). Also, genomic imprinting can be an extremely important factor influencing gene expression. It occurs when one parent contributes copies of genes that are silenced through a process known as methylation. Imprinting then can result in offspring who express either the maternal or paternal copy of the gene, but not both (see Figure 12.13).

3. What would you predict would be the outcome of natural selection on a mutation that increases fertility early in life but increases susceptibility to cancerous growths later in life?

a. Natural selection would favor individuals with the mutation because the fitness effects early in life would be bigger than the harm the mutation causes in old age.

Correct. Because of antagonistic pleiotropy, a mutation can result in beneficial effects for one trait, such as increased fertility, and detrimental effects on other traits, such as increased susceptibility to cancerous growths. Natural selection should favor the optimal trade-off that maximizes the number of offspring surviving to maturity over the course of an organism's entire life. So according to life history theory, the fitness benefits of an allele that increases fertility early in life may overcome the increased probability of death from cancerous growths later in life, especially if the detrimental effects of the allele manifest after the organism has finished reproducing. Individuals with the mutation would have relatively greater reproductive success, leaving more offspring even if they might die a little sooner than individuals without the mutation (see Section 12.1).

b. Natural selection would not favor individuals with the mutation because individuals susceptible to cancerous growths would die sooner and have lower lifetime reproductive success than individuals without the mutation.

Incorrect. Many genes are pleiotropic—they affect the expression of many different phenotypic traits, so a mutation to a pleiotropic gene can have beneficial effects for one trait and detrimental effects on other traits. Because of antagonistic pleiotropy, a mutation can result in beneficial effects for one trait, such as increased fertility, and detrimental effects on other traits, such as increased susceptibility to cancerous growths. Natural selection should favor the optimal trade-off that maximizes the number of offspring surviving to maturity over the course of an organism's entire life. So according to life history theory, even though an allele may cause an increased probability of death from cancerous growths later in life, the fitness benefits of increased fertility can be enough to maximize lifetime reproductive success, especially if the extrinsic mortality rate is high. Individuals with the mutation would have relatively greater reproductive success, leaving more offspring even if they might die a little sooner than individuals without the mutation (see Section 12.1).

c. Natural selection would favor a reaction norm that balanced reproductive success and susceptibility to cancerous growths.

   Incorrect. Reaction norms represent the pattern of phenotypic expression of a single genotype across a range of environments. In contrast, pleiotropy occurs when a single gene affects the expression of many different phenotypic traits. A mutation to a pleiotropic gene can have beneficial effects for one trait and detrimental effects on other traits—a phenomenon known as antagonistic pleiotropy. So, a mutation to a single gene can result in increased fertility and increased susceptibility to cancerous growths. Natural selection should favor the optimal trade-off that maximizes the number of offspring surviving to maturity over the course of an organism's entire life. According to life history theory, the fitness benefits of an allele that increases fertility early in life may overcome the increased probability of death from cancerous growths later in life, especially if the detrimental effects of the allele manifest after the organism has finished reproducing. Individuals with the mutation would have relatively greater reproductive success, leaving more offspring even if they might die a little sooner than individuals without the mutation (see Section 12.1).

d. Natural selection would remove individuals with the mutation from the population because mutations are detrimental.

   Incorrect. Mutations can be beneficial, neutral, or detrimental (see sections 5.1 and 5.2). More importantly, many genes are pleiotropic—they affect the expression of many different phenotypic traits, so a mutation to a pleiotropic gene can have beneficial effects for one trait and detrimental effects on other traits. Because of antagonistic pleiotropy, a mutation can result in beneficial effects for one trait, such as increased fertility, and detrimental effects on other traits, such as increased susceptibility to cancerous growths. Natural selection should favor the optimal trade-off that maximizes the number of offspring surviving to maturity over the course of an organism's entire life. According to life history theory, the fitness benefits of an allele that increases fertility early in life may overcome the increased probability of death from cancerous growths later in life, especially if the detrimental effects of the allele manifest after the organism has finished reproducing. Individuals with the mutation would have relatively greater reproductive success, leaving more offspring even if they might die a little sooner than individuals without the mutation (see Section 12.1).

e. A single mutation cannot be both beneficial and detrimental.

   Incorrect. Pleiotropy occurs when a single gene affects the expression of many different phenotypic traits, so a single mutation to a pleiotropic gene can have beneficial effects for one trait and detrimental effects on other traits—a phenomenon known as antagonistic pleiotropy. Because of antagonistic pleiotropy, a mutation can result in increased fertility and increased susceptibility to cancerous growths. Natural selection should favor the optimal trade-off that maximizes the number of offspring surviving to maturity over the course of an organism's entire life. According to life history theory, the fitness benefits of an allele that increases fertility early in life may overcome the increased probability of death from cancerous growths later in life, especially if the detrimental effects of the allele manifest after the organism has finished reproducing. Individuals with the mutation would have relatively greater reproductive success, leaving more offspring even if they might die a little sooner than individuals without the mutation (see Section 12.1).

Identify Key Terms

1. a; 2. g; 3. f; 4. h; 5. k; 6. b; 7. d; 8. i; 9. j; 10. c; 11. e; 12. l

Link Concepts

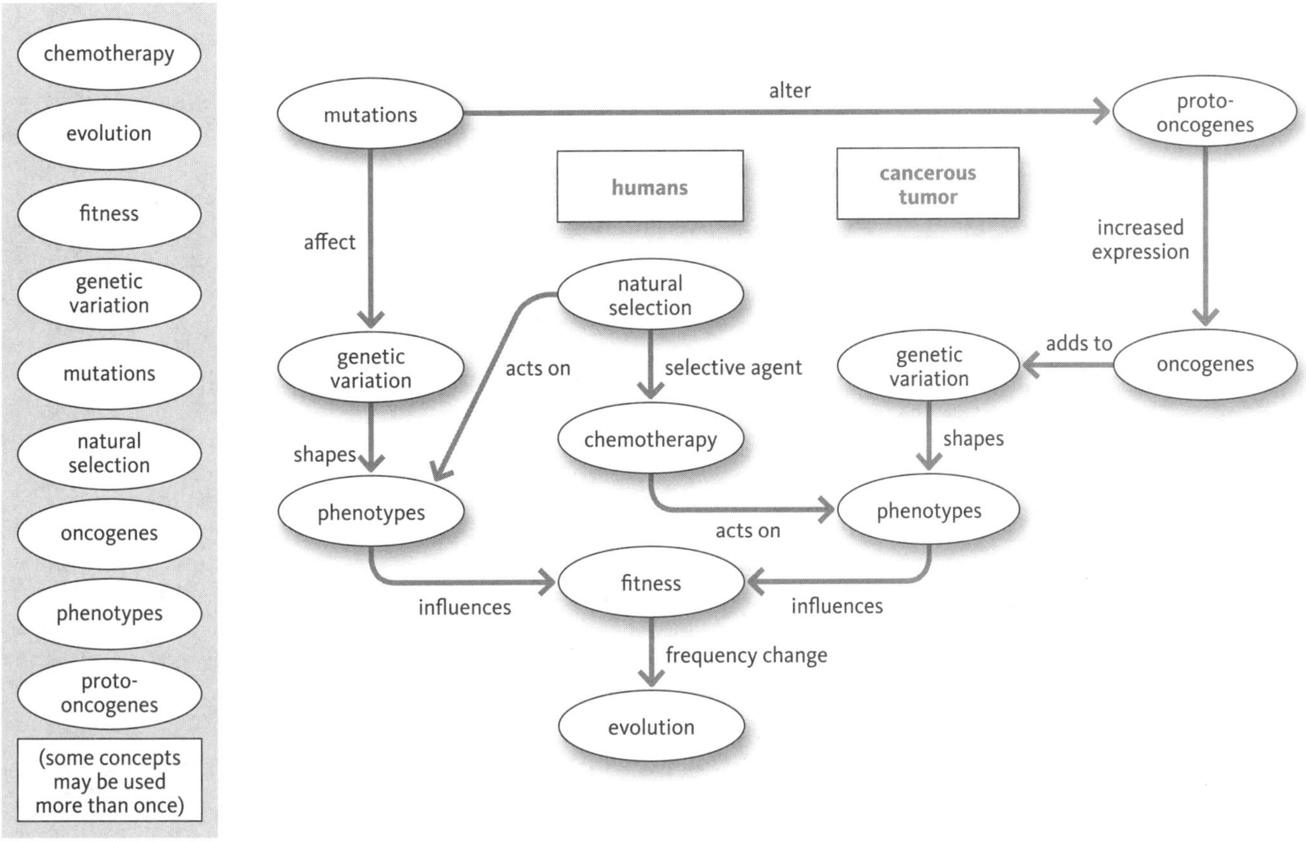

Interpret the Data

- What happened to the cluster 3 cancer cell lineage (dark gray hatched area on the left side marked by 5.10%) after chemotherapy?

  The lineage survived chemotherapy.

- What happened to the cluster 2 cancer cells after chemotherapy?

  The lineage went extinct.

- Why did the cluster 4 lineage increase so dramatically after chemotherapy?

  Because they had mutations that allowed them to be resistant to chemotherapy, they were able to survive and reproduce at a high rate, especially since other cancer lineages were eliminated.

- Was the cluster 5 lineage resistant to chemotherapy?

  Yes. The lineage arose from mutations to the cluster 4 lineage, a lineage that survived chemotherapy because of a resistance trait. The cluster 5 lineage came to dominate the population, causing the first cancer relapse.

Delve Deeper

1. How might repeated infections function to enable host shifting in some pathogens?

   A pathogen infecting a new host faces strong selective pressure. If individuals survive, genetic variation in the new population will be relatively low. Beneficial mutations within this new population may not arise quickly enough to overcome the new host's immune response. However, multiple infections may bring additional individuals with mutations beneficial for surviving on the new host, and horizontal transfer may bring new combinations of alleles together. Genetic variation in the pathogen population on the new host would increase with each new infection, and this relatively rapid increase in genetic variation may lead to rapid increases in fitness in the new population. As a result, repeated infections might provide the essential genetic variation to the pathogen population to enable host shifting.

2. Why do devastating diseases, such as Huntington's disease, continue to plague humans?

   Not all diseases affect fitness as soon as an organism is born. In fact, some diseases may only become issues because we now have longer lifespans than in the past. A disease such as Huntington's usually doesn't show signs until individuals are past breeding age. Individual with Huntington's can pass on the alleles to their offspring before they even know they carry the disorder.

   Many diseases may involve a similar trade-off. Natural selection may favor some traits early in life, but those traits may be detrimental later in life. For example, production of the p53 tumor suppressor protein may prevent cancer in young individuals, allowing them to survive and reproduce. However, the p53 tumor suppressor protein may negatively interact with surrounding tissues. As we age, these negative interactions accumulate, and once we've passed breeding age, natural selection will have little effect.

3. What are the three conditions necessary for evolution by natural selection to occur?

   First, individuals must differ in the characteristics of a trait. Individual variation in the expression or characteristics of a trait is the raw material for evolution by natural selection. The ultimate source of that variation is mutations to the genetic architecture within an individual. Some mutations are detrimental, some are beneficial, and some are neutral (although additional mutations can change that). How the genotype (the genetic makeup of an individual) affects the phenotype (the observable, measurable characteristic) can be complex, but the phenotype is the manifestation of individual differences on which natural selection can act. Second, the differences among individuals in a trait must be at least partially heritable. Heritable variation is integral to natural selection for most organisms because it is an effective mechanism by which genotypic variation may be transmitted. Third, some individuals survive and reproduce more successfully than others because of differences in a trait. Natural selection acts on the relative success of individuals within a population. Individuals with heritable phenotypic traits that confer some advantage or disadvantage within the population will, on average, have higher or lower survival and/or reproductive rates than other individuals who possess different versions of those traits. Natural selection acts to alter the abundances of those traits, leading to evolution of the population.

Test Yourself

1. c; 2. b; 3. d; 4. a; 5. b; 6. f; 7. d; 8. f; 9. b; 10. c; 11. d